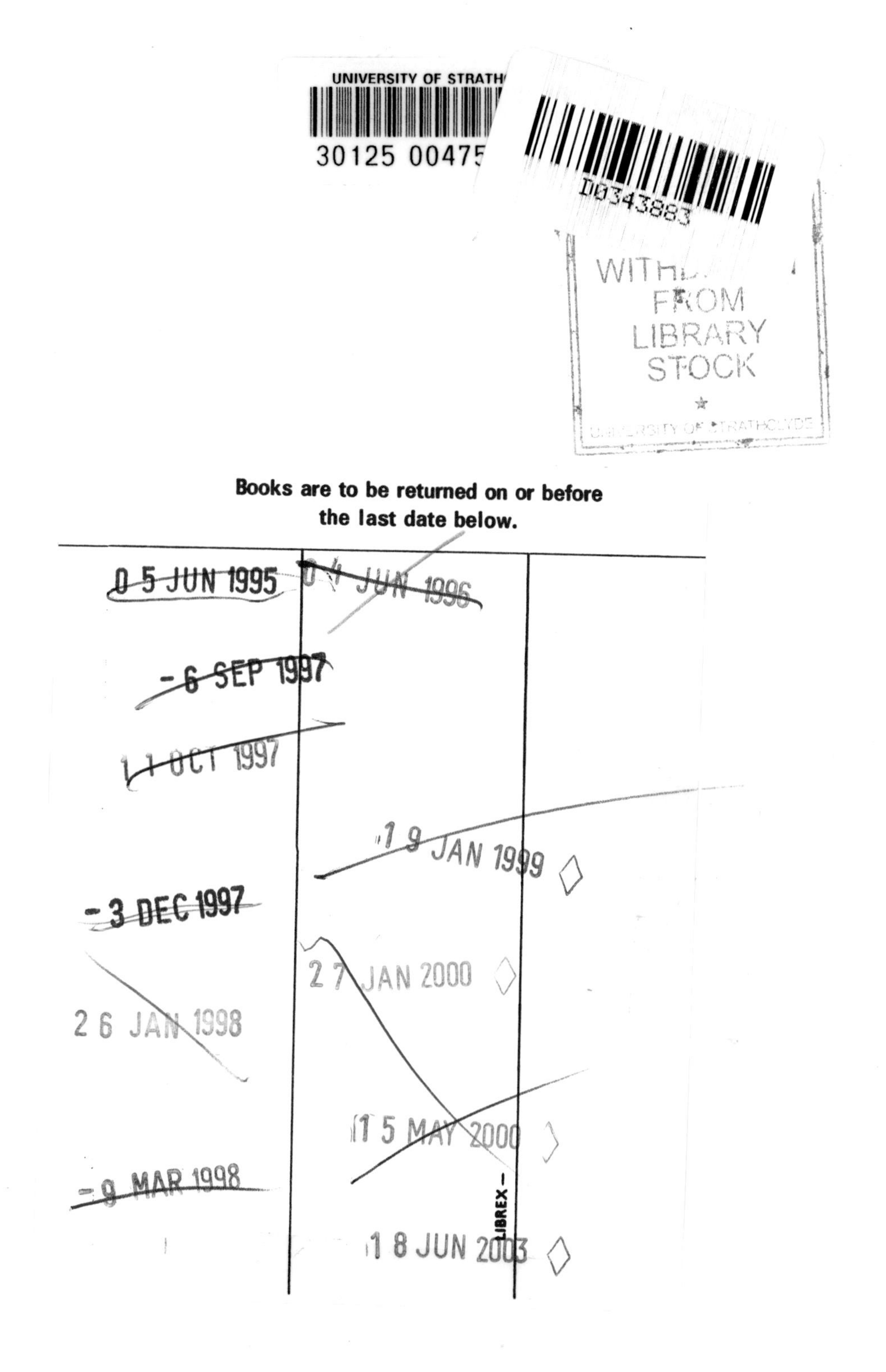

UNIVERSITY OF STRATH
30125 00475
D0343883
WITH
FROM
LIBRARY
STOCK
Books are to be returned on or before
the last date below.
0 5 JUN 1995
- 6 SEP 1997
1 1 OCT 1997
- 3 DEC 1997
2 6 JAN 1998
- 9 MAR 1998
1 9 JAN 1999
2 7 JAN 2000
1 5 MAY 2000
1 8 JUN 2003
LIBREX

Molecular Genetics of Yeast

The Practical Approach Series

Affinity Chromatography
Anaerobic Microbiology
Animal Cell Culture (2nd Edition)
Animal Virus Pathogenesis
Antibodies I and II
Behavioural Neuroscience
Biochemical Toxicology
Biological Data Analysis
Biological Membranes
Biomechanics—Materials
Biomechanics—Structures and Systems
Biosensors
Carbohydrate Analysis
Cell–Cell Interactions
The Cell Cycle
Cell Growth and Division
Cellular Calcium
Cellular Interactions in Development
Cellular Neurobiology
Centrifugation (2nd Edition)
Clinical Immunology
Computers in Microbiology
Crystallization of Nucleic Acids and Proteins
Cytokines
The Cytoskeleton
Diagnostic Molecular Pathology I and II
Directed Mutagenesis
DNA Cloning I, II, and III
Drosophila
Electron Microscopy in Biology
Electron Microscopy in Molecular Biology
Electrophysiology
Enzyme Assays
Essential Developmental Biology
Essential Molecular Biology I and II
Experimental Neuroanatomy
Extracellular Matrix
Fermentation
Flow Cytometry
Gas Chromatography

Gel Electrophoresis of Nucleic Acids (2nd Edition)
Gel Electrophoresis of Proteins (2nd Edition)
Gene Targeting
Gene Transcription
Genome Analysis
Glycobiology
Growth Factors
Haemopoiesis
Histocompatibility Testing
HPLC of Macromolecules
HPLC of Small Molecules
Human Cytogenetics I and II (2nd Edition)
Human Genetic Disease Analysis
Immobilised Cells and Enzymes
Immunocytochemistry
In Situ Hybridization
Iodinated Density Gradient Media
Light Microscopy in Biology
Lipid Analysis
Lipid Modification of Proteins
Lipoprotein Analysis
Liposomes
Lymphocytes
Mammalian Cell Biotechnology
Mammalian Development
Medical Bacteriology
Medical Mycology
Microcomputers in Biochemistry
Microcomputers in Biology
Microcomputers in Physiology
Mitochondria
Molecular Genetic Analysis of Populations
Molecular Genetics of Yeast
Molecular Imaging in Neuroscience
Molecular Neurobiology
Molecular Plant Pathology I and II
Molecular Virology
Monitoring Neuronal Activity
Mutagenicity Testing
Neural Transplantation
Neurochemistry
Neuronal Cell Lines
NMR of Biological Macromolecules
Nucleic Acid Hybridisation
Nucleic Acid and Protein Sequence Analysis
Nucleic Acids Sequencing
Oligonucleotides and Analogues
Oligonucleotide Synthesis
PCR
Peptide Hormone Action
Peptide Hormone Secretion
Photosynthesis: Energy Transduction
Plant Cell Biology
Plant Cell Culture
Plant Molecular Biology
Plasmids (2nd Edition)
Pollination Ecology
Postimplantation Mammalian Embryos
Preparative Centrifugation
Prostaglandins and Related Substances

Protein Architecture
Protein Blotting
Protein Engineering
Protein Function
Protein Phosphorylation
Protein Purification Applications
Protein Purification Methods
Protein Sequencing
Protein Structure
Protein Targeting
Proteolytic Enzymes
Radioisotopes in Biology
Receptor Biochemistry
Receptor–Effector Coupling
Receptor–Ligand Interactions
Ribosomes and Protein Synthesis
RNA Processing I and II
Signal Transduction
Solid Phase Peptide Synthesis
Spectrophotometry and Spectrofluorimetry
Steroid Hormones
Teratocarcinomas and Embryonic Stem Cells
Transcription Factors
Transcription and Translation
Tumour Immunobiology
Virology
Yeast

Molecular Genetics of Yeast

A Practical Approach

Edited by

JOHN R. JOHNSTON

Department of Bioscience and Biotechnology,
University of Strathclyde, Royal College Building,
204 George Street, Glasgow G1 1XW, UK

at

OXFORD UNIVERSITY PRESS

Oxford New York Tokyo

Oxford University Press, Walton Street, Oxford OX2 6DP
Oxford New York Toronto
Delhi Bombay Calcutta Madras Karachi
Kuala Lumpur Singapore Hong Kong Tokyo
Nairobi Dar es Salaam Cape Town
Melbourne Auckland Madrid
and associated companies in
Berlin Ibadan

Oxford is a trade mark of Oxford University Press

A Practical Approach is a registered trade mark
of the Chancellor, Masters, and Scholars of the University of Oxford
trading as Oxford University Press

Published in the United States
by Oxford University Press Inc., New York

A catalogue record for this book is available from the British Library

Library of Congress Cataloging in Publication Data
Molecular genetics of yeast : a practical approach / edited by John R. Johnston.
(The Practical approach series ; 141)
Includes bibliographical references and index.
1. Saccharomyces cerevisiae–Genetics. 2. Molecular genetics–Technique. 3. Genetics, Experimental. I. Johnston, John R. II. Series.
QH470.S23M65 1994 589.2'330487328—dc20 93–33833
ISBN 0 19 963430 0 (Hbk.)
ISBN 0 19 963429 7 (Pbk.)

Typeset by Footnote Graphics, Warminster, Wilts
Printed in Great Britain by Information Press Ltd, Eynsham, Oxon.

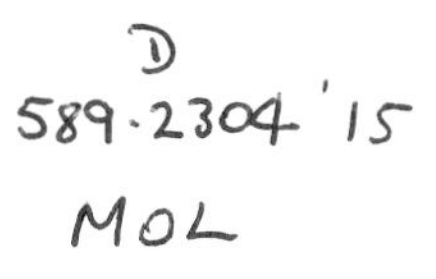

Preface

Our knowledge of the genetics and molecular biology of yeast continues to advance at an encouraging rate and elicits considerable excitement. To a large extent, the term yeast in this respect is synonymous with the budding species *Saccharomyces cerevisiae* and the overwhelming emphasis within this book is upon this organism. We should remember, however, that very substantial progress is being made with other yeasts, in particular the fission species *Schizosaccharomyces pombe*. In many studies yeast has come to occupy the role of a model eukaryotic cell and many recent findings assume importance for cells of higher eukaryotes, including human. At the same time, yeast is a microbe of major economic and social significance. Not only does it provide the fermentations for most breads and alcoholic beverages but increasingly it has been recruited to the 'new biotechnology' to produce heterologous proteins and other molecules mainly, but not exclusively, for pharmaceutical use.

The primary aim of this book is to bring together many of the current experimental methods underlying recent progress in a range of areas within yeast molecular genetics. Limitation of space has necessitated selection and therefore the contents are representative rather than comprehensive. In particular, the intention to include specialized chapters on YAC vectors, gene disruption and replacement, secretion, and the Yeast Genome Sequencing Project was not realized. However, aspects of these topics are included within certain other chapters. Similarly, inclusion of more cellular aspects of yeast molecular biology was severely restricted. The molecular and cellular levels of yeast biology are becoming increasingly integrated and Chapter 15 is an example of this rewarding trend. My thanks are due to the authors for their excellent contributions and professional cooperation. My task was also made the lighter by the efficient and friendly service provided by the publisher's editorial staff.

Glasgow JOHN R. JOHNSTON
December 1993

Contents

Contributors

SALLY E. ADAMS
Immunotherapeutics Division, British Biotechnology Ltd, Watlington Road, Cowley, Oxford OX4 5LY, UK.

JEF D. BOEKE
Department of Molecular Biology and Genetics, Johns Hopkins University School of Medicine, 725 N. Wolfe Street, Baltimore, MD 21205, USA.

ALISTAIR J. P. BROWN
Department of Molecular and Cell Biology, University of Aberdeen, Marischal College, Aberdeen AB9 1AS, UK.

JEFFREY L. BROWN
Department of Biology, McGill University, 1205 Dr Penfield Avenue, Montreal, Quebec, Canada H3A 1B1.

NIGEL R. BURNS
Immunotherapeutics Division, British Biotechnology Ltd, Watlington Road, Cowley, Oxford OX4 5LY, UK.

HOWARD BUSSEY
Department of Biology, McGill University, 1205 Dr Penfield Avenue, Montreal, Quebec, Canada H3A 1B1.

MICHAEL W. CLARK
Department of Biology, McGill University, 1205 Dr Penfield Avenue, Montreal PQ, Canada H3A 1B1.

STEPHEN J. ELLEDGE
Department of Biochemistry and Institute for Molecular Genetics, Baylor College of Medicine, Houston, TX 77030, USA.

HORST FELDMANN
Institut für Physiologische Chemie, Physikalische Biochemie und Zellbiologie, Ludwig-Maximilians-Universität München, Schillerstrasse 44, 80336 München, Germany.

D. J. GARFINKEL
Laboratory of Eukaryotic Gene Expression, NCI-Frederick Cancer Research and Development Center, ABL-Basic Research Program, PO Box B, Frederick, MD 21702-1201, USA.

R. DANIEL GIETZ
Department of Human Genetics, Faculty of Medicine, University of Manitoba, 770 Bannatyne Avenue, Room 250, Winnipeg, Manitoba, Canada R3E 0W3.

Contributors

CHRISTOPHER HADFIELD
Department of Genetics, University of Leicester, Leicester LE1 7RH, UK.

ALAN D. HARTLEY
Biological Laboratory, University of Kent, Canterbury, Kent CT2 7NJ, UK.

DOUGLAS W. HURD
Department of Biochemistry, University of Oxford, South Parks Road, Oxford OX1 3QU, UK.

JOHN R. JOHNSTON
Department of Bioscience and Biotechnology, University of Strathclyde, Royal College Building, 204 George Street, Glasgow G1 1XW, UK.

JILL B. KEENEY
Department of Molecular Biology and Genetics, Johns Hopkins University School of Medicine, 725 N. Wolfe Street, Baltimore, MD 21205, USA.

MORTEN C. KIELLAND-BRANDT
Carlsberg Laboratory, Department of Yeast Genetics, Gamle Carlsberg Vej 10, DK-2500 Copenhagen Valby, Denmark.

ALAN J. KINGSMAN
Department of Biochemistry, University of Oxford, South Parks Road, Oxford OX1 3QU, UK.

SUSAN M. KINGSMAN
Department of Biochemistry, University of Oxford, South Parks Road, Oxford OX1 3QU, UK.

GÉRARD LOISON
Sanofi-Elf-Biorecherches, Labège-Innopole BP137, 31676 Labège Cédex, France.

MARC LUSSIER
Department of Biology, McGill University, 1205 Dr Penfield Avenue, Montreal, Quebec, Canada H3A 1B1.

GEMMA MARFANY
Department of Biochemistry, University of Oxford, South Parks Road, Oxford OX1 3QU, UK.

ENCARNA MARTIN-RENDON
Department of Biochemistry, University of Oxford, South Parks Road, Oxford OX1 3QU, UK.

M. MASTRANGELO
Department of Chemistry, Allegany Community College, Willowbrook Road, Cumberland, MD 21502, USA.

JOHN T. MULLIGAN
Department of Genetics, Stanford University School of Medicine, Stanford, CA 94305, USA.

PETER PHILIPPSEN
Institute of Applied Microbiology, Biozentrum, University of Basel, Klingelbergstrasse 70, CH-4056 Basel, Switzerland.

HORST PICK
Institute of Applied Microbiology, Biozentrum, University of Basel, Klingelbergstrasse 70, CH-4056 Basel, Switzerland.

PETER W. PIPER
Biochemistry and Molecular Biology, University College London, London WC1E 6BT, UK.

L. A. RINCKEL
Laboratory of Eukaryotic Gene Expression, NCI-Frederick Cancer Research and Development Center, ABL-Basic Research Program, PO Box B, Frederick, MD 21702-1201, USA.

TERRY ROEMER
Department of Biology, McGill University, 1205 Dr Penfield Avenue, Montreal, Quebec, Canada H3A 1B1.

ANNE-MARIE SDICU
Department of Biology, McGill University, 1205 Dr Penfield Avenue, Montreal, Quebec, Canada H3A 1B1.

ROBERT S. SIKORSKI
Department of Molecular Biology and Genetics, Johns Hopkins University School of Medicine, 725 N. Wolfe Street, Baltimore, MD 21205, USA.

J. N. STRATHERN
Laboratory of Eukaryotic Gene Expression, NCI-Frederick Cancer Research and Development Center, ABL-Basic Research Program, PO Box B, Frederick, MD 21702-1201, USA.

ROLF STUCKA
Institut für Physiologische Chemie, Physikalische Biochemie und Zellbiologie, Ludwig-Maximilians-Universität München, Schillerstrasse 44, 80336 München, Germany.

MICK F. TUITE
Biological Laboratory, University of Kent, Canterbury, Kent CT2 7NJ, UK.

ACHIM WACH
Institute of Applied Microbiology, Biozentrum, University of Basel, Klingelbergstrasse 70, CH-4056 Basel, Switzerland.

ROBIN A. WOODS
Department of Biology, University of Winnipeg, 515 Portage Avenue, Winnipeg, Manitoba, Canada R3B 2E9.

DELIN ZHU
Biological Laboratory, University of Kent, Canterbury, Kent CT2 7NJ, UK.

Abbreviations

A	absorbance
A	adenine
ADH	alcohol dehydrogenase (gene, promoter)
AMV	avian myeloblastosis virus
Ap	ampicillin
ARS	autonomously replicating sequence
ATP	adenosine triphosphate
BSA	bovine serum albumin
C	cytosine
CAA	casamino acids
CAT	chloramphenicol acetyltransferase
CEN	centromere
CLP	chromosome length polymorphism
Cm	chloramphenicol
dA	2′-deoxyribosyladenine
DAPI	4,6-diamidino-2-phenylindol dihydrochloride
dATP	2′-deoxyadenosine triphosphate
dCTP	2′-deoxycytidine triphosphate
DEPC	diethyl pyrocarbonate
DMSO	dimethyl sulphoxide
d.p.m.	disintegrations per minute
ds	double stranded
dT	2′-deoxyribosylthymine (thymidine)
DTE	dithioerythreitol
DTT	dithiothreitol
dTTP	2′-deoxythymidine triphosphate
EC	European Community
EDTA	ethylenediaminetetracetic acid
EEO	electroendosmosis
EM	electron microscope/micrograph
EMBO	European Molecular Biology Organisation
EthBr	ethidium bromide
EtOH	ethanol
5-FO, 5-FOA	5-fluoroorotic acid
5-FU, 5-FUA	5-fluorouracil
FTL	freeze-thaw lysate
G	guanine
GAL	galactose
GRAP	galactose-regulated artificial promoter sequence

GUT	glycerol utilization (gene, promoter)
HS	heat shock
K	killer
kb	kilobase pairs
kDa	kilodalton
KT	killer toxin
LGTA	low gelling temperature agarose
LMPA	low melting point agarose
MF	mating factor
NMR	nuclear magnetic resonance
OD	optical density
Ori	origin of replication sequence
PCR	polymerase chain reaction
PEG	polyethylene glycol
PFG	pulsed field gel
PFGE	pulsed field gel electrophoresis
p.f.u.	plaque forming units
PGK	3-phosphoglycerate kinase (gene, promoter)
PHO	phosphatase (gene, promoter)
PMSF	phenylmethylsulphonyl fluoride
PRB	proteinase B (gene, promoter)
SDS	sodium dodecyl sulphate
SDS–PAGE	SDS–polyacrylamide gel electrophoresis
SE	sonicated extracts
SLB	sphaeroplast lysis buffer
ss	single stranded
TAE	tris acetate EDTA
TBE	tris boric EDTA
TCA	trichloroacetic acid
TM	trade mark
Ty	transposon (yeast)
UAS	upstream activator sequence
VLP	virus like particle
YAC	yeast artificial chromosome
YCp	yeast centromere plasmid
YEp	yeast episomal plasmid
YES	yeast–*E. coli* shuttle (vector)
YIp	yeast integrating vector
YRp	yeast replication plasmid

1

Procedures for isolating yeast DNA for different purposes

ACHIM WACH, HORST PICK, and PETER PHILIPPSEN

1. Introduction

The DNA isolation procedures described in this chapter have been extensively used with *Saccharomyces cerevisiae* strains. However, they can be adapted for other *Saccharomyces* species and for other yeast genera. The demands with respect to purity, size, and amount of yeast DNA can be very different and mainly depend on the type of experiment to be performed, e.g. DNA hybridization for screening many strain isolates for rearrangements, to verify the status of transforming DNA, or to search for homologous sequences; re-isolation of recombinant yeast plasmids for transformation into *Escherichia coli*; PCR-type reactions; construction of gene banks with partially cleaved and size fractionated DNA; or DNA ligations on a preparative scale. (The isolation of chromosomal sized DNA for electrophoretic karyotyping is discussed in Chapter 5.)

We present three different methods of preparing total DNA from yeast strains and comment on a commercially available yeast DNA isolation kit. These methods differ in overall yield, in speed, and in the final purity of the DNA. Since it is not necessary to isolate very pure DNA for routine DNA cleavages and hybridizations, the fast small scale procedure is very often used and allows preparation of partially purified DNA from at least 24 strains within 5–6 h. However, when gene banks have to be constructed involving size fractionation of partially cleaved DNA, we highly recommend the isolation of preparative amounts of clean DNA via CsCl density gradients even if this takes up to a week. In cases where a series of DNA reactions have to be performed with a small number of strains, we propose the use of the medium scale procedure which yields 50–100 μg of DNA from up to 12 strains within a day.

The basic scheme of the three procedures is very similar:

(a) collection and washing of cells followed by enzymatic degradation of their cell wall

(b) lysis of protoplast and deproteination of DNA and RNA at higher temperature with sodium dodecyl sulphate (SDS) in the presence or absence of proteinase K
(c) precipitation of dodecyl sulphate with potassium ions
(d) separation of the solubilized DNA and RNA from all denatured and insoluble molecules by centrifugation
(e) further purification of total nucleic acids from most soluble cellular components by precipitation with alcohol
(f) resolution of nucleic acids and separation of DNA from RNA. Since DNA represents only 2% of total yeast nucleic acids, complete separation of DNA from RNA is not a trivial task. Complete or at least partial separation can be achieved in different ways as described in the protocols (see also *Chapter 3*)

Although it is not difficult to isolate yeast DNA by strictly following the procedures provided, it will be quite informative to study the comments and figures directly following the description of each procedure. When troubleshooting, these sections may also be valuable for detecting possible sources of problems.

2. Large scale preparation of clean DNA by CsCl density gradient equilibrium centrifugation

For construction of gene banks it is highly recommended that large amounts of very clean genomic DNA are prepared. The method of choice is still CsCl density gradient equilibrium centrifugation. Although this procedure is time-consuming, it is worthwhile because of the overall yield obtained and the reproducible performance of all subsequent DNA-manipulating steps, e.g. controlled partial digestion with restriction nucleases and ligation of size-fractionated yeast DNA to dephosphorylated vector DNAs at different relative concentrations.

Protocol 1. Large scale preparation of clean DNA

1. Grow yeast cells in 1 litre of liquid medium (e.g. YPD, see Appendix 1) at 30°C to a cell density of $1\text{–}2 \times 10^8$ cells/ml.[a]
2. Collect cells by centrifugation for 5 min at 5000 r.p.m. (e.g. four 250 ml aliquots, Sorvall GSA rotor), decant supernatant, wash cells by resuspending each aliquot in 100 ml of 10 mM EDTA, pH 8.0, and centrifuge again.
3. Resuspend each aliquot in 30 ml of 1.0 M sorbitol, 20 mM dithiothreitol, 10 mM EDTA, pH 8.0, and incubate for 10 min at room temperature.

4. Collect cells by centrifugation, resuspend each aliquot in 5.0 ml of 1.0 M sorbitol, 20 mM dithiothreitol, 10 mM EDTA, pH 8.0. Combine aliquots into one tube, add 20 mg of Zymolyase 20 000 dissolved in 1.0 ml of 10 mM EDTA, mix well, and incubate at 37°C. At 10–15 min intervals observe protoplast formation by light microscopy until all cells have converted into round protoplasts. This may take 30–90 min. (Also see footnote *a*.)
5. Collect protoplasts by centrifugation (10 min, 4000 r.p.m., Sorvall GSA rotor), remove supernatant, and carefully resuspend protoplasts in 5.0 ml of 1.0 M sorbitol, 10 mM Tris–HCl, 10 mM EDTA, pH 8.0.
6. Dissolve 10 mg of proteinase K in 75 ml 1% SDS, 10 mM Tris–HCl, 10 mM EDTA, pH 8.0, mix with the resuspended protoplasts, and incubate for 40–60 min at 60°C with occasional shaking[a].
7. Add 80 ml of 5.0 M potassium acetate, pH 5.4, let sit on ice for 60 min, and centrifuge for 20 min at 12 000 r.p.m. (Sorvall GSA rotor).
8. If the supernatant is still turbid or contains a white surface layer, it should be poured through several layers of medical wound gauze into a graduated glass cylinder.
9. Mix clear supernatant with one volume of isopropanol and collect white precipitate (RNA and DNA) with a glass hook or by centrifugation.
10. Remove remnants of salt in the white precipitate by washing with 70% ethanol. Dry for a short time in a vacuum desiccator, just to remove the alcohol.
11. Resolve nucleic acids in 7.9 ml of 50 mM Tris–HCl, 10 mM EDTA, pH 8.0. After the nucleic acids have dissolved completely (a short incubation at 42°C helps), add 8.0 g of CsCl and 200 μl of 10 mg/ml DAPI solution. Stir to dissolve CsCl. A precipitate may form which can be removed by a brief centrifugation. Prior to the addition of CsCl, a small aliquot should be checked by agarose gel electrophoresis to estimate the quality of the preparation as shown in *Figure 1A*.
12. Transfer into ultracentrifuge tubes, e.g. those for Beckman Ti60 rotor, and run at 21°C for 48 h at 42 000 r.p.m.[a].
13. Illuminate tubes with long wavelength UV light (*caution*: eye protection required) and harvest DNA (1–2 ml) by puncturing the side of the tube with a large diameter syringe needle (the lower intense blue band is nuclear DNA, while the upper diffuse band is mitochondrial DNA). When inspected by agarose gel electrophoresis (*Figure 1B*), it will be seen that most but not all of the RNA has separated from the nuclear DNA. (Also see footnote *a*.)
14. Prepare buffered CsCl solution as in step **12** (but with one-tenth the concentration of DAPI), mix with the DNA fraction, and repeat density gradient equilibrium centrifugation.

Protocol 1. *Continued*

15. Isolate DNA again by puncturing the side of the tube with a hyper-dermic needle.

16. Dilute the high-molar CsCl fraction with 2 volumes of TE buffer (10 mM Tris, 1 mM EDTA, pH 8.0). Precipitate DNA with 2 volumes of ethanol. The dilution step prevents precipitation of the CsCl by ethanol.

17. Collect DNA precipitate by centrifugation or with the aid of a glass hook, wash well with 70% ethanol, dry briefly, dissolve in 1–2 ml of TE buffer, and store at +4°C, *not* at −20°C[a].

18. Check DNA preparation by agarose gel electrophoresis. This DNA is practically RNA-free (*Figure 1B*, lane 2). Use a UV absorption spectrum between 320 nm and 240 nm to determine the concentration and the purity of the DNA preparation (for details see step **14** of *Protocol 2*). The yield from 1 litre of culture should be at least 1 mg DNA.

[a] See Section 2.1 for notes on this protocol.

2.1 Comments on *Protocol 1*

Step **1**. The growth temperature for heat-sensitive strains is 20–23°C and that for cold-sensitive strains is 30–35°C. When cells have to be grown in minimal medium or selective medium the cell density often does not reach 10^8 cells/ml in shaking flasks. However, good yields are obtained when 50 mM sodium phosphate, pH 6.4, is added to the minimal medium. Cell densities can, in principle, be determined by measuring the OD_{600} of the culture. However, there is no universal correlation between cell density and OD_{600} since this correlation depends on the strain used and on the growth conditions.

Step **4**. Thiol reagents, like dithiothreitol or 2-mercaptoethanol, are not essential for cell wall degradation. They increase the rate of protoplast formation by reducing disulphide bridges in the cell wall, making it more accessible for glucanase action. It is of importance for the efficiency of the cell wall removal by Zymolyase, that the cell culture has not entered late stationary growth phase, since resistance to cell wall degradation progressively increases with advancing entry into the stationary phase. Yeast grown in low phosphate medium can also be resistant to cell wall degradation. The time required for sufficient cell wall degradation depends, in addition, on the strain in use. The period given in the protocol is satisfactory for most of the yeast strains we have tested, including commonly used laboratory strains derived from S288C or Σ1278. Insufficient conversion to protoplasts will decrease the overall yield of DNA and sometimes leads to an overrepresentation of the 2 μm plasmid. Commercial glucanase preparations (Zymolyase from Seikagaku Kogyo, Ltd, or Miles Corp.; Lyticase or yeast lytic enzyme from ICN Biomedicals Inc.)

can contain protease activity and traces of phosphatase activity but no DNase or RNase. Glucanase from *Arthrobacter luteus*, which is used for most commercial preparations, is active between pH 6 to 9 with a pH optimum of 7.5 and it can be inactivated by a 5 min treatment at 60°C. The glucan cell wall of many yeasts, including *Ashbya, Candida, Hansenula, Kluyveromyces, Lipomyces, Pichia, Torulopsis, Saccharomycopsis*, and *Saccharomyces* can be degraded. Completion of the protoplast conversion can be inspected by different means:

(a) in a light microscope intact cells appear to be oval in shape, while protoplasts are round;
(b) OD_{800} readings taken at 10 min intervals allow quantitative monitoring of protoplast formation: to do so, 50 μl sample aliquots are mixed with 950 μl of water and the OD_{800} is measured after 20 sec; the conversion is complete when the OD_{800} has decreased by 80–90%;
(c) samples can be examined visually by checking the immediate clarification of 100 μl of the turbid sample suspension after addition of 100 μl of 1% SDS.

Step **6**. Proteinase K is omitted in several protocols. When added to help cell lysis and deproteination of nucleic acids, care has to be taken to ensure that it is completely removed at a later stage. Otherwise, enzymes added to the final DNA preparation could be degraded. Since proteins band at lower density than DNA in CsCl density gradients, complete removal is achieved in step **12**. Older procedures (1) employ phenol/chloroform to extract proteins from nucleic acid solutions (see also comment to step **7** of *Protocol* 2).

Step **12**. In theory, RNA can be completely separated from DNA in CsCl density gradients since RNA bands at higher density than DNA. However, after one equilibrium density gradient, there is still a substantial amount of small RNA molecules in the DNA band because of the vast excess of RNA and because small RNA molecules do not band well. Therefore, it is *essential* not to use RNase when CsCl gradients are employed and it is recommended that a second CsCl gradient is run if complete separation of DNA and RNA is desired (see *Figure 1B*).

Step **13**. *S. cerevisiae* nuclear DNA (58% A+T) and mitochondrial DNA (80% A+T) can be separated by density gradient centrifugation (CsCl or NaI) in the presence of dyes which preferentially bind to (A+T)-rich DNA, such as 4,6-diamidino-2-phenylindol (DAPI) or bisbenzimide (2, 3). Removal of these dyes is possible via extraction of the non-diluted CsCl fraction with isopropanol. Upon dilution of the DNA-containing CsCl fraction, precipitation of the DNA with ethanol and washing of the precipitated DNA with 70% ethanol (steps **16** and **17**), most of the dye is also removed. The mitochondrial DNA can be collected separately from nuclear DNA or together with nuclear DNA. When isolated separately, 2 μm plasmid DNA can be present in the

Figure 1. Agarose gel electrophoresis of nucleic acids at different steps of the DNA isolation procedure. (A) Total nucleic acids prior to CsCl density gradient centrifugation. 10 μl (lane 1), 5 μl (lane 2), or 1 μl (lane 3) from step **12** were loaded next to a λ DNA size marker. **Note:** the chromosomal DNA band migrates more slowly than the largest λ DNA fragments (24 and 21 kb) indicating that the DNA molecules are at least 30 kb, probably much larger. The 4.6 kb double-stranded, viral RNA, present in many *S. cerevisiae* strains, is seen as a prominent band. Sometimes the 6.3 kb circular 2 μm DNA (see *Figure 4*) is visible as a faint band in front of the viral DNA. The two discrete bands in the lower part of lane 3 are 25S (3.4 kb) and 18S (1.75 kb) ribosomal RNA. When electrophoresis is performed with ethidium bromide already present in the gel, it is advisable to restain the gels because the excess of rapidly migrating RNA molecules will bind so much ethidium bromide that the DNA band becomes barely detectable. (B) Lane 1: nuclear DNA collected from the first CsCl density gradient centrifugation (step **14**, here with vertical rotor not fixed angle rotor). Lane 2: nuclear DNA after the second CsCl density gradient centrifugation (step **16**, here again with vertical rotor). In both cases 4 μl were loaded per slot. **Note:** it may look as if the DNA preparation after the first CsCl gradient contains only little viral dsRNA and traces of small RNA (bottom lane 1). However, ethidium bromide intercalates only with the small fraction of RNA molecules which form stable, double-stranded structures. Thus, this method does not detect the total RNA content in the sample and often leads to an underestimation of the remaining contamination with RNA. (C) Complete clevage of CsCl-purified DNA with *EcoRI* (lane 1) or *Hin*dIII (lane 2). **Note:** the strong bands in the restriction pattern originate from ribosomal RNA genes (see *Figure 4*). Fragment sizes are 2.65, 2.46, 1.95, 0.66, 0.59, 0.36, and 0.28 kb for the *EcoR*I digestion (the two smallest fragments are not seen on the gel) and 1.2 and 2.7 kb for the *Hin*dIII digestion. Traces of viral double-stranded RNA are still visible as a band at 4.6 kb. (D) Partial cleavage of CsCl-purified DNA with *Sau*3AI. The same amount of DNA (~0.5 μg) was used in each assay but the amount of *Sau*3AI was doubled from assay to assay (from 2 mU, lane 1, to 1 U, lane 10). Each sample was incubated at 37 °C for 30 min. **Note:** the DNA preparation still contained traces of 4.6 kb viral double-stranded RNA. Complete digestion of rDNA with *Sau*3AI generates, besides many fragments of low molecular weight, also larger DNA fragments of 1.4, 0.98, and 0.64 kb which are visible as discrete bands in lane 10. The molecular size marker used in the gels is a mixture of λ DNA digested with *Hin*dIII and λ digested with *Hin*dIII + *Eco*RI. Fragment sizes, from top to bottom, are: 23.7 and 21.2 kb (one band), 9.42 kb, 6.56 kb, 5.15 and 4.97 kb (one band), 4.36 and 4.27 kb (one band), 3.53 kb, 2.97 kb (faint partial band), 2.32 kb, 2.03 kb, 1.90 kb, 1.58 kb, 1.37 kb, 0.95 kb, 0.83 kb, and 0.56 kb.

mitochondrial DNA fraction. The method of choice for preparative preparations of pure 2 μm plasmid or other circular recombinant plasmids is still CsCl density gradient in the presence of ethidium bromide (0.5 mg/ml), with the CsCl solution adjusted to 1.55 g/ml. Centrifuge in swinging bucket rotor at 150 000 *g* for 60 h, isolate the supercoiled plasmids (lower band) by puncturing the side of the tube with a syringe needle, and precipitate the DNA with ethanol after extraction with chloroform and 3-fold dilution with 10 mM Tris, 10 mM EDTA, pH 8.0. DNA is completely separated from RNA when this second centrifugation step is performed.

Step **17**. The freezing point of a saturated salt solution is near −20°C. It is therefore *not* recommended that the DNA solution is stored in freezers with a cycling temperature around −20 °C because this can result in frequent freezing and thawing of the DNA (4).

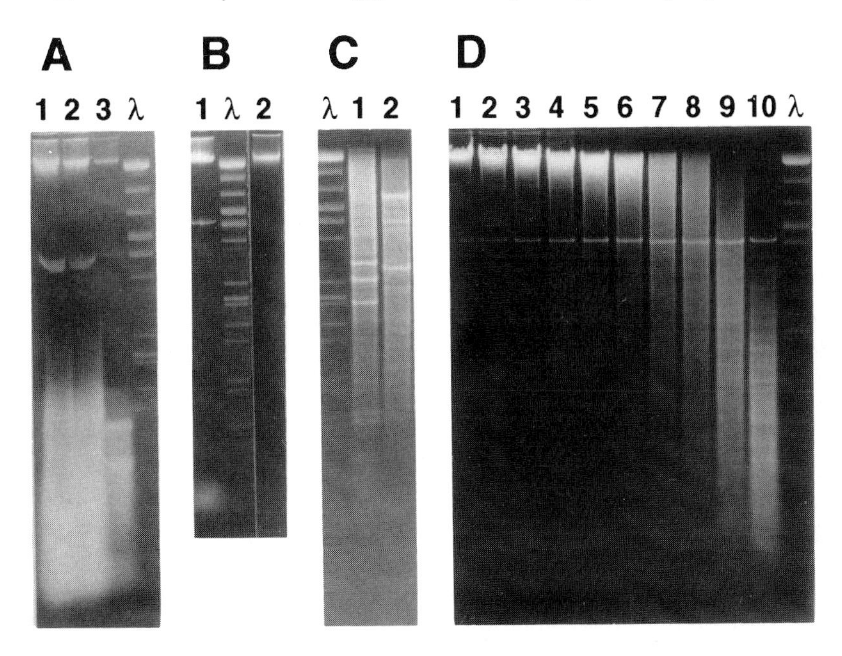

3. Isolation of yeast DNA from 30 ml cultures

For many experiments it is not necessary to prepare DNA of highest purity. The following procedure (5) yields 50 μg or more of yeast total DNA which can be completely cleaved with most restriction nucleases. It can be utilized in DNA ligation reactions and, after additional shearing, in PCR-type experiments. It is the method of choice when DNA of selected strains is required over a longer time period for several series of analytical experiments.

Protocol 2. Medium scale isolation of DNA[a]

1. Grow cells in 30 ml of YPD (Appendix 1) at 30°C to a cell density of approximately 2×10^8 cells/ml (early stationary phase)[a].
2. Spin cells at 5000 r.p.m. for 5 min in a Sorvall SS34 or similar rotor. Use 30 ml polypropylene tubes with screw caps.
3. Resuspend cells in 10 ml of 10 mM EDTA, pH 8.0, and spin at 5000 r.p.m. for 5 min.
4. Resuspend cells in 3.0 ml of 1.0 M sorbitol, 0.1 M EDTA, 50 mM dithiothreitol, pH 7.5. Use the same tube until step **8**. Add 0.5 mg Zymolyase 20 000 dissolved in 200 μl of 1.0 M sorbitol and incubate with occasional shaking at 37°C. The conversion to protoplasts is complete when no more oval cells are visible in the microscope. This

Protocol 2. *Continued*

takes 15–120 min depending on the strain used, the growth medium, and the growth phase. (Also see footnote *a*.)

5. Spin protoplasts at 5000 r.p.m. for 5 min in a Sorvall SS34 rotor and carefully discard the supernatant.
6. Resuspend protoplasts in 3.0 ml of 50 mM Tris–HCl, 50 mM EDTA, pH 8.0 by slowly and repeatedly drawing the protoplasts into a pipette and releasing them into the tube. Then mix with 0.3 ml of 10% SDS and incubate at 65°C for 30 min. (Also see footnote *a*.)
7. Add 1.0 ml of 5 M potassium acetate, mix, and let sit on ice for 60 min or longer[a].
8. Spin at 15 000 r.p.m. for 30 min and transfer the supernatant (~4 ml) to a 10 ml or 15 ml Corex tube or to a polypropylene tube.
9. Add 4.0 ml of ice-cold absolute ethanol. On mixing, the nucleic acids and some residual proteins will immediately precipitate at room temperature[a].
10. Spin in a Sorvall SM-24 or similar rotor at 10 000 r.p.m. for 10 min and discard the supernatant. Allow the salt to diffuse out of the pellet by incubation with 4.0 ml of 50% ethanol for at least 5 min (vortex to break up the pellet). Spin at 10 000 r.p.m. for 10 min. (Also see footnote *a*.)
11. Discard the supernatant, shortly dry the pellet under reduced pressure, and redissolve the DNA and RNA in 3.0 ml of 10 mM Tris, 1 mM EDTA, pH 7.5. This may take more than 1 h. A 10 min incubation at 42°C can help. If a small pellet remains, spin at 10 000 r.p.m. for 10 min and transfer the supernatant to a new tube. (Also see footnote *a*.)
12. Add 150 ml of 1 mg/ml DNase-free pancreatic RNase (Boehringer Mannheim) and incubate at 37°C for 30 min[a].
13. Add 3.0 ml of isopropanol, mix, and remove the precipitated DNA with small plastic forceps or a glass hook, or by centrifugation. Wash the DNA with 50% isopropanol and dry under reduced pressure.
14. Redissolve the DNA in 0.5 ml of TE buffer. The yield should be 60–90 μg of total DNA from a haploid strain. For quantification, dilute 50 μl with 0.95 μl of 10 mM Tris–HCl, 10 mM NaCl, pH 8.0, and measure the optical density at 250, 260, 280, and 300 nm. An OD_{260} of 0.2 corresponds to 10 μg DNA per ml, if the ratio of OD_{260}/OD_{280} is 1.9–2.0, the OD_{300} is close to zero, and the OD_{250} is less than the OD_{260}. The average fragment size of the DNA is 100 kb. The DNA should be stored at +4°C or −70°C, *not* at −20°C. (Also see footnote *a*.)

[a] See Section 3.1 for comments on this protocol.

3.1 Comments on *Protocol 2*

Steps **1** and **4**. See comments on steps **1** and **4** of *Protocol 1*.

Step **6**. Good homogenization and a not-too-concentrated suspension of protoplasts are critical before the SDS is added.

Step **7**. Besides nucleic acids, all soluble proteins, polysaccharides, lipids, and small molecular weight metabolites are still present in the slightly turbid solution that forms after potassium acetate addition. The white precipitate that forms consists mainly of insoluble potassium dodecyl sulphate and denatured proteins. Several procedures employ a treatment with 100 μg proteinase K at this step followed by repeated extractions with phenol/chloroform/isoamyl alcohol (25:24:1 by volume). These extractions remove more material, including proteinase K, from the DNA solution than does the precipitation with potassium acetate. However, the latter is much more convenient, and results in, for most purposes, sufficiently clean DNA.

Step **9**. The concentration of nucleic acids in solution is so high that complete precipitation of DNA and RNA is achieved at room temperature even with a final ethanol concentration of only 33%. We recommend this lowered amount of added alcohol and the incubation at room temperature because unwanted material which may precipitate with 66% ethanol or at low temperatures will stay in solution.

Step **10**. If the nucleic acid pellet is contaminated with salt, the nucleic acids will take longer to dissolve. In addition, double-stranded RNA may not be completely degraded by RNase (step **13**).

Step **11**. Analysis of the nucleic acids on an agarose gel (see *Figure 1*) before treatment with RNase allows conclusions about the presence of the very abundant 4.6 kb linear double-stranded viral RNA (6, 7) and the 6.3 kb circular 2 μm plasmid (8) in the DNA/RNA preparation.

Step **12**. It is often assumed that RNase treatment removes the RNA from the DNA/RNA solution. However, it only converts the large RNA molecules into mono- and oligonucleotides by cleavage at pyrimidines and it is actually the subsequent precipitation with isopropanol that causes partial separation of these mono- and oligonucleotides from the DNA. Alternatively, small RNA molecules can be also separated from large DNA molecules by passage of the nucleic acid solution through BioGel A15 (Bio-Rad,) or using commercially available columns (see Section 3.2).

Step **14**. In standard 0.7% agarose gel electrophoresis, the uncleaved DNA, containing DNA fragments from 30 to over 500 kb, should run as a single band of slightly lower mobility than the 23.7 kb marker fragment of λ DNA cleaved with *Hin*dIII (*Figure 1*).

3.2 Alternative medium scale procedure using a commercially available kit

This medium scale procedure using Quiagen columns permits separation of RNA from DNA. The kit (Diagen) allows the preparation of 60 μg of easily cleavable total yeast DNA from 20 ml cultures. Protoplast formation, lysis, and proteinase K treatment follow the standard protocol except that a different detergent is used instead of SDS and that RNase treatment precedes the incubation with proteinase K. The mixture of DNA and degraded RNA is passed over a column which binds only DNA. Upon washing and elution, essentially RNA-free DNA is obtained. We have tested this procedure only once in our laboratory and found it easy to perform, yielding the expected amount and quality of DNA. It seems possible that up to 12 preparations can be made in one day.

4. Fast small scale isolation of yeast total DNA

The procedure described in this section was designed for the simultaneous and rapid preparation of yeast genomic DNA from many samples. Up to 24 samples can easily be handled in parallel and the DNA preparation is sufficiently clean that most restriction enzymes, used to fragment this DNA, work satisfactorily well. The prepared DNA can be used in analytical or preparative methods such as Southern blot analysis, polymerase chain reaction (PCR) after additional shearing, or transformation of *Escherichia coli* (provided that a shuttle vector is present in the preparation).

Protocol 3. Rapid small scale isolation of DNA[a]

1. Grow a yeast culture overnight in 2 ml of YPD (Appendix 1) at 30 °C to early stationary growth phase (~ 2×10^8 cells/ml)[a].
2. Transfer 1.3 ml of the culture to a 1.5 ml microcentrifuge tube, harvest the cells by centrifugation in a table top microcentrifuge at 6000 r.p.m. for 3 min, and decant the supernatant.
3. Resuspend the cells in 1 ml of sterile water by vortexing and proceed as in step **2**.
4. Resuspend the cells in 0.2 ml of protoplasting buffer (10 μl/ml 2-mercaptoethanol, 100 mM Tris–HCl, pH 7.5, 10 mM EDTA, pH 7.5, 0.2 mg/ml Zymolyase (20 000 units/mg)) by vortexing and incubate at 37 °C for 1–2 h with occasional inversion of the tube to prevent sedimentation of the cells. It is of particular importance to have a homogeneous suspension prior to step **5**. (Also see footnote *a*.)
5. Add 0.2 ml of lysis solution (0.2 M NaOH, 10 g/l SDS) and mix gently by inversion[a].

6. Incubate the sample at 65°C for 20 min and then cool it rapidly on ice. This treatment remarkably reduces the turbidity of the suspension if protoplast formation was correct.
7. Add 0.2 ml of 5 M potassium acetate (pH 5.4), mix gently by inversion, and incubate on ice for 15 min. Insoluble potassium dodecylsulphate and denatured proteins precipitate from the solution and will be removed in the next step.
8. Centrifuge for 3 min in a microcentrifuge at room temperature, and transfer the supernatant to a fresh tube without distorting the pellet[a].
9. Add 0.6 vol. of isopropanol to the tube, mix gently by inversion, and let stand at room temperature for 5 min to allow complete precipitation of DNA[a].
10. Centrifuge for 30 sec in a microcentrifuge and discard the supernatant.
11. Add 1 ml of 70% ethanol to the pellet, mix by inversion of the microcentrifuge tube, and let stand at room temperature for at least 10 min. This interval should be respected so that there is enough time for salt ions that are entrapped in the DNA aggregate to diffuse out.
12. Repeat the centrifugation as in step **10**.
13. Dry the pellet briefly and dissolve DNA in 20–50 μl of TE buffer (10 mM Tris–HCl, pH 7.5, 1 mM EDTA). Complete solution of the DNA can take some time. The process should be promoted by occasional gentle flipping of the microcentrifuge tubes with a finger. (Also see footnote *a*.)
14. The prepared DNA can then stored at 4°C.

[a] See Section 4.1 for comments on this protocol.

4.1 Comments on *Protocol 3*

Step **1**. When experimental constraints require the cultivation of yeast in minimal medium, it is advisable to increase the initial culture medium by a factor of four and to include 50 mM sodium phosphate, pH 6.4, in the medium. Harvest the cells at a density of ~ 5×10^7 cells/ml from the total culture volume. The resulting pellet is then treated as described in step **3**.

Step **4**. See comment on step **4** of *Protocol 1*.

Step **5**. Sorbitol, usually included in the protoplasting buffer as an osmotic stabilizer at 1.0 M concentration, is omitted from our buffer to foster spontaneous lysis when protoplasts have formed. The released cellular DNA is, however, sensitive to mechanical forces like vortexing or shearing forces. To

Figure 2. Agarose gel electrophoresis and Southern blot analysis of yeast DNA, isolated by the fast small scale procedure described in *Protocol 3*. 10 μl of genomic DNA per sample were subjected to *Hin*dIII/*Pst*I double digestion (10 units each) in the presence of 2 μg of RNase A for 8 h. (A) Ethidium bromide-stained restriction fragments separated in a 0.9% (w/v) agarose gel. λ lane: λ DNA cleaved with *Hin*dIII and *Eco*RI. Lane 1: YPH499, haploid wild-type strain. Lane 2: YAW1m, isogenic mutant strain of YPH499 whose genomic copy of *PMA1* was modified by an integrative transformation yielding a *PMA1* allele with partially duplicated *PMA1* sequences and a central *LEU2* marker (as illustrated in panel C). Lanes 3–5: YAWm1 transformed with centromeric plasmid (*CEN6, ARSH4, URA3*), containing Bal31-deleted *PMA1* fragments of different length in a *Pst*I site. Lanes 6–10: samples as in lanes 1–5, but from yeast strains of the opposite mating type. **Note:** strongly stained bands originate from 4.6 kb viral, double-stranded RNA, from *Hin*dIII–*Hin*dIII fragments of ribosomal RNA genes (6.2 and 2.7 kb, see *Figure 4*). (B) Southern blot analysis after blotting of the separated DNA fragments (A) on to a nylon membrane (GeneScreen Plus) by capillary forces (6 h), hybridized with a [^{32}P]CTP-labelled probe consisting of the 5101 bp fragment of the *PMA1* gene (11) and the 5067 bp fragment of its isogene, *PMA2* (12). Lanes 1–10 are as in (A). **Note:** intensities of the individual bands correlate well with the length of the complementary sequences between the particular DNA fragments and the probe. (C) Schematic illustration of *PMA1* and *PMA2* wild-type alleles and the modified *PMA1* allele obtained after integrative transformation. Restriction enzymes are abbreviated as follows: H, *Hin*dIII; Ps, *Pst*I; S, *Sal*I.

prevent elevated DNA damage, it is therefore of *crucial* importance to handle the suspension as gently as possible after Zymolyase has been added.

Step **8**. In some cases, the precipitate of insoluble potassium dodecyl sulphate and denatured proteins that forms after the addition of potassium acetate does not produce a tight pellet after centrifugation. Therefore, great care should be taken during transfer of the supernatant to a fresh tube not to contaminate it with particles originating from the pellet. Rather it is preferable to leave some supernatant or, if it does become contaminated, to centrifuge the sample once again.

Step **9**. The precipitation of large DNA molecules is a dynamic process that requires some period to complete. Similarly, incubation at room temperature is preferable to low temperature incubation since low temperature enhances the precipitation of salts.

Step **13**. Excessive drying of the pellet will lead to strong dehydration of the DNA, resulting in insoluble DNA aggregates. One should therefore *not* dry the pellet excessively (e.g. no longer than 3 min of vacuum in a Speed-Vac concentrator (Savant Instruments)). A good indicator for dehydration is the colour of the resulting genomic DNA pellet. A white, non-transparent pellet indicates immoderate and irreversible dehydration.

5. Recommended volumes for further experimental utilization of DNA

(a) Use 10–20 μl in Southern blot analysis. 10 units of restriction enzyme(s)

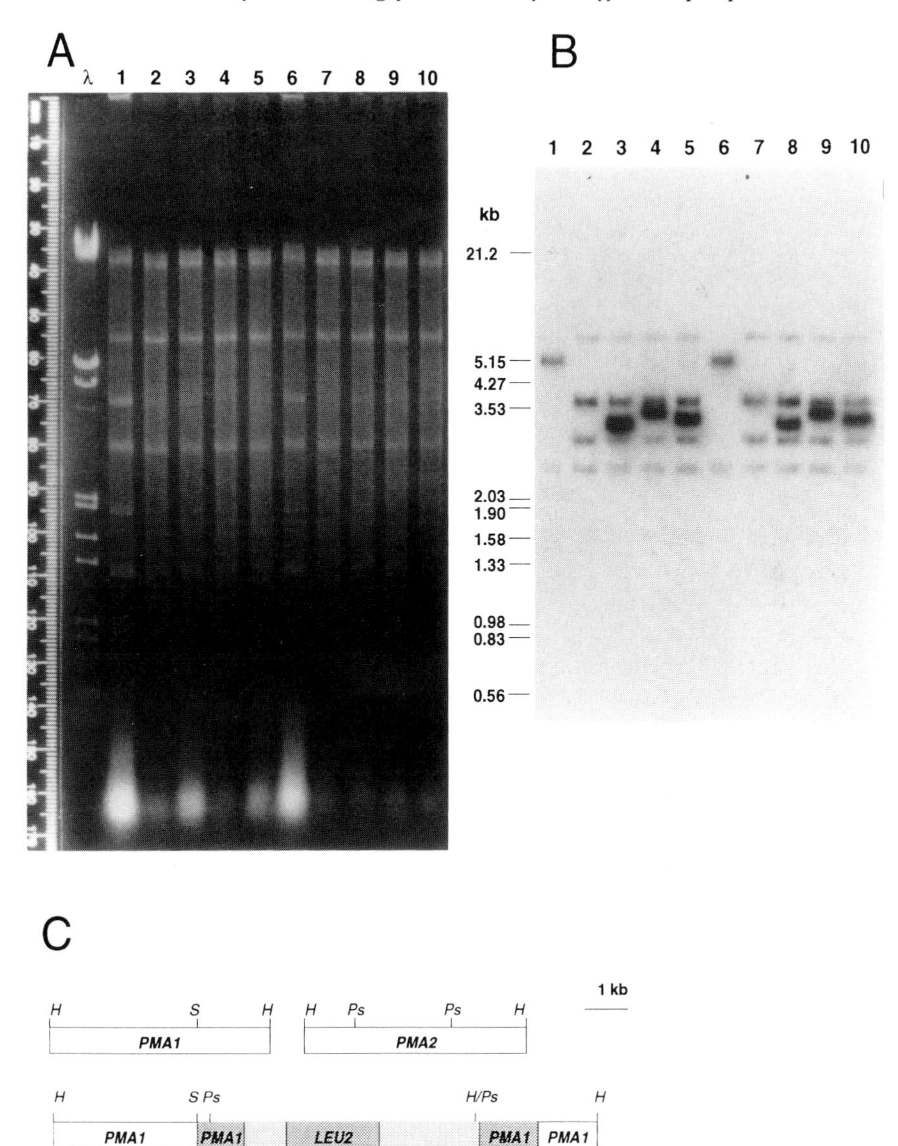

are usually sufficient to cut the yeast DNA completely within 8–12 h. Moreover, RNase A activity in different restriction enzyme buffers is high enough to ensure complete cleavage of the sample RNA during this time. Thus, simultaneous incubation of the sample with restriction enzyme(s) and RNase A (routinely 2 μg are added) is encouraged. Examples of Southern blot analyses of samples prepared by the small scale procedure are shown in *Figures 2* and *3*.

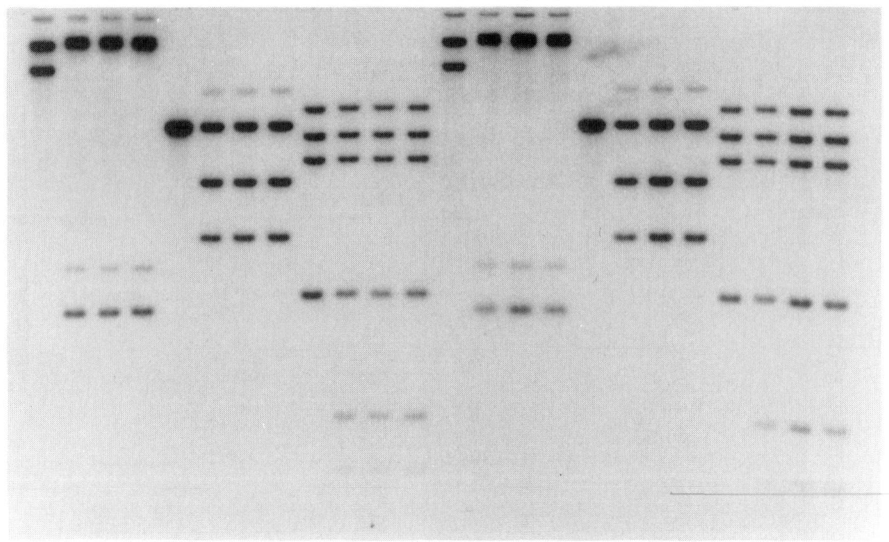

Figure 3. Screening of yeast transformants, obtained after integration of a modified *PMA1* allele by homologous recombination, by Southern blot analysis with *PMA1* and *PMA2* probes (as described in *Figure 2*). As a result of the integration, the *PMA1* allele was modified (see *Figure 2C*) such that additional restriction sites (*Bst*EII, *Hin*dIII, and *Xba*I) were placed directly downstream of the *PMA1* stop codon (position of the *Pst*I site in *Figure 2C*). DNA was isolated from wild-type and from six independent transformants. The following restriction enzymes were used to digest 10 μl of sample DNA: lanes 1–4 and 13–16, *Bst*EII; lanes 5–8 and 17–20, *Hin*dIII; lanes 9–12 and 21–24, *Xba*I. **Note**: the reproducibility of the preparation is fine and at least those enzymes employed here clleavage the isolated DNA to completion.

(b) Employ 1–5 μl and serial dilutions thereof for PCR amplification of specific DNA regions. Additional shearing of the DNA by mechanical forces before the PCR reaction is recommended. PCR can, for example, be used as a rapid means of cloning yeast genes or for determining the mating type of cells. In inspecting the latter, one might also consider the utilization of intact yeast cells in the PCR reaction (9).

(c) Apply 5–10 μl to transform *E. coli* when transfer of yeast–*E. coli* shuttle vectors is desired.

6. Interpretation of band patterns in restriction spectra of S. *cerevisiae* DNA

The 16 linear chromosomes of a haploid *S. cerevisiae* genome contain a total of 15 000 kb (10). Approximately 15% of this DNA belongs to several classes

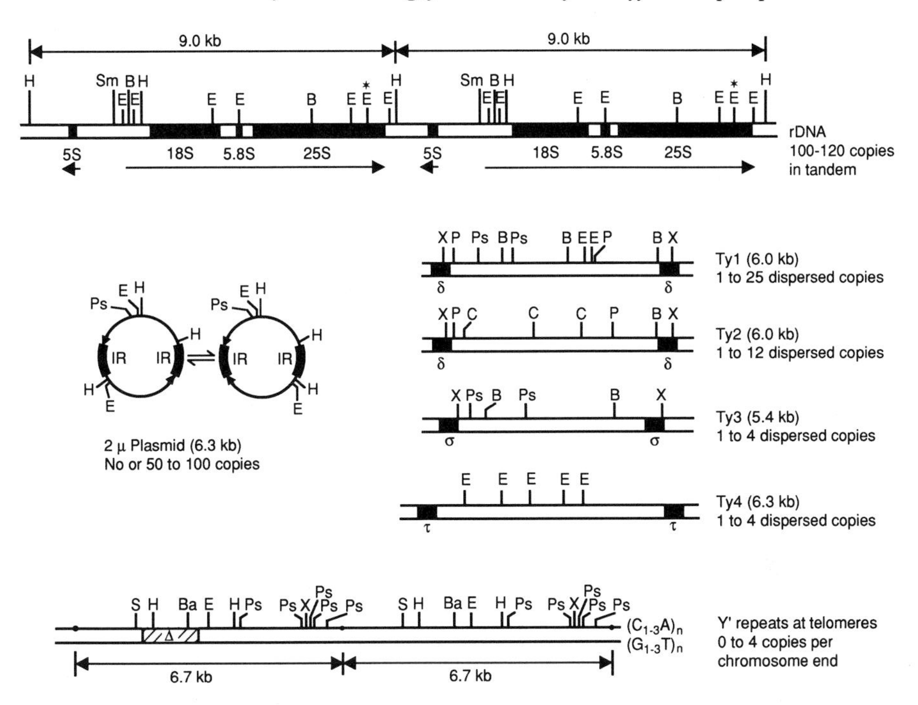

Figure 4. Repeated sequences in nuclear DNA of *S. cerevisiae*. Repeat unit lengths and positions of representative restriction sites are shown for the cluster of ribosomal DNA (13), for the dispersed copies of transposable elements (14), for the single or tandem Y′ telomere repeats (15), and for the 2 μm plasmid (8). The arrows underneath the rDNA units represent the primary transcription products. The short direct terminal repeats of Ty elements (δ, γ, τ) are present at many locations as so-called solo-copies. The 0.6 kb inverted repeats (IR) of the 2 μm plasmid act as efficient recombination sites leading to the formation of two isomeres and also to dimers and trimers. The region marked Δ in the Y′ repeat is deleted in some copies. Abbreviations for restriction sites are as follows: B, *Bgl*II; Ba, *Bam*HI; C, *Cla*I; E, *Eco*RI; H, *Hin*dIII; P, *Pvu*II; Ps, *Pst*I; and X, *Xho*I. E* marks an *Eco*RI site in ribosomal DNA which is cleaved inefficiently.

of repeated DNA. Non-chromosomal repeated DNA molecules are mitochondrial DNA (3) (75–80 kb with 20–40 copies per cell) and 2 μm plasmids present in many *S. cerevisiae strains* (8) (6.3 kb, 50–100 copies). When total DNA is cleaved with restriction nucleases and separated by gel electrophoresis, fragments originating from repeated DNA can be seen as discrete bands in the background of several thousand fragments (see *Figures 1C* and *D* and *2*). A summary of the classes of tandemly or dispersed repeated DNA units in nuclear DNA of *S. cerevisiae* is shown in *Figure 4*. The pattern of ribosomal DNA fragments (100 copies each) can in many cases be used to decide whether cleavage is complete or not, since the pattern of completely cleaved rDNA is predictable from the known sequence. This may save a lot of time.

The DNA sequences of *S. cerevisiae* ribosomal DNA are accessible in the EMBL/GenBank/DDBJ nucleotide sequence databases (5S rRNA, accession number (AC) X00486; 18S rRNA, AC M35588 and V01335; 5.8S rRNA and non-transcribed spacer, AC M38503; 25S rRNA, AC J01355). A more complete discussion of restriction digestion patterns including weak discrete bands and a comparison of the ribosomal DNA cleavage pattern of different yeasts has recently been published (5).

References

1. Cryer, D. R., Eccleshall, R., and Marmur, J. (1975). *Methods Cell Biol.*, **12**, 39.
2. Williamson, D. H. and Fennell, D. J. (1975). *Methods Cell Biol.*, **12**, 335.
3. Fox, T. D., Folley, L. S., Mulero, J. J., McMullin, T. W., Thorsness, P. E., Hedin, L. O., and Costanzo, M. C. (1991). In *Methods in Enzymology*, Vol. 194, (ed. R. W. Davis, D. Botstein, and J. R. Roth), pp. 149–65. Academic Press, San Diego.
4. Davis, R. W., Botstein, D., and Roth, J. R. (1980). In *Advanced bacterial genetics* (ed. R. W. Davis, D. Botstein, and J. R. Roth), pp. 211–14. Cold Spring Harbor Laboratory Press, Cold Spring Harbor, NY.
5. Philippsen, P., Stotz, A., and Scherf, C. (1991). In *Methods in Enzymology*, Vol. 194, (ed. R. W. Davis, D. Botstein, and J. R. Roth), pp. 169–82. Academic Press, San Diego.
6. Icho, T. and Wickner, R. B. (1989). *J. Biol. Chem.*, **264**, 6716.
7. Wickner, R. B. (1991). In *The molecular and cellular biology of the yeast Saccharomyces* (ed. J. R. Broach, J. R. Pringle, and E. W. Jones), pp. 263–96. Cold Spring Harbor Laboratory Press, Cold Spring Harbor, NY.
8. Broach, J. R. and Volkert, F. C. (1991). In *The molecular and cellular biology of the yeast Saccharomyces*, (ed. J. R. Broach, J. R. Pringle, and E. W. Jones), pp. 297–331. Cold Spring Harbor Laboratory Press, Cold Spring Harbor, NY.
9. Huxley, C., Green, E. D., and Dunbaum, I. (1990). *Trends Genet.*, **6**, 236.
10. Carle, G. F. and Olson, M. V. (1985). *Proc. Natl Acad. Sci. USA*, **83**, 3756.
11. Serrano, R., Kielland-Brandt, M. C., and Fink, G. R. (1986). *Nature*, **319**, 689.
12. Schlesser, A., Ulaszewki, S., Ghislain, M., and Goffeau, A. (1988). *J. Biol. Chem.*, **263**, 19480.
13. Warner, J. (1989). *Microbiol. Rev.*, **53**, 256.
14. Boeke, J. D. and Sandmeyer, S. B. (1991) In *The molecular and cellular biology of the yeast Saccharomyces*. (ed. J. R. Broach, J. R. Pringle, and E. W. Jones), pp. 193–261. Cold Spring Harbor Laboratory Press, Cold Spring Harbor, NY.
15. Louis, E. J. and Haber, J. E. (1992). *Genetics*, **1331**, 559.

2

Construction of cloning and expression vectors

CHRISTOPHER HADFIELD

1. Introduction

Numerous different types of cloning and expression vectors have been made for yeast and the majority are widely distributed amongst the international yeast research community. In the main, these are shuttle vectors that are able to propagate, and be selected for, in both yeast and *Escherichia coli*. This dual host capability has enabled yeast vectors to be constructed on the back of existing *E. coli* vectors and, in addition, has enabled recombinant derivatives to be made using efficient and powerful *E. coli* host-vector cloning strategies before introduction of the finished product into yeast.

Although there exists a wide selection of yeast vectors, it is not uncommon that the exact vector required for a particular purpose has not been made and so a tailor-made vector has to be constructed. Perhaps, for example, this will involve introducing a new marker or promoter, or conversion from a multi-copy vector to a single-copy type. Thus, besides knowing how to employ existing vectors, it is also necessary to know how to make new ones, when circumstances dictate.

In this chapter, the composition and properties of different types of yeast vectors will be considered in relation to performance and choice. Additionally, how different component elements of vectors may be manipulated will be examined. One of the most common types of vector manipulation undertaken is to create a heterologous gene construction to obtain efficient expression in yeast. The options available for making such constructions will be considered, along with the circumstances under which they may be employed.

2. Types and properties of yeast vectors

Although there are increasingly large numbers of yeast vectors being produced, the majority of these can be regarded as variations of a basic set of vector types. Many developments in yeast vectors do not involve the yeast elements themselves, but, since they are chimeric plasmids, simply reflect improvements in cloning technology found in *E. coli* vectors. In broad terms, yeast vectors fall into two classes: those that do not replicate themselves in yeast, but integrate into a chromosome and are passively inherited, and those that replicate autonomously. Generally, integrating vectors are inherited with very high stability, whereas autonomous plasmids tend to be unstable to varying degrees, depending upon composition.

An *E. coli* plasmid carrying a yeast auxotrophic marker constitutes a basic yeast integrating plasmid, which, upon transformation into yeast, will integrate via homology of the yeast plasmid marker with its chromosomal counterpart. Such integrative transformation is rare (1–10 transformants per μg of DNA), but can be made a thousand-fold more frequent by cutting the plasmid within the yeast DNA *in vitro* prior to transformation, as the ends produced are highly recombinogenic (1). An integrative vector can also be derived from a cloned yeast chromosomal DNA fragment, which can be liberated from the plasmid DNA by *in vitro* endonuclease digestion and used directly to transform yeast. The ends of such fragments are likewise highly recombinogenic; the transforming DNA will transplace its chromosomal homologue and, by inserting a cloned gene within it, can thereby serve as a vector.

Introduction of an origin of replication into an integrating-type plasmid also increases transformation efficiency. Such replicating plasmids may carry an autonomous replication sequence (*ARS*), or an *ARS* plus a centromere, or all or part of the endogenous *Saccharomyces cerevisiae* 2 μm plasmid. Their performance as vectors is fundamentally influenced by the differing nature of these elements.

These simple vector types form the basis of the majority of yeast vectors. They have been described in various respects in several articles (e.g. 2–7). The characteristics of the basic yeast vectors are briefly considered below.

2.1 2 μm-based plasmid vectors

These are the general work-horse vectors for yeast, being multicopy, highly (but not completely) mitotically stable, and freely replicating. Such vectors transform efficiently and, due to their episomal nature, can be recovered again from the yeast cell, if required. This latter property makes them useful for library screening; however, one potential drawback in this regard is that they can result in the isolation of multicopy suppressors as well as true complementing clones.

Early vectors, such as pJDB248 (8), utilized the whole 2 μm plasmid, with the yeast marker and *E. coli* plasmid elements inserted at the single *Pst*I site located in the *RAF* (or *D*) gene (*Figure 1*). Loss of RAF function does not impair 2 μm stability to much extent, so such vectors are still highly stable. However, in order to make smaller vectors, with more usable cloning sites and more efficient for cloning, smaller fragments of 2 μm were employed in subsequent vectors. The minimal portion of 2 μm consists of just the *ORI-STB* region. The origin (*ORI*) provides replication, whilst *STB* is the *cis-*

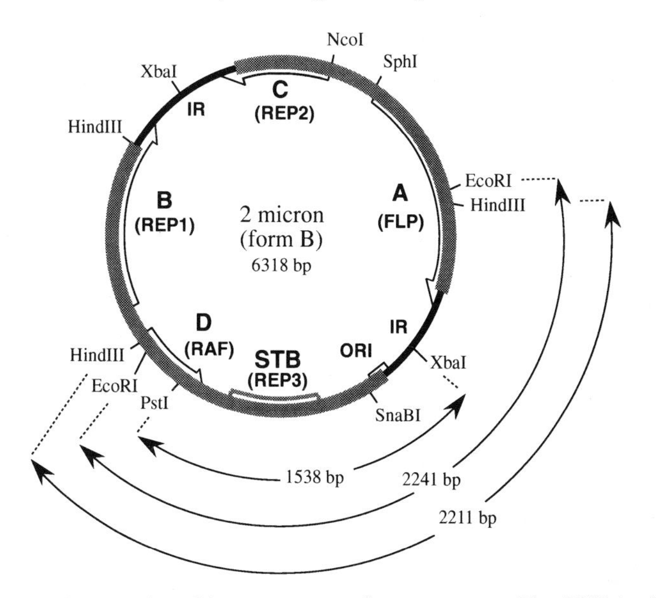

Figure 1. Origin of 2 μm plasmid components of yeast vectors. The 2241 bp *Eco*RI and 2211 bp *Hin*dIII *ORI-STB* fragments can only be derived from the B-form of the plasmid. (A-form not shown.)

acting element necessary for partitioning. With these vectors, the REP1, REP2, and RAF functions are supplied *in trans* by endogenous 2 μm plasmids, to facilitate heritable stability. Absence of the two inverted repeats, which are the site of action of FLP, results in the absence of copy number amplification. Typically, there is about a 1% non-selective loss per generation of such plasmids, and copy number is usually in the range of 20–40 per cell (9–11).

More recent developments with these vectors have focused on the desirability of high stability over a long period of non-selective growth. As a consequence, vectors containing the whole 2 μm plasmid have re-emerged into fashion. With all elements intact, these can not only exhibit higher stability, but can be used in 2 μm-free host cells, so that the vector can occupy all of the cell's plasmid-bearing potential. The 2 μm plasmid has a very densely packed genome and the choice of non-disruptive insertion sites is very limited. A unique *Sna*BI site between *ORI* and *STB* (*Figure 1*), a non-transcribed region, has been successfully used for inserting cloned DNA (12), although the inserted foreign DNA can reduce heritable stability (13).

The problem of incomplete plasmid vector stability has also been addressed in a different way by the development of autoselection systems, in which retention of the plasmid is essential for the continued growth and survival of the cells. For example, Bussey and Meaden (14) employed a plasmid-borne gene that encoded yeast killer toxin and immunity. Cells that lose the plasmid also lose immunity and are then killed by the toxin secreted by the plasmid-containing cells. Another example utilizes *ura3 fur1* strains as hosts for plasmids carrying the *URA3* gene, as described by Loison (Chapter 11).

2.2 *ARS* plasmids

The replication origins of these plasmids occur within some cloned yeast chromosomal DNA fragments; for example, *ARS1* occurs at the 3′ end of the *TRP1* gene of *S. cerevisiae*. The inheritance of *ARS* plasmids is usually highly unstable, either with or without selection; typically in the order of 10% loss per generation. Plasmid-containing cells may harbour hundreds of plasmids, reflecting the fact that the defect in inheritance is not in replication, but in segregation to daughter cells in budding yeasts. However, the plasmids can become stably inherited by integrating into the chromosome via the homology with its auxotrophic marker, and such integrations can be multiple copy. With consideration, *ARS* plasmids can be used adequately as general purpose vectors, including the construction and screening of genomic libraries (15).

2.3 Centromere-containing vectors

Presence of a centromere ensures low copy number, typically one (or sometimes two) plasmid(s) per cell. Such plasmids show high mitotic stability (<1% loss per generation), which appears to increase with greater size (16, 17). In these vectors, replication is provided by an *ARS*, whilst stability and low copy number are mediated by the centromere (*CEN*). Circular *CEN* shuttle plasmids are particularly useful as vectors for libraries (genomic or cDNA), enabling complementing clones to be isolated without the problem of multicopy suppression (18). Since they are independent replicons, *CEN-ARS* plasmid clones can be readily recovered from yeast cells.

Linear *CEN-ARS* plasmids having telomeres at either end form artificial chromosomes (or YACs), which may accommodate very large pieces of cloned chromosomal DNA (e.g. 19). These provide a means of cloning large fragments from large genomes.

2.4 Integrative vectors

Integrative vectors have the merit of high heritable stability without selection, but usually integrate in single-copy. This is a disadvantage if high-level expression is required. However, multiple integration of integrative plasmids can occur at a target locus and, despite the duplication of the flanking yeast DNA that occurs with such integrants (1), they are highly stable (<0.1% non-selective loss per generation). For efficient transformation, such vectors usually contain a unique restriction site within the yeast chromosomal fragment DNA, used for targeting the integration.

Integrative fragments replace the homologous counterpart in the chromosome (termed transplacement), typically as a single copy, and are fully stably maintained without loss in non-selecting conditions. Positioning the marker and cloned gene outside of any genes within the yeast chromosomal fragment ensures that the transplacing fragment does not impair a yeast function. Integrative fragment vectors can also be targeted for multicopy integration, thereby combining stability with high copy number. This is achieved by targeting integrations to DNA that is multiply reiterated in the genome, such as the ribosomal DNA repeats (21), transposable elements (Ty) (21–23), or δ sequences (24).

3. Basic vector components

3.1 *E. coli* plasmid component

The 'backbone' of yeast shuttle vectors is an *E. coli* plasmid replicon and selectable marker. Typically, this is usually pBR322, or sometimes its antecedents in some early vectors, or its descendants, such as pUC or pIC series plasmids, in more recent vectors. One advantage of the pUC-type of plasmid is that it replicates to a higher copy number in *E. coli* than pBR322, so that plasmid DNA yields are higher. Common to these plasmids is the presence of ampicillin-resistance as a selectable marker for *E. coli*.

3.2 Yeast auxotrophic markers

The most commonly used yeast selectable markers are auxotrophic markers, and the most frequently used of these are *LEU2, HIS3, TRP1*, and *URA3*. These can be subcloned and manipulated in the following restriction fragments: *LEU2*, as a 2218 bp *Sal*I–*Xho*I fragment; *HIS3*, as a 1770 bp *Bam*HI fragment; *TRP1*, as a 1459 bp *Eco*RI fragment also containing *ARS1*, or as a smaller *Eco*RI–*Pst*I (or –*Stu*I or –*Bgl*II) fragment lacking *ARS1* and transcription terminator activity; and *URA3* as a 1170 bp *Hin*dIII fragment. Sequence details can be obtained from the EMBL or GenBank databases.

Each of these genes produces a homologue product that can functionally substitute for the corresponding *E. coli* enzyme. Furthermore, they are sufficiently well expressed in *E. coli* to enable clones to be directly selected by complementation (see ref. 3). This is a highly useful feature for vector construct manipulation.

The defective leucine marker *leu2*-d is a useful variant, present in many vectors, that can be used to provide ultra-high copy number. Lacking the *LEU2* promoter, this marker is expressed at a very low level and, in order to provide enough β-isopropylmalate dehydrogenase (the *LEU2* product) to facilitate growth, vectors carrying *leu2*-d for selection must be present at very high copy number (≥200 copies per cell). However, the selection pressure also results in increased instability of episomal plasmids (9), as a subpopulation of cells is constantly being selected. The *leu2*-d allele was cloned as a randomly sheared 1.2 kb fragment using poly(dA/dT) homopolymer tailing, inserted into the *Pst*I site of 2 μm in pJDB219 (8) and subsequent derivatives. An alternative, similar marker is *ura3*-d, whose use is described by Loison (Chapter 11). It can be difficult to obtain primary transformants with such defective markers and some vectors include a second, non-defective marker for primary selection to avoid this problem.

3.3 Yeast dominant markers

Dominant markers are also becoming more widely used. Of these, antibiotic-resistance determinants are easy to work with as they can function in both yeast and *E. coli*. Chloramphenicol-resistance (conferred by a 1.8 kb *Sal*I fragment (10)) can be directly selected in both yeast and *E. coli* (see *Table 1*). G418-resistance (11) can be directly selected in yeast and as a secondary resistance phenotype in *E. coli*. Transformants of *E. coli* are first selected as ampicillin-resistant using the pBR322/pUC plasmid determinant and then replica plated on to kanamycin or neomycin plates (for cost-saving) to select a sub-population of cells that carry the G418-resistance determinant (see *Table 1*). Such markers can be introduced into polylinker sites to increase the number of sub-cloning options, as exemplified in *Figure 2*.

Table 1. Selection conditions for shuttle antibiotic-resistance markers[a]

	Concentration of antibiotic used for selection	
Marker	**Yeast**[b]	***E. coli***[c]
Cm^R	haploid strains: usually 2 mg/ml polyploid and industrial strains: usually > 2 mg/ml	10 μg/ml
$G418^R$	haploid strains: usually 0.5 mg/ml polyploid and industrial strain: usually < 0.5 mg/ml	25 μg/ml[d]

[a] References: Hadfield *et al.* (38, 11).
[b] Medium used is YPGE (see Appendix 1). Yeast strains vary in their sensitivities to antibiotics, so inhibitory concentrations need to be individually determined.
[c] Medium used is L broth (see Appendix 1).
[d] Less expensive kanamycin or neomycin can be substituted for G418 for use with *E. coli.*

3.4 Yeast replicons

3.4.1 2 μm derived

The minimum-sized 2 μm plasmid *ORI-STB* fragment used for incorporation into replicative vectors is generally the 1538 bp *Pst*I–*Xba*I fragment. Alternatively, in many vectors the slightly larger *Eco*RI or *Hin*dIII fragments are employed (*Figure 1*).

3.4.2 *ARS*

ARS1, occurring in the *TRP1*-containing *Eco*RI fragment, is the most commonly utilized *ARS* element. Another useful *ARS* element is *ARS2*, which can be sub-cloned on a 0.6 kb *Xho*I fragment (26). Other *ARS*s may be used, but they are not in such a convenient form.

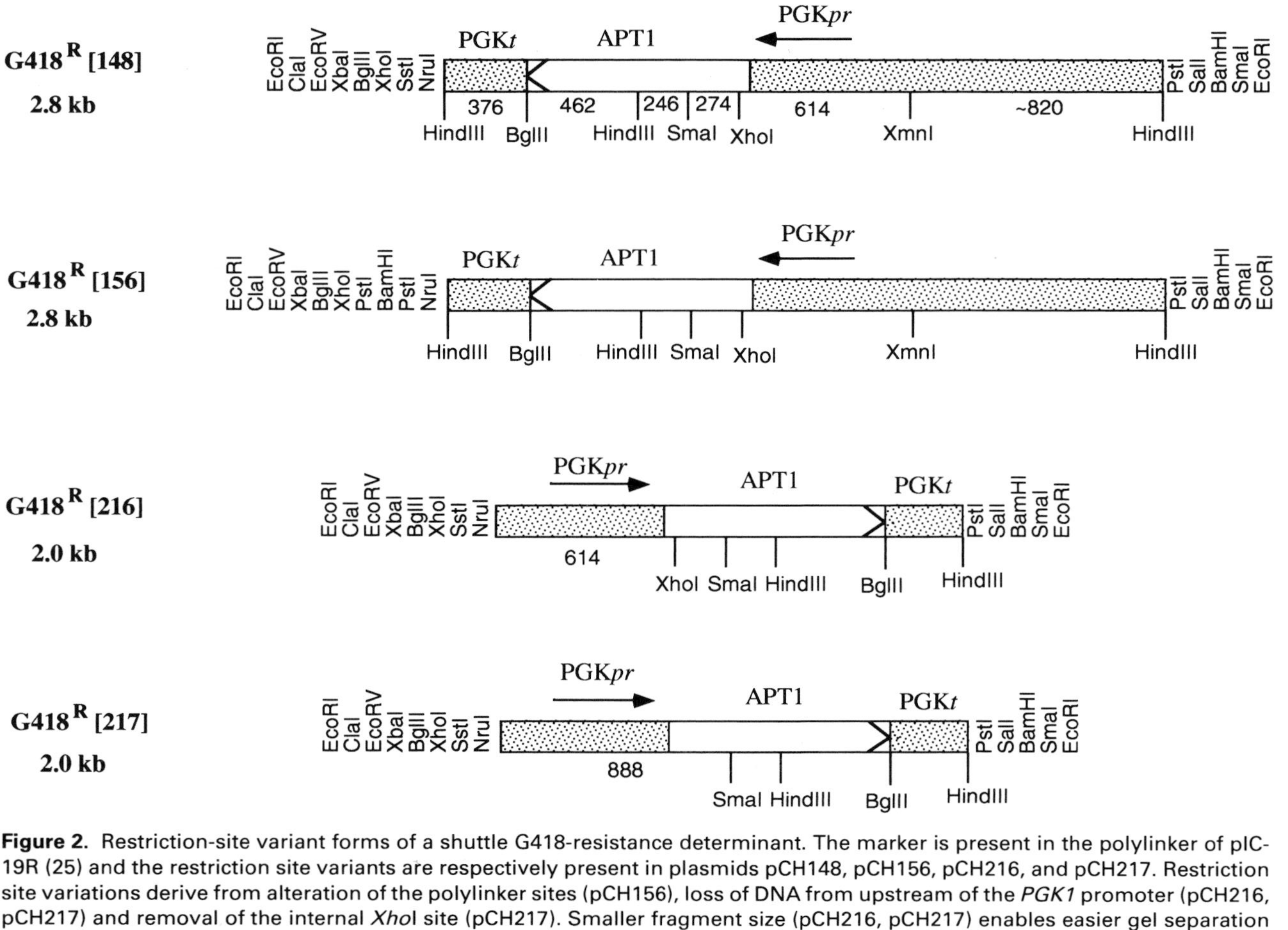

Figure 2. Restriction-site variant forms of a shuttle G418-resistance determinant. The marker is present in the polylinker of pIC-19R (25) and the restriction site variants are respectively present in plasmids pCH148, pCH156, pCH216, and pCH217. Restriction site variations derive from alteration of the polylinker sites (pCH156), loss of DNA from upstream of the *PGK1* promoter (pCH216, pCH217) and removal of the internal *Xho*I site (pCH217). Smaller fragment size (pCH216, pCH217) enables easier gel separation from the pIC-19R band for purification.

3.4.3 Centromeres

Like *ARS* elements, centromeres are small elements that can be sub-cloned as small fragments or within larger ones. *CEN3* and *CEN11* are commonly used centromeres that can be manipulated as 628 bp and 858 bp *Sau*3A fragments, respectively (27, 4). These conveniently sized fragments can be derived from larger ones present in plasmids like pYe(CEN3)11 or pYe(CEN11)12 (27).

Centromeres provide stability and low copy number, but do not provide replication, and so must be used in conjunction with a replication origin. Thus, for example, vector YCp50 contains *CEN4* and *ARS1* (28).

4. Basic vector construction methodology

The manipulation of vectors—whether it be in initial construction or later modification—requires precise and calculated steps. Starting with a precursor plasmid, a new component DNA fragment may be ligated in, the new recombinant plasmid identified, and then reisolated as DNA for the next step. The final product is thereby arrived at in a stepwise manner. The major procedures involved in this process are the isolation of specific DNA fragments from gels and the cloning of them into the burgeoning vector construction.

4.1 The isolation of DNA fragments from gels

For agarose gels, Tris–acetate electrophoresis buffer is used, the gel is stained with ethidium bromide, and the DNA bands are viewed under UV illumination in the standard way (29). Using a scalpel, the appropriate fragment is then excised in a minimum-sized gel slice, and this is then processed to extract the DNA. There are numerous methods for doing this. Two generally reliable procedures for isolating DNA from agarose gel slices are commercially available from BIO 101 and Promega (see *Protocol 1*A and B). Such fragments will ligate with plasmid vector with reasonable efficiency, although care must be taken to follow the clean-up steps scrupulously, as agarose is notorious for inhibiting ligation. In cases where two gel-isolated fragments are to be ligated together, this is usually very difficult to achieve with fragments isolated from agarose gels. For this purpose, it is better to use fragments isolated from polyacrylamide gels (see *Protocol 2*), as traces of polyacrylamide do not inhibit ligation. Although polyacrylamide gels are usually used for the separation of small fragments (<400 bp), it is possible to isolate fragments of several kilobases from them by using the minimum concentration, 3.5%, and simply running them for longer.

4.2 Cloning manipulations

Successful cloning of DNA fragments needs an understanding of the factors governing the molecular events involved. Indeed, use of optimal calculated cloning conditions is essential for the successful production of recombinant

Protocol 1. Purification of agarose gel-separated DNA fragments

1. Excise DNA band from agarose gel in minimum-sized gel slice and transfer to a preweighed microcentrifuge tube.
2. Weigh tube plus contents and calculate weight of gel slice. Assuming that the gel has a density of 1 mg/μl, add 3 volumes of 6 M sodium iodide (final volume should be 300–500 μl if intending to process by the Magic™ system). Incubate at 37–55 °C for 5–15 min until the agarose has melted.
3. Process with GENECLEAN™ or Magic™ as described below.

A. *GENECLEAN™ (BIO 101)*

1. Resuspend the GLASSMILK suspension by vortexing vigorously, and add 10 μl per μg of DNA. Vortex and incubate at room temperature for 10 min, mixing occasionally.
2. Spin at 13 000 r.p.m. in a microcentrifuge for 30 sec. Discard the supernatant.
3. Wash the pellet by adding 500 μl of NEW wash (supplied with the kit), vortexing, and pelleting. Repeat three times.
4. Remove all traces of NEW wash. Resuspend pellet in 10 μl of TE buffer per μg of DNA. Incubate at 55 °C for 10 min.
5. Microcentrifuge at 13 000 r.p.m. for 30 sec. Remove supernatant, which contains eluted DNA, and transfer to a microcentrifuge tube.

B. *Magic™ (Promega)*

1. Add 1 ml of 'Magic DNA Clean Up' resin and mix by inversion.
2. Attach a 3 ml disposable syringe barrel to the Luer-lok extension of a 'Magic Minipreps' mini-column.
3. Pipette resin–DNA mix into the syringe barrel attached to the mini-column and push the mixture into the column using the syringe plunger.
4. Wash with 2 ml of Column Wash Solution (made up to 50% (v/v) with ethanol prior to use) by removing the mini-column from the syringe and taking up the solution in the syringe. Reattach the syringe to the mini-column and gently push the solution through the mini-column with the syringe plunger.
5. Transfer the mini-column to a microcentrifuge tube, and spin in a microcentrifuge for 20 sec to dry the resin.
6. Transfer the mini-column to a new microcentrifuge tube and apply 50 μl TE buffer (see Appendix 1) at 65 °C. Spin in a microcentrifuge for 20 sec to elute the DNA.

Protocol 2. Purification of a polyacrylamide gel-separated DNA fragment[a]

1. Incubate the gel slice (containing 1 μg of fragment DNA) in 2–3 volumes of elution buffer (0.5 M ammonium acetate, 1 mM EDTA (pH 8.0)) for 1 h at 37 °C.
2. Remove the elution buffer with a Gilson pipette and transfer to a polypropylene centrifuge tube, with a sealing cap, 10–50 ml capacity, depending upon total elution volume for ethanol precipitation, as described below.
3. Repeat steps **1** and **2** to give a total of six serial elutions.
4. To the pooled eluate, add 0.1 volume of 3 M sodium acetate pH 5.5 and 2 volumes of ethanol. Place the sealing cap on the centrifuge tube and mix the contents thoroughly by several inversions. Chill down in a dry ice/ethanol bath for 10 min.
5. Centrifuge at 43 000 *g*, 4°C for 30 min to pellet the precipitated DNA.
6. Pour off the supernatant and drain well; then add a similar volume of 70% ethanol. Replace the tube cap and gently invert a couple of times to ensure the pellet is washed.
7. Centrifuge at 43 000 *g*, 4°C for 10 min.
8. Pour off supernatant and drain well, blotting the rim of the inverted tube with clean tissue to remove non-draining 70% ethanol.
9. Dry completely under vacuum and resuspend the DNA in 400 μl of TE buffer.

[a] After Maxam and Gilbert (30).

clones, and minimizes the occurrence of double inserts or scrambles. It is essential to set up the cloning ligation with the correct concentration, and relative molar ratio, of vector and insert fragment. How to calculate these conditions is outside the scope of this article, but has been described elsewhere (31, 32). After selecting the appropriate concentration conditions for the ligation, it can be set up and undertaken as described in *Protocol 3*.

5. Expression constructs

Heterologous gene constructions, designed for efficiently expressing foreign gene products in yeast, utilize a yeast promoter, to provide transcription initiation and any required regulation, and a yeast terminator, which provides

Protocol 3. Cloning ligations

1. A typical volume for cloning a fragment into a plasmid vector is 20 μl; for a monomeric circularization the volume is often at least 100 μl; for a concatameric linear ligation the volume may be only 5 μl. If either of the input DNAs is too dilute, concentrate it by ethanol precipitation, wash with 70% ethanol, vacuum dry (as described in *Protocol 7*, for example) and resuspend in the required volume.
2. Set up the ligation in a microfuge tube on ice:
   (a) add vector and fragment DNAs (diluted in 5 mM Tris–HCl pH 7.6)
   (b) add ligation buffer to 1 × final concentration (0.1 volume of 10 × stock)
   (c) add T4 DNA ligase (5–7 Weiss units in up to 20 μl final volume is sufficient for blunt-end ligations, and may be reduced by half for sticky ends)
3. Incubate at 15 °C (cold-room waterbath) overnight (≥17 h).
4. Transform *E. coli* directly with an aliquot (e.g. half) [a]. Store the remainder at −20 °C.

[a] Also include a control (e.g. uncut vector DNA) to test that the competent cells are transforming efficiently. Failure to recover clones can often be due to poor cell competence rather than the clones not being present in the ligation mix.

discretely-sized mRNA and also can improve expression level by up to 10-fold (33). Generally, a yeast gene expresses an mRNA that encodes only one polypeptide, and the correct 'hook-up' between transcription and translation of the message is necessary to obtain efficient expression.

5.1 Translational fusion constructs

In this type of gene construction, the yeast promoter fragment provides both the transcription initiation and translation initiation sites, plus some amino-terminal codons. Fusion with the foreign protein-encoding sequence therefore generates a hybrid protein. This may create problems if the foreign protein requires a specific amino-terminal domain for activity. The *E. coli lacZ*-encoded β-galactosidase is highly tolerant of such fusion and has provided a mainstay approach for analysing promoter activity under different regimes. A useful vector for such analysis is shown in *Figure 3*. This type of fusion has also been well exploited for the isolation of new promoters; for example, using fusion to sequences encoding herpes simplex thymidine kinase (35) or phleomycin-resistance (36) as means of selecting for cloned fragments with promoter activity.

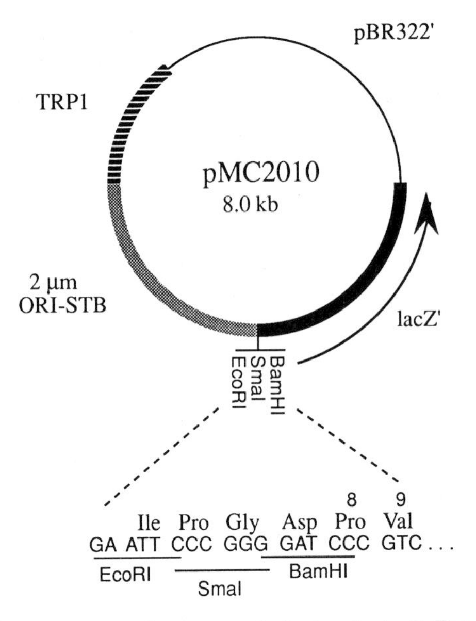

Figure 3. Analysis of promoter performance via fusion to *lacZ*. The 5′ end of the *E. coli lacZ* gene has been modified to provide cloning sites for translational fusion to β-galactosidase (34).

Another very important use of this type of construction is in the creation of fusions to the *MFα1* leader sequence to create constructs that facilitate secretion of foreign proteins. An excellent series of vectors for this purpose are pDP314, pDP315, and pDP316 (*Figure 4a*). Plasmid pDP314 contains the natural *Hin*dIII site occurring in the *MFα1* leader sequence. Fusion at this site results in the expression of a fusion protein with the full *MFα1 pre-pro*-EA_2 leader. The *pre* sequence encodes the secretion signal peptide; the two carboxy-terminal Lys-Arg residues of the *pro* sequence are recognized by the Golgi-located *KEX2* endopeptidase, which cleaves carboxy-terminally to the Arg residue, liberating an EA_2-mature protein product fusion. The two Glu-Ala dipeptides (EA_2) are then serially removed by STE13 exopeptidase to generate the mature product. However, it has been well documented that this latter processing is inefficient. This can be avoided by direct carboxy-terminal fusion to the KEX2 cleavage site. This has been provided in pDP315, via *in vitro* mutagenesis to introduce a *Stu*I site. Alternatively, if the *pro* sequence is not required, fusion directly to the *pre* peptide is possible with pDP316, in which an *Sph*I site has been introduced at the fusion junction by site-directed mutagenesis.

The creation of such precise in-frame junctions requires precise tailoring of the end of a DNA fragment encoding the protein sequence. There are two major methods available for doing this: PCR (polymerase chain reaction) and

Bal31 exonuclease digestion. PCR is the most modern method and, for precise tailoring, with ease and efficiency, is the method of choice. PCR requires the availability of a DNA thermocycler and primer oligonucleotides, based upon a known DNA sequence. Since two primers are needed to generate a PCR fragment, it is often convenient to have the substrate DNA fragment cloned into a pUC-type vector so that one of the primers can be a 'universal' M13 sequencing primer that will prime from the *lacZ* DNA in the vector. Thus, only the specific internal primer needs to be synthesized. This enables a PCR strategy as shown in *Figure 5a*. The internal PCR primer can be designed to generate a precise 5′ end for fusion (e.g. to the *Stu*I site in pDP315), whilst the 3′ end of the fragment can be generated by endonuclease cleavage within the polylinker sequence of the vector present in the fragment due to the external priming of the universal primer (e.g. at a *Bam*HI site, so that the 3′ fragment end will ligate to the *Bam*HI site in pDP315). The basic procedure for generating such fragments is described in *Protocol 4*.

If it is not possible to undertake PCR (e.g. because the sequence is not known), Bal31 provides a double-stranded exonuclease activity that will progressively remove terminal nucleotides in a time-course, so that the amount of terminal DNA removed can be quite finely judged (see *Figure 6b* and *Protocol 5*). Such digestion is not precise, however, so a spread of termini will be obtained and about 50% of them would be expected to be blunt-ended. Treatment with Klenow large-fragment DNA polymerase in the presence of deoxynucleotides will fill in any 5′-recessed ends, to increase the frequency of blunt ends to 75% (see *Protocol 6*). Alternatively, a higher frequency of flush ends can be generated by treatment with mung bean nuclease (see *Protocol 7*). Linkers can then be added to introduce terminal restriction sites (see *Protocol 8*). These are short double-stranded oligonucleotides with flush ends that encode a restriction endonuclease site. Ligating them to the ends of a blunt-ended DNA fragment therefore adds the restriction site to the termini. Subsequent cleavage with the appropriate restriction enzyme then generates an artificial restriction fragment with the appropriate ends.

Phosphorylated linkers ligate considerably more efficiently than non-phosphorylated linkers and so a method designed for their use is described. Ideally, linkers should be bought that have been chemically phosphorylated by the manufacturer, as enzyme-mediated phosphorylation is not as efficient. However, phosphorylated linkers are usually substantially more expensive than non-phosphorylated ones. If necessary, non-phosphorylated linkers can be phosphorylated by kinase-treatment as described in *Protocol 9*.

The Bal31-generated termini of the clones obtained will subsequently have to be analysed to find one with a suitable end-point. Initially, some clones may be discarded by fine restriction mapping. Ultimately, they will have to be sequenced across the deletion end-points. Likewise, it would be advisable to sequence a PCR fragment, although this could be a daunting task if it is large. For such cases, it would be better to use a less error-prone DNA polymerase

Protocol 4. Generating precisely defined DNA fragments by PCR[a]

1. (a) To a 0.5 ml microcentrifuge tube on ice add:
   - 26.5 μl water (double-distilled)
   - 5 μl 10 × *Taq* buffer (0.5 M KCl, 0.1 M Tris–HCl pH 9, 1% Triton X-100 ± $MgCl_2$ up to 100 mM[b])
   - 5 μl mixed dNTPs, each at 2 mM (dCTP + dGTP + dATP + dTTP)
   - 6 μl primer 1 (50 μg/ml[c])
   - 6 μl primer 2 (50 μg/ml)
   - 1 μl substrate DNA (1 ng/μl)
   - 0.5 μl *Taq* (or similar) DNA polymerase (5 units/μl)

   (b) Mix by flicking the bottom of the tube with a finger or by briefly vortexing. Then add approximately 50 μl of liquid paraffin, which will form an overlay. Microcentrifuge for 15 sec to bring all material to the bottom of the tube.
2. Place the tube in the thermal cycler and initiate the PCR. For the 17-mer universal primers of the pUC or pIC plasmids, typical PCR conditions are: denature at 94°C for 20 sec, anneal at 55°C for 20 sec, and extend at 72°C for 30 sec for a total of 25–30 cycles.
3. Remove 5 μl and run on a gel to check if the PCR has worked. Also run appropriate molecular weight markers in an adjacent gel track, in order to estimate the size of the PCR product. A single band of the expected size should be obtained from the PCR, as shown in *Figure 3* for example.

[a] Further details from ref. 37.
[b] $MgCl_2$ can influence the priming specificity, with different primers being influenced differently by changes in concentration. Start with a modest concentration of $MgCl_2$, such as 2.5 mM, and alter if non-specific primings occur.
[c] If using a 'universal' primer, these are available from Gibco-BRL, Pharmacia, or New England Biolabs.

Protocol 5. Fragment trimming by Bal31 exonuclease digestion

1. Bal31 will trim both ends of a linear molecule; so adopt a suitable strategy, as exemplified in *Figure 6a*.
2. The activity of the Bal31 will have been predetermined by the manufacturer. For example, with enzyme from New England Biolabs, 1 unit

deletes 200 bp from each end of Φ174 DNA (5.4 kb) at 650 μg/ml in 10 min at 30 °C (i.e. 20 bp deleted/min). Utilize this information to calculate the approximate amount of digestion required, keeping the molar concentration of the DNA molecules the same. Thus, for example, a 2.7 kb DNA molecule will have to be present at 325 μg/ml to be exo-digested to the same extent as the Φ174 DNA standard.

3. (a) Set up a number of timed Bal31 incubations, each in a constant volume, on ice as follows:
   - 4 μl DNA (equivalent of 6.5 μg of 5.4 kb DNA)
   - 5 μl 2 × Bal31 buffer (1.2 M NaCl, 24 mM $CaCl_2$, 24 mM $MgCl_2$, 40 mM Tris–HCl pH 8.0, 2 mM EDTA)
   - 1 μl Bal31 (1 unit per μl or 0.1 unit μl) (10 μl total volume)

   (b) Mix well. Microcentrifuge for a few seconds to bring down any droplets to the bottom of the tube. Incubate at 30 °C for a range of times either side of the number of minutes calculated to produce the required amount of exonuclease digestion.

4. To terminate the digestions, add 20 μl of phenol–chloroform (1:1 (v/v), equilibrated against TE buffer) and mix well for 30 sec (this denatures the protein).

5. Microcentrifuge for 2 min. Remove the upper aqueous phase with a Gilson pipette and transfer to a new microfuge tube.

6. Calculate the volume by taking it up again in the Gilson pipette tip; then remove a volume containing 1 μg of DNA. This is to be used for gel electrophoresis to check the extent of digestion.

7. Run an agarose gel with the timed digestions, including a zero minute control and a suitable size marker ladder. This is then used to assess the exonuclease digestions, as indicated in the example shown in *Figure 6(b)*.

8. Repeat the timed digests, focusing on a narrower time range, to obtain the precise extent of deletion required.

9. Clean up phenol–chloroform extracted DNAs by ethanol precipitation and washing, as follows:

   (a) Add 600 μl of phenol–chloroform and 300 μl of TE buffer; mix well for 30 sec.

   (b) Repeat step **5**.

   (c) Add 50 μl of 3 M sodium acetate pH 5.2 and 900 μl of ethanol. Mix together well.

   (d) Process as in *Protocol 8A*, steps **7–10**.

   (e) Resuspend in a suitable small volume of TE buffer (e.g. 1 μg/μl).

Protocol 6. Filling-in 3′-recessed ends with Klenow (large fragment) DNA polymerase to create blunt ends

1. Mix together in a microcentrifuge tube on the bench:
   - 16 μl of DNA fragment (1 μg made up to volume with 10 mM Tris–HCl pH 7.5)
   - 2 μl of 10 × Klenow buffer (0.1 M Tris–HCl pH 7.5, 50 mM $MgCl_2$, 75 mM dithiothreitol)
   - 1 μl of dNTPs mix (dCTP + dGTP + dATP + dTTP, each at 1 mM)
   - 1 μl of Klenow DNA polymerase (1–5 units/μl)
2. Incubate at 25–30 °C for 30 min.
3. Terminate the reaction by either heat inactivation (15 min in a 75 °C waterbath) or phenol–chloroform extraction (followed by ethanol precipitation to clean up, as described in *Protocol 5*, step **9**).

Protocol 7. Blunt-ending DNA with mung bean nuclease

1. Mix together in a microcentrifuge tube, on ice:
   - 10 μl of DNA fragment (1 μg made up to volume with water)
   - 1.2 μl of 10 × mung bean nuclease buffer (0.5 M sodium acetate pH 5.0, 0.3 M NaCl, 10 mM $ZnSO_4$)
   - 1 μl of mung bean nuclease (0.25 units/μl)
2. Incubate at 30 °C for 30 min.
3. Terminate the reaction by phenol–chloroform extraction (followed by ethanol precipitation to clean up, as described in *Protocol* 5, step **9**).

Protocol 8. Linker addition of a terminal restriction site

The linker ligation will be set up with a ratio of at least 10 linker molecules to every fragment molecule. One microgram of DNA fragment is used to give a convenient amount for subsequent handling procedures. The excess linkers act to prevent the blunt ends of the fragment DNA molecules from ligating either to themselves (circularizing) or to other fragment molecules (concatamerization).

A. *Linker ligation*

1. (a) Add to a 0.5 ml microcentrifuge tube on ice:
   - 1 μg blunt-ended DNA fragment[a]
   - 0.1 μg of phosphorylated linkers (8–12 bp)

- water to 25.5 μl
- 3 μl of 10 × ligation buffer (0.5 M Tris–HCl pH 7.6, 0.1 M $MgCl_2$, 0.1 M dithiothreitol, 10 mM ATP)
  1.5 μl T4 DNA ligase (5–7 Weiss units/μl)
  (total volume 30 μl)

(b) Ensure that the contents are mixed together, and then microcentrifuge for 3 sec to bring down droplets to bottom of tube.

2. Incubate at 15°C overnight.
3. Add 300 μl of phenol–chloroform (1:1 (v/v) equilibrated against TE buffer) and 120 μl of TE buffer. Mix thoroughly by shaking the tube for 1 min.
4. Microcentrifuge for 2 min at 13 000 r.p.m.
5. Remove the upper aqueous layer and transfer to a 1.5 ml microcentrifuge tube.
6. Re-extract the phenol–chloroform phase by adding 150 μl of TE buffer, mixing together by shaking, and repeating steps **4** and **5**; pool the aqueous phases.
7. Add 30 μl of 3 M sodium acetate pH 5.2 and 660 μl of ethanol. Mix together well, then place the tube in a dry ice/ethanol bath for 5 min.
8. Microcentrifuge at 13 000 r.p.m. for 15 min.
9. Pour off the supernatant, draining the tube and blotting the rim with clean tissue whilst holding the tube upside down. Add I ml of 70% ethanol and gently rock the tube back and forth a few times to wash the pellet.
10. Microcentrifuge at 13 000 r.p.m. for 5 min. Pour off liquid, drain, and blot as before. Dry under vacuum.

B. *Restriction of the linkers to yield fragments with one cleaved linker at each end*

1. Resuspend in 24 μl of TE buffer.
2. Add 3 μl of 10 × restriction buffer (as specified by the manufacturer) and 3 μl of restriction enzyme (10 units/μl). Incubate for 5 h at the recommended temperature (usually 37°C) to ensure that all linker restriction sites are digested.
3. Add 600 μl of phenol–chloroform and 270 μl of TE buffer. Mix thoroughly by shaking the tube for 1 min.
4. Microcentrifuge for 2 min at 13 000 r.p.m.
5. Remove the upper aqueous layer and transfer to a 1.5 ml microcentrifuge tube.

Protocol 8. *Continued*

C. *Removal of contaminating free linker oligonucleotides*

Small oligonucleotides are precipitated inefficiently by ethanol, whereas larger DNA fragments can be recovered at high efficiency by ethanol precipitation. This difference can be exploited to remove the linkers that did not ligate to the fragment DNAs.

1. Add 50 μl of 3 M sodium acetate pH 5.2 and 1 ml of ethanol. Mix together well, then place the tube in a dry ice/ethanol bath for 5 min.
2. Repeat steps A**8–10**.
3. Add 450 μl of TE buffer to the DNA pellet. Agitate the tube and then allow 5 min for the DNA pellet to redissolve.
4. Add 50 μl of 3 M sodium acetate pH 5.5 and 1 ml of ethanol. Mix together well, then place the tube in a dry ice/ethanol bath for 5 min.
5. Repeat steps A**8–10**.
6. Dry the pellet under vacuum. Resuspend in 10 μl of TE buffer.

[a] If the fragment is shorter than 120 bp, increase the quantity of linkers present to ensure a minimum 10-fold molar excess.

Protocol 9. Phosphorylation of linkers by kinase treatment

1. Dissolve 50 μg (1 absorbance unit at 260 nm) of non-phosphorylated linkers in 100 μl of TE (7.6) buffer (10 mM Tris–HCl pH 7.6, 1 mM EDTA pH 8).
2. (a) Mix in a microcentrifuge tube:
   - 3 μl of water
   - 4 μl of linkers (2 μg in total)
   - 1 μl of 10 × linker kinase buffer (0.7 M Tris–HCl pH 7.6, 0.1 M $MgCl_2$, 50 mM dithiothreitol)
   - 1 μl of 10 mM ATP
   - 1 μl of polynucleotide kinase (~10 units/μl)

   (b) Incubate at 37 °C for 30 min.
3. Then add:
   - 7 μl of water
   - 1 μl of 10 × linker kinase buffer
   - 1 μl of 10 mM ATP
   - 1 μl of polynucleotide kinase (~10 units/μl)

   Incubate at 37 °C for 30 min.
4. Store at −20 °C. Use aliquots directly in cloning ligations.

than *Taq*, such as Vent™ (New England Biolabs) or *Pfu* (Stratagene). Detection of a protein product of the expected size and having the expected activity, would be a good indication of success. When PCR works correctly, it gives rise to a single, clean DNA band visible on a gel. However, if the primers are not specific enough, several bands may be obtained on a gel (*Figure 5b*). With luck, non-specific bands may be eliminated by increasing the annealing temperature (*Figure 5b*) or altering the Mg^{2+} concentration. Such problems are outside of the scope of this article, but have been reviewed elsewhere (e.g. 37).

The creation of successful fusion constructs to the *MFα1* leader variants in pDP314, pDP315, and pDP316 are demonstrated in *Figure 4b*, in which production of elastase inhibitor, elafin, secreted in the culture medium, can be seen with each fusion.

5.2 Transcriptional fusion constructs

In this type of heterologous gene construction the coding sequence of a yeast gene is, in effect, replaced with that of a foreign gene. The yeast promoter provides the transcription initiation, whilst translation initiation is provided by the foreign coding sequence. The protein product will therefore be full-length and resemble the native form of the foreign protein.

A variety of expression plasmids have been constructed that contain an isolated promoter fragment that has a restriction site positioned immediately downstream of the transcription initiation site, with no associated translation initiation codon; this is then followed by a terminator-containing fragment. A foreign coding sequence can be inserted at the restriction site to create a heterologous gene; the fusion junction being in the untranslated leader sequence. As the foreign coding sequence provides the translation initiator codon, for correct coupling to give efficient initiation, this needs to be an AUG codon and be the most 5′ such codon in the resultant mRNA. Creation of this type of fusion construct is facilitated by the fact that the untranslated leader contains no specifically required sequence and may vary in length without unduly affecting translation efficiency. The main feature to avoid is the creation of secondary structure loops within the mRNA leader that will act to inhibit translation.

Such constructs are therefore suitable for the cloning and expression of cDNAs. A typical example is pYcDE-2 and its derivatives, as shown in *Figure 7*, which utilize the *ADH1* promoter. Similarly, pBEJ15 contains a *PGK1* promoter with a *Bgl*II cloning/expression site, followed by a *PGK1* terminator element (*Figure 7*). In both of these cases, the cloning/expression site was introduced via Bal31 deletion initiated from within the natural coding sequence of the gene, deleting the ATG translation start codon, but stopping short of the transcription start-point. A linker was then added to insert the restriction site, as described above.

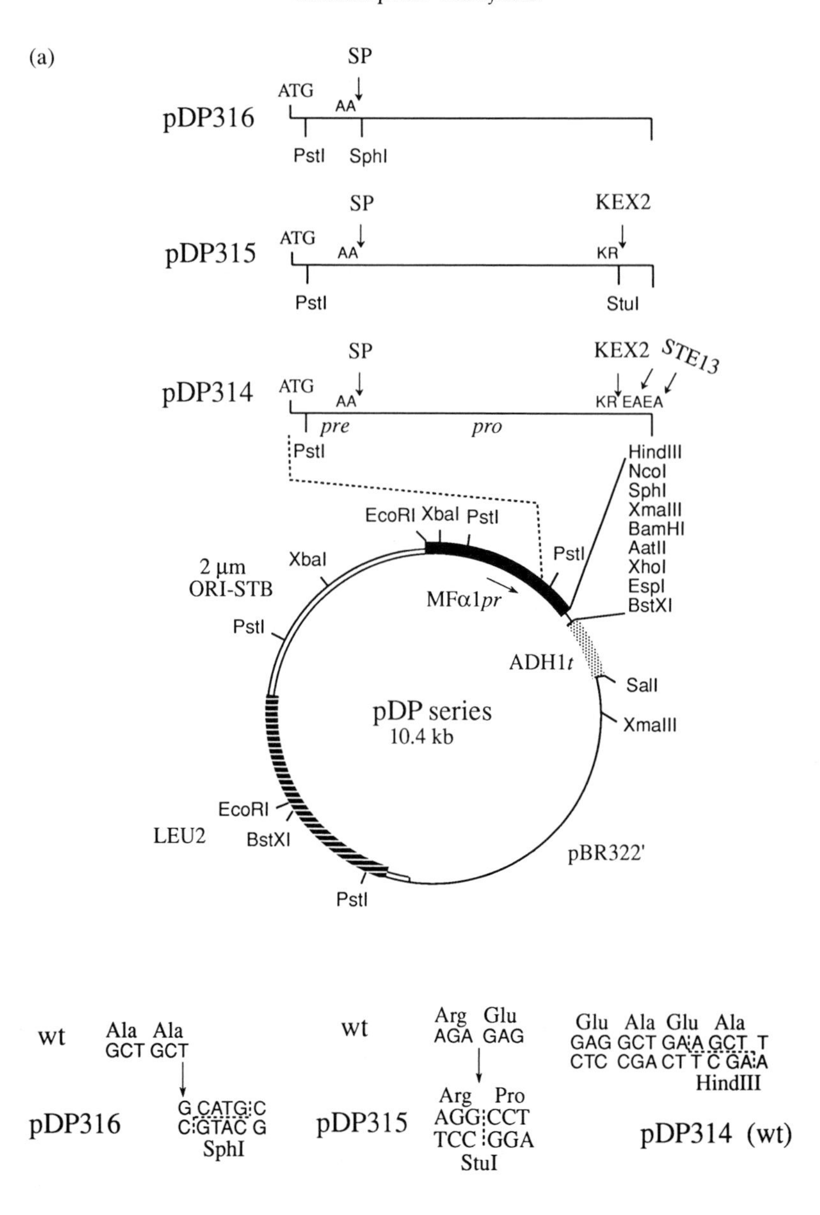

As already considered, the current method of choice for creating promoter fragments for such vectors is via PCR. This is exemplified by the incorporation of the *HSP12* (39) promoter into pDAD1, a construct analogous to pYcDE-2. Starting with a fragment of the *HSP12* gene cloned into pIC-19R, a PCR promoter fragment was isolated using an 'internal' oligonucleotide primer that would hybridize just upstream of the translation start codon of the gene

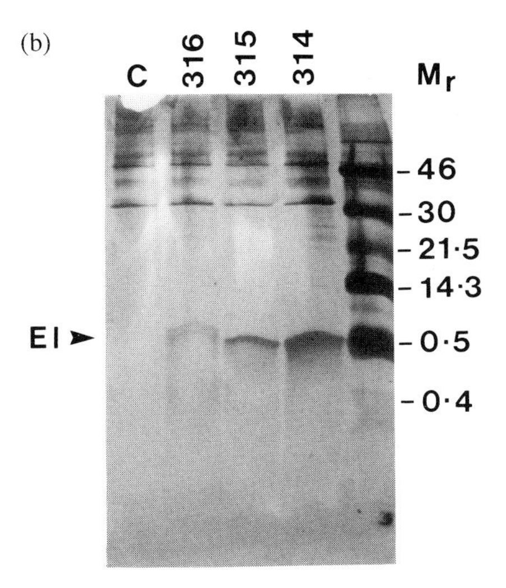

Figure 4. Fusion to the *MFα1* leader sequence for protein secretion. (a) The pDP series of vectors. (Constructed by D. Pioli, Zeneca Pharmaceuticals.) The *MFα1 pre-pro* leader is unmodified in pDP314, which utilizes the naturally occurring *Hin*dIII site as the fusion cloning site. In pDP315 a *Stu*I site has been introduced to provide a blunt cloning site for fusion immediately carboxy-terminal of the KEX2 cleavage site. Fusion to the *pre* (signal peptide) sequence in pDP316 has been facilitated by the introduction of an *Sph*I site, which can be cleaved and the 3′ projecting single-stranded DNA trimmed back with mung bean nuclease, to make a blunt end which terminates in the last base of the Ala19 codon. Signal peptidase (SP) cuts carboxy-terminally to this alanine residue. (b) Protein gel showing elastase inhibitor produced from the alternative *MFα1* leader peptide fusions in pDP314, pDP315, and pDP316. Culture supernatants were concentrated and aliquots run on an SDS–polyacrylamide gel. All three fusions give rise to an elastase inhibitor (EI) band. (Courtesy of D. Pioli.)

and prime DNA synthesis in a direction towards the promoter, whilst simultaneously introducing an *Eco*RI site. A 'universal reverse' primer was used to prime DNA synthesis in the opposite direction from outside of the promoter (*Figure 8*). After PCR, the resultant fragment was digested with *Sal*I, which cut in the vector polylinker sequence upstream of the *HSP12* promoter, and *Eco*RI, which cut at the new primer-introduced site downstream of the transcription start-site. This was then cloned into pCJ17 (a derivative of pYcDE-2, in which the *ADH1* promoter had been replaced with a polylinker), creating pDAD1.

HSP12 displays stationary-phase-induced gene expression. This is conferred on heterologous constructs based on pDAD1, as demonstrated by monitoring expression of wheat α-amylase cDNA in pDAD1 compared with pYcDE-2, whose *ADH1* promoter shows 'constitutive' regulation, as shown in *Figure 8*.

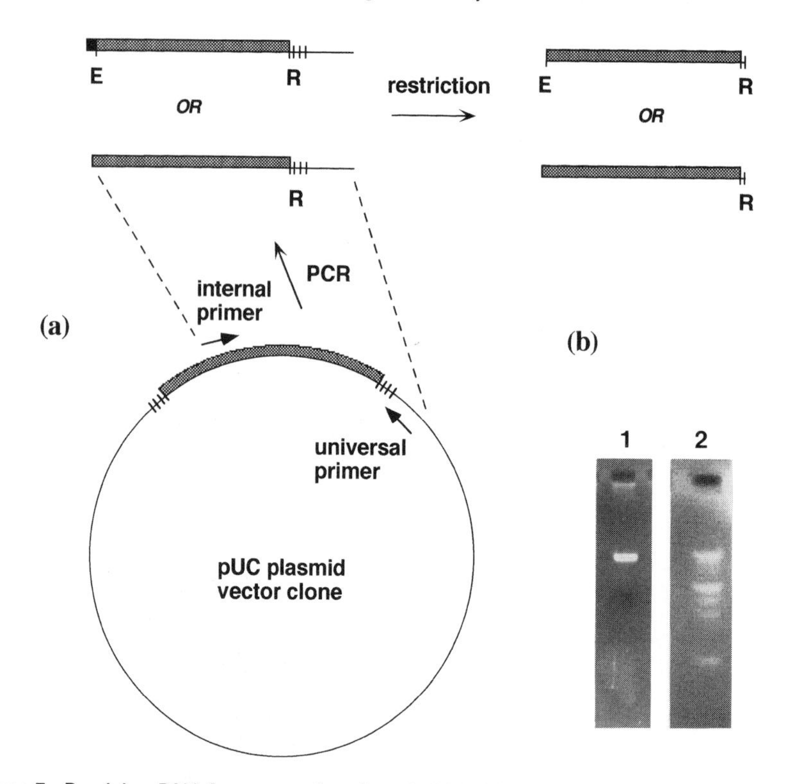

Figure 5. Precision DNA fragment trimming via PCR. (a) A useful general strategy using a fragment cloned into the polylinker (| | | |) of a pUC-type plasmid. If the trimmed end is to be blunt, then the 5′ terminal base of the internal primer will represent that of the fragment generated. Alternatively, if a restriction site (E) is to be introduced, the recognition site bases are introduced at the 5′ end of the primer. It is also advisable to add two to four additional bases at the 5′ end, as some restriction enzymes fail to cut inefficiently if the site is right at the end. However, addition of bases reduces the initial specificity of the primer, so not too many should be added. Having the fragment cloned into the pUC polylinker enables one of the universal sequencing primers to be used at the non-deletion end. The fragment obtained after PCR is cleaved by a restriction enzyme that cuts at a chosen site of the polylinker (R) at the non-deletion end. (b) Agarose gel analysis of the PCR product. Lane 1: a single discrete-sized fragment resulting from a successful PCR reaction. Lane 2: too many bands, due to incorrect priming (annealing temperature too low in this example).

5.3 Transcription terminators

Transcription terminators incorporated into expression constructs are typically present as simple restriction fragments, which are derived from the 3′ ends of yeast genes. Such fragments do not need to be as precisely defined as promoters. It is adequate that they are derived from cleavages upstream of the terminator (within the coding sequence) and downstream of the poly-

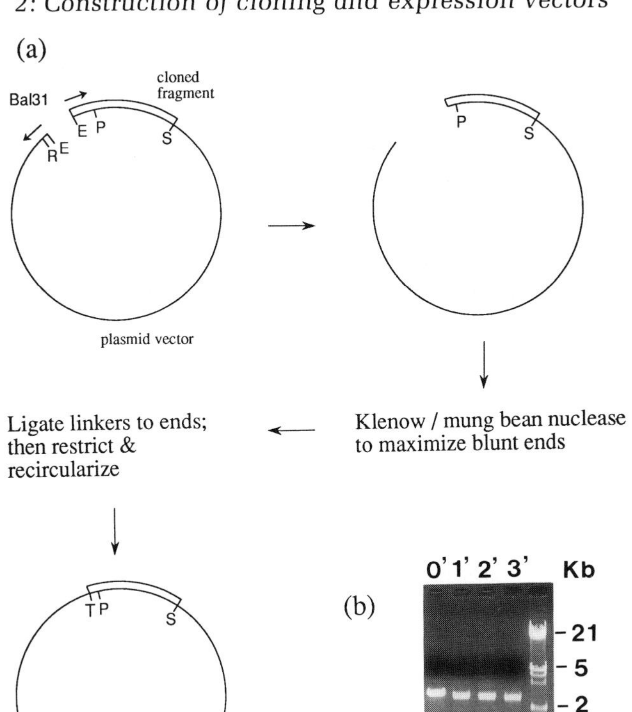

Figure 6. Deleting DNA by Bal31 exonuclease digestion and introducing a terminal restriction site. (a) A useful general strategy with a plasmid-cloned DNA fragment. The plasmid is cut at the site where the deletion is to be introduced, and a series of timed Bal31 deletions is made to produce a range of deletions that include the required size of delection. (As the deleting activity is bidirectional, the plasmid construct must be arranged so that exonuclease digestion does not eliminate an essential plasmid function.) The deletions are then analysed by gel electrophoresis: besides a change in fragment size, the deletion may also be analysed by loss of certain strategically positioned restriction sites. The appropriate Bal31-deleted DNA is then treated to maximize the number of blunt ends, before linkers are ligated to the ends. Subsequent endonuclease digestion of the linker restriction sites generates new termini, which are joined together via ligation under conditions which favour recircularization. These plasmids are recovered by transformation of *E. coli* and individual clones screened for the desired deletion. (b) Bal31 deletion time series. Agarose gel separation of DNA fragments after 0, 1, 2, or 3 min digestion.

adenylation site(s); the resulting fragment will therefore contain all necessary terminator sequence information. This is the case, for example, with the *CYC1* terminator present in pYcDE-2 (*Figure 7*), the *PGK1* terminator in pBEJ15 (*Figure 7*), and the *ADH1* terminator in the pDP plasmids (*Figure*

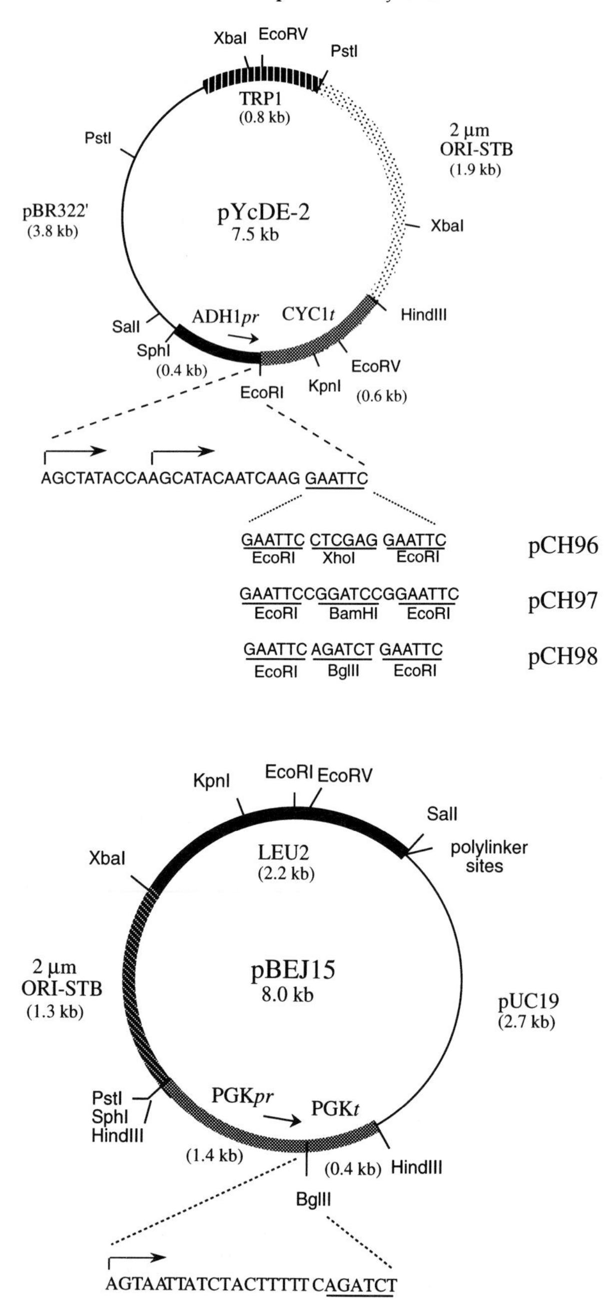

Figure 7. Expression vectors. pYcDE-2 (see ref. 38) and pBEJ15 (11) contain promoter fragments in which the associated ATG translation start codon has been removed by exonuclease digestion and a linker restriction site inserted. In pCH96, pCH97, and pCH98 additional linkers have been inserted to introduce alternative cloning sites.

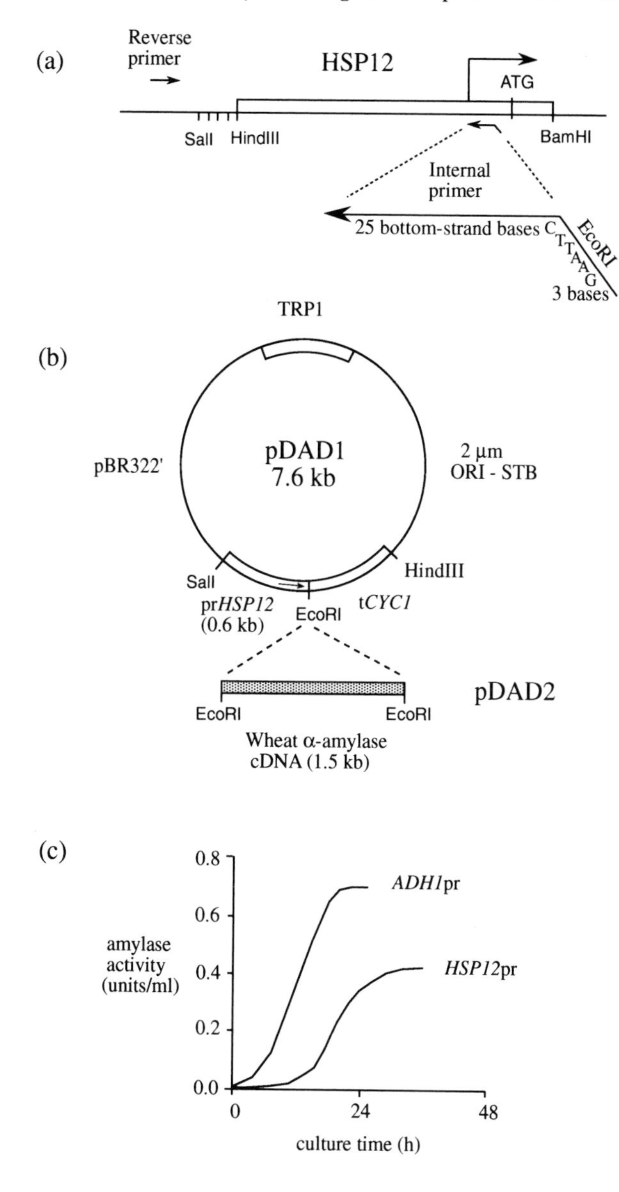

Figure 8. Creation and performance of a new promoter expression construct. (a) Isolation of a promoter fragment from an HSP12 subclone by PCR. Compositional features of the internal primer used are indicated. (b) pDAD1: created by substitution of the *ADH1* promoter fragment in pYcDE-2 (see *Figure 6*) with the *HSP12* promoter PCR fragment. (c) Comparison of secretory wheat α-amylase production: from the 'constitutive' *ADH1* promoter (YEpJK1-αw) and the stationary-phase-induced *HSP12* promoter (pDAD2). YPD (2% glucose) was inoculated with 10^6 selectively grown cells per ml; the cells were grown at 30°C and the secreted amylase activity present in the medium was monitored.

4a). It should also be noted that although some terminators function bidirectionally, many are unidirectional; therefore it is prudent to incorporate the terminator fragments into a heterologous gene construction in the same orientation as occurs in the donor yeast gene.

6. Vector purpose and design

A diversity of cloning vectors for yeast continue to be developed that show increased ease, efficiency, and effectiveness of use. Some of those currently on offer are considered below.

6.1 Efficient cDNA cloning

The construction of cDNA libraries presents many problems: full-length cDNAs must be synthesized and then efficiently cloned. One solution is provided by λ YES, as described in detail by Mulligan and Elledge (Chapter 4). This vector strategy contains a number of elegant touches. Firstly, recombinant vector–cDNA clones are created by efficient linear concatenate ligation and also recovered efficiently by λ packaging. Non-recombinants are eliminated by an adaptor/filling-in strategy. Then, after recovery of the phage clones, growth on an appropriate *E. coli* host results in *in vivo* excision of a plasmid form, which can then be isolated for transformation into yeast. The cDNA cloning site in the vector is directly downstream of a *GAL* promoter for inducible expression, and the vector contains a centromere for low copy number, so that multicopy suppressors should not be obtained if screening for complementation of a yeast host mutation.

An alternative vector strategy was utilized by Minet and Lacroute (40), employing the shuttle vector plasmid pFL60, which facilitates directional cloning, so that all recovered clones have the cDNA insert in the expression orientation (*Figure 9*). The plasmid was linearized by *Eco*RI and ligated with *Eco*RI/poly(dT) adaptors. Subsequent digestion with *Xho*I and removal of the small oligonucleotide fragment yielded a linearized vector that could serve as a unidirectional primer for cDNA synthesis in the presence of poly(A)$^+$ RNA. After second-strand synthesis, the linear plasmid molecules joined with cDNAs were size-selected by electrophoresis on agarose gels, and then circularized by ligation of their blunt ends. These are then recovered by high efficiency transformation, or electroporation, of *E. coli*.

6.2 Genomic library construction

Several *S. cerevisiae* genomic libraries already exist, with some of these in yeast shuttle vectors, such as one based on YCp50 by Rose *et al.* (18). This contains random *Sau*3A fragments inserted into the *Bam*HI site in the Tet^R gene of the pBR322 component of the plasmid. As the vector is centromeric, transforming plasmids from this library persist in single copy, favouring the

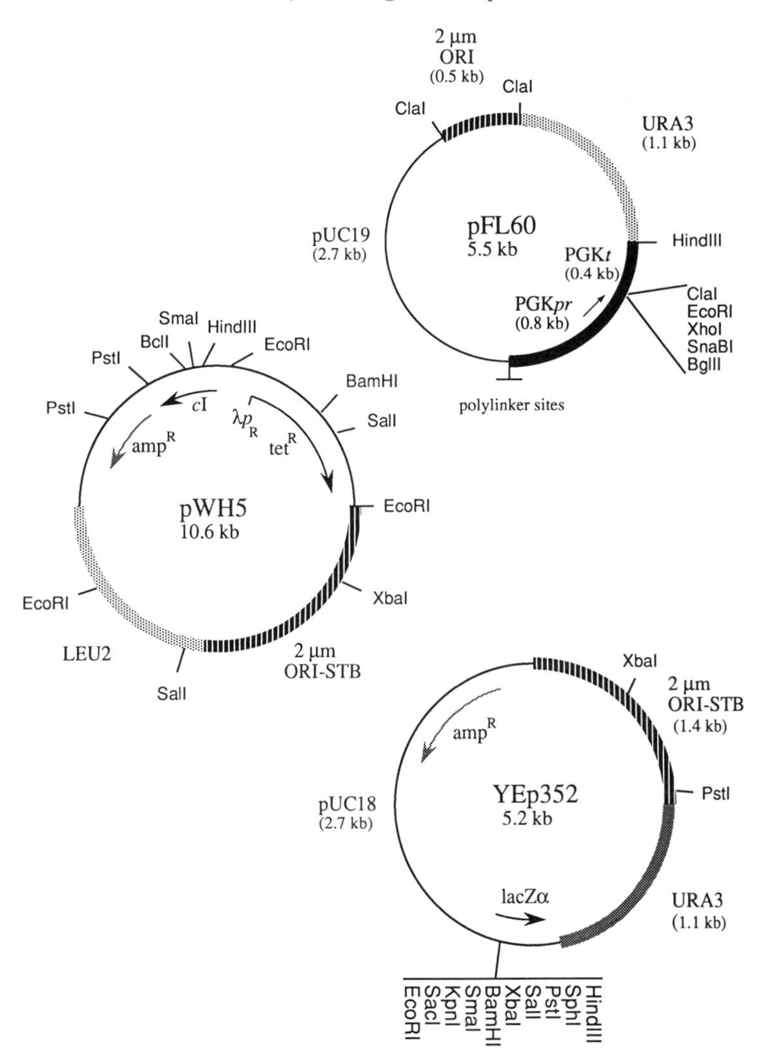

Figure 9. Yeast shuttle plasmid vector designs for cDNA library construction and expression in yeast (pFL60), for cloning with positive selection for inserts in *E. coli* (pWH5), and for *lacZ* polylinker sites, enabling blue/white X-gal screening for inserts in *E. coli* (YEp352).

isolation of clones complementing host mutations rather than multicopy suppressors. For the creation of new genomic libraries, vectors offering more sophisticated cloning methods than YCp50 are now available. Thus, for example, λ YES (see Chapter 4) could also be employed for this purpose; maximum insert size is approximately 8.4 kb. For larger insert sizes, and hence fewer random clones needed per library to represent the genome fully, cosmids can be used. Stucka and Feldmann describe in detail how cosmids can be employed to clone genomic DNA fragments in a size range of 35–45 kb (see Chapter 3).

For those who prefer not to get involved in λ-based technology, genomes the size of yeast are small enough to be cloned in plasmid vectors. The method of choice would then be to use a positive-selection shuttle vector plasmid, like pWH5 (41) shown in *Figure 9*. For library construction, random *Sau*3A fragments can be cloned into the *Bcl*I site. Resultant inactivation of the *c*I repressor gene enables the p_R^{λ} promoter to be active and cause expression of tetracycline-resistance, by which recombinants can be selected in *E. coli*.

6.3 Sub-cloning

There are very many yeast vectors that could be utilized for sub-cloning; however, relatively few of these have any helpful features to facilitate this purpose. Positive selection vectors are obvious exceptions, but they tend to have few cloning site alternatives. More generally useful for this purpose are vectors like YEp352 (42), which contains the multisite polylinker and *lacZ* inactivation screening system from pUC18 (*Figure 9*).

6.4 Expression alternatives

There exists a wide choice of possible promoters to drive heterologous gene expression. Of those commonly used, the *PGK1* and *ADH1* promoters generally give a good level of constitutive expression. Other promoters, however, have also been found to function well in this regard, with some, such as *PRB1* and *GUT1*, showing higher levels of heterologous gene expression (43).

Of the regulated promoters, the *GAL1*, *GAL10*, and *GAL7* promoters are very tightly repressed by glucose and very powerfully induced by galactose (>1000-fold). The *GAL1* and *GAL10* genes share a common UAS, transcribe in opposite orientations, and can be used to express two products simultaneously in approximately equivalent amounts. Thus, for example, Carlson (44) co-expressed heavy- and light-chain immunoglobulins in yeast to produce recombinant antibodies, using vector pBM150, which carries the *GAL1-10* divergent promoters on a *CEN4-ARS1-URA3* plasmid (28).

Other useful regulated promoters are: *ADH2*, which is induced 100-fold by loss of glucose repression (45); *PHO5*, which is induced 10-fold by low phosphate concentration (46); *CUP1*, which is induced 20-fold by addition of copper ions to the medium (47). UAS elements can be transferred to other promoters, conferring the regulation to them. Using this approach, heterologous promoters have been created that contain the response elements for mammalian steroid hormones (48, 49). Addition of the hormone to the culture medium induces expression up to several hundred-fold.

The obtaining of very high levels of expression depends upon many factors besides the strength of the promoter (see *Chapter 11* and ref. 7). The positive

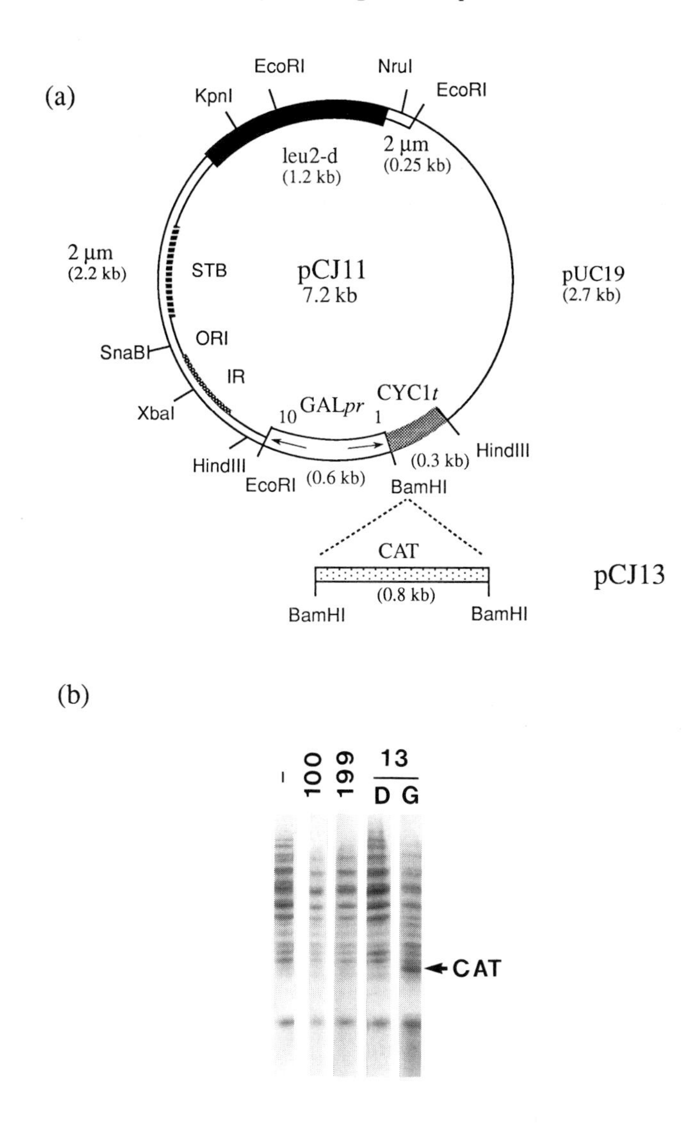

Figure 10. High-level expression. (a) *GAL1 leu2*-d expression vector pCJ11. Derivative pCJ13 contains the CAT-encoding sequence cloned into the expression site. (Constructed by C. Jones.) (b) Protein gel comparison of CAT protein yield from different expression constructs. The following yeast soluble cell protein extracts were analysed by SDS–PAGE: untransformed (–), pCH100 (100), pCH199 (199), and pCJ13 (13). pCH100 and pCH199 are CAT-expressing constructs containing the *ADH1* or *PGK1* promoter, respectively, derived from pYcDE-2 (*Figure 6*). All strains were grown on glucose (D), with the exception of the track marked G (galactose).

contributions of a strong inducible *GAL1* promoter, and ultra-high copy number provided by *leu2*-d, can be demonstrated in the expression of chloramphenicol acetyl transferase (CAT). When expressed from the *ADH1* promoter in pCH98 (*Figure 7*) (pCH100 (11)), or an analogous construction containing the *PGK1* promoter (pCH199), both present on a 2 μm *ORI-STB* vector at about 20 copies per cell, approximately 1 and 2 CAT units per mg of cell protein, respectively, were obtained in YPD medium (see Appendix 1), following inoculation from cultures grown on selective minimal medium. However, when CAT was expressed from an induced *GAL* promoter on a similar plasmid containing *leu2*-d as the selectable marker (pCJ13), 66 CAT units per mg of cell protein were obtained. This larger amount of product could be seen as a prominent band on a protein gel, whereas no such band was evident with the other two constructs (*Figure 10*).

Acknowledgements

Thanks are due to CEC BRIDGE (BIOT-CT90–0165) for support of B. E. Jordan and D. A. Dawson; to C. Jones and P. A. Meacock for provision of pCJ11 and 13; and to D. Pioli for provision of pDP314, 315, 316 plasmids and elafin data.

References

1. Orr-Weaver, T. L., Szostak, J. W., and Rothstein, R. S. (1983). In *Methods in enzymology*, Vol. 101 (ed. R. Wu, L. Grossman, and K. Moldave), pp. 228–45. Academic Press, New York.
2. Beggs, J. D. (1981). In *Genetic engineering 2* (ed. R. Williamson), pp. 175–203. Academic Press, London.
3. Rothstein, R. (1985). In *DNA cloning: a practical approach*, Volume II (ed. D. M. Glover), pp. 45–66. IRL Press, Oxford.
4. Parent, S. A., Fenimore, C. M., and Bostian, K. A. (1985). *Yeast*, **1**, 83.
5. Armstrong, K. A., Som, T., Volkert, F. C., Rose, A., and Broach, J. R. (1989). In *Yeast genetic engineering* (ed. P. J. Barr, A. J. Brake, and P. Valenzuela), pp. 165–92. Butterworths, Boston.
6. Romanoss, M. A., Scorer, C. A., and Clare, J. J. (1992). *Yeast*, **8**, 423.
7. Hadfield, C., Raina, K. K., Shashi-Menon, K., and Mount, R. C. (1993). *Mycol. Res.*, **97**, 897.
8. Beggs, J. D. (1978). *Nature*, **275**, 104.
9. Erhart, E. and Hollenberg, C. P. (1983). *J. Bacteriol.*, **156**, 625.
10. Hadfield, C., Cashmore, A. M., and Meacock, P. A. (1987). *Gene*, **52**, 59.
11. Hadfield, C., Jordan, B. E., Mount, R. C., Pretorius, G. H. J., and Burak, E. (1990). *Curr. Genet.*, **18**, 303.
12. Chinery, S. A. and Hinchliffe, E. (1989). *Curr. Genet.*, **16**, 21.
13. Bijvoet, J. F. M., van der Zanden, A. L., Goosen, N., Brouwer, J., and van de Putte, P. (1991). *Yeast*, **7**, 347.

14. Bussey, H. and Meaden, P. (1985). *Curr. Genet.*, **9**, 285.
15. Nasmyth, K. and Reed, S. I. (1980). *Proc. Natl Acad. Sci. USA*, **77**, 2119.
16. Clarke, L. and Carbon, J. (1980). *Nature*, **287**, 504.
17. Hieter, P. C., Mann, C., Snyder, M., and Davis, R. W. (1985). *Cell*, **40**, 381.
18. Rose, M., Novick, P., Thomas, J. H., Botstein, D., and Fink, G. R. (1987). *Gene*, **60**, 237.
19. Burke, D. T. and Olsen, M. V. (1991). In *Methods in enzymology*, Vol. 194 (ed. C. Guthrie and G. R. Fink), pp. 251–70. Academic Press, New York.
20. Lopes, T. S., Klootwijk, J., Veestra, A. E., Vanderaar, P. C., Heerikhuizen, H. V., Raue, H. A., and Planta, R. J. (1989). *Gene*, **79**, 199.
21. Kingsman, S. M., Kingsman, A. J., Dobson, M. J., Mellor, J., and Roberts, N. A. (1985). In *Biotechnology and genetic engineering reviews*, Vol. 3 (ed. G. E. Russel), pp. 377–416. Intercept, Newcastle upon Tyne.
22. Boeke, J. D., Xu, H., and Fink, G. R. (1988). *Science*, **239**, 280.
23. Jacobs, E., Dererchin, M., and Boeke, J. E. (1988). *Gene*, **67**, 259.
24. Sakai, A., Shimizu, Y., and Hishinuma, F. (1990). *Appl. Microbiol. Biotech.*, **33**, 302.
25. Marsh, J. L., Erfle, M., and Wykes, E. J. (1984). *Gene*, **32**, 481.
26. Tschumper, G. and Carbon, J. (1982). *J. Mol. Biol.*, **156**, 293.
27. Fitzgerald-Hayes, M., Clarke, L., and Carbon, J. (1982). *Cell*, **29**, 235.
28. Johnston, M. and Davies, R. W. (1984). *Mol. Cell. Biol.*, **4**, 1440.
29. Sealey, P. G. and Southern, E. (1982). In *Gel electrophoresis of nucleic acid: a practical approach* (ed. D. Rickwood and B. D. Hames), pp. 39–76. IRL Press, Oxford.
30. Maxam, A. M. and Gilbert, W. (1977). *Proc. Natl Acad. Sci. USA*, **74**, 560.
31. Sambrook, J., Fritsch, E. F., and Maniatis, T. (1989). *Molecular cloning. A laboratory manual*, 2nd edn, Vol. 1, pp. 1.53–1.73. Cold Spring Harbor Laboratory Press, Cold Spring Harbor, NY.
32. Hadfield, C. (1987). In *Gene cloning and analysis, a laboratory guide* (ed. G. J. Boulnois), pp. 61–106. Blackwell Scientific Publications, Oxford.
33. Zaret, K. S. and Sherman, F. (1982). *Cell*, **28**, 563.
34. Casadaban, M. J., Martinez-Arias, A., Shapira, S. K., and Chou, J. (1983). In *Methods in enzymology*, Vol. 100 (ed. R. Wu, L. Grossman, and K. Moldave), pp. 293–308. Academic Press, New York.
35. Goodey, A. R., Doel, S. M., Piggott, J. R., Watson, M. E. E., Zealey, G. R., Cafferkey, R., and Carter, B. L. A. (1986). *Mol. Gen. Genet.*, **204**, 505.
36. Gatignol, A., Dassain, M., and Tiraby, G. (1990). *Gene*, **91**, 35.
37. Erlich, H. A. (ed.) (1989). *PCR technology*. M. Stockton Press, New York.
38. Hadfield, C., Cashmore, A. M., and Meacock, P. A. (1986). *Gene*, **45**, 149.
39. Praekelt, U. M. and Meacock, P. A. (1990). *Mol. Gen. Genet.*, **223**, 97.
40. Minet, M. and Lacroute, F. (1990). *Curr. Genet.*, **18**, 287.
41. Wright, A., Maundrell, K., Heyer, W.-D., Beach, D., and Nurse, P. (1986). *Plasmid*, **15**, 156.
42. Hill, J. E., Myers, A. M., Koener, T. J., and Tzagoloff, A. (1986). *Yeast*, **2**, 163.
43. Sleep, D., Belfield, G. P., Ballance, D. J., Steven, J., Jones, S., Evans, L. R., Moir, P. D., and Goodey, A. R. (1991). *Biotechnology*, **9**, 183.
44. Carlson, J. R. (1988). *Mol. Cell. Biol.*, **8**, 2638.
45. Price, V. L., Taylor, W. E., Clevenger, W., Worthington, M., and Young, E. T.

(1990). In *Methods in enzymology*, Vol. 185 (ed. S. P. Colwick and N. O. Kaplan), pp. 308–18. Academic Press, New York.
46. Hinnen, A., Meyhack, B., and Heim, J. (1989). In *Yeast genetic engineering* (ed. P. J. Barr, A. J. Brake, and P. Valenzuela), pp. 193–213. Butterworths, Boston.
47. Etcheverry, T. (1990). In *Methods in enzymology*, Vol. 185 (ed. S. P. Colwick and N. O. Kaplan), pp. 319–29. Academic Press, New York.
48. Purvis, I. J., Chotai, D., Dykes, C. W., Lubahn, D. B., French, F. S., Wilson, E. M., and Hobden, A. N. (1991). *Gene*, **106**, 35.
49. Schena, M., Picard, D., and Yamamoto, K. R. (1991). In *Methods in enzymology*, Vol. 194 (ed. C. Guthrie and G. R. Fink), pp. 389–98. Academic Press, New York.

3

Cosmid cloning of yeast DNA

ROLF STUCKA and HORST FELDMANN

1. Introduction

Cosmids are small plasmid cloning vectors that contain

- a phage λ *cos* site
- particular plasmid DNA sequences (origin of replication, selection markers)
- one or more unique restriction sites for cloning
- optionally, further markers or sequences for specific applications (1, 2)

After *in vitro* packaging into premature phage heads, recombinant cosmids can easily be introduced into *Eschericha coli* cells and propagated with high efficiency like plasmids. Depending on the vector, genomic DNA fragments in the size range 35–45 kb can be accommodated: 78–105% of the length of wild-type λ DNA (38–55 kb) can be packaged efficiently. The *in vitro* packaging systems give an approximately 10-fold higher yield of transformants per μg of DNA than the standard $CaCl_2$ transformation procedure.

These features make cosmids very convenient tools for the construction and handling of genomic libraries. Obvious advantages of cloning DNA segments in cosmids are:

- larger genes can be obtained on a single recombinant clone
- several linked genes can be isolated together with their intergenic regions
- fewer colonies need to be maintained and screened to isolate a clone of interest

Additionally, the isolation of sequentially overlapping cosmid clones has enabled physical linkage and extensive characterization of genes from various organisms. This approach has also been successfully applied to the yeast, *Saccharomyces cerevisiae*, in the framework of the EC yeast genome project (3).

2. Construction of yeast cosmid libraries

To construct a library with as complete coverage as possible with as few clones as possible, the cloned DNA fragments should be randomly distributed on the DNA. Under these conditions, the number of clones (N) in a library representing each genomic segment with a given probability (P) is

$$N = \ln(1 - P)/\ln(1 - f)$$

where f is the insert length expressed as fraction of the genome size (4).

With the size of 14 000 kb for the yeast genome and assuming an average insert length of 35 kb, a cosmid library containing 1840 random clones represents the yeast genome at $P = 99\%$ (i.e. about five times the genome equivalent). Collections of 2760 ($P = 99.9\%$) or 4600 ($P = 99.99\%$) such clones correspond to seven or 12 times, respectively, the yeast genome equivalent. The actual number of cosmid clones obtained by the procedures outlined below is very high (> 200 000/μg DNA). *Table 1* lists the number of independent cosmid clones needed to cover about four or 14 times, respectively, the equivalent of each single yeast chromosome.

These figures are of interest in setting up ordered yeast cosmid libraries or sorting out and mapping chromosome-specific sublibraries. For example, a chromosome XI-specific sublibrary composed of 138 clones has been sorted out from an unordered cosmid library by colony hybridization using PFG-purified chromosome XI as a probe. The 'nested chromosomal fragmentation' method was then applied to rapid sorting of these clones (5). Finally, a set of some 30 overlapping cosmids was sufficient to build a contig of chromosome XI (5). We have selected 3000 independent clones from a total yeast cosmid library to prepare the DNA of each single cosmid from mini-lysates of

Table 1. Numer of cosmids representing single yeast chromosomes

Chromosome	Size (kb)	Number of cosmids	
		$P = 99\%$	$P = 99.999\%$
I	220	26	80
II	840	108	324
III	320	39	119
IV	1600	208	624
V	610	78	233
VI	280	35	103
VII	1200	155	601
VIII	560	71	214
IX	450	57	170
X	760	98	293
XI	670	79	258
XII	2200	288	861
XIII	920	119	356
XIV	800	103	309
XV	110	142	427
XVI	960	124	372

5 ml cultures (6). These samples were numbered and the corresponding cultures kept as glycerol stocks (LB medium, 15% glycerol) at −70°C. By chromosomal walking, some 40 of these clones have been sorted out to cover yeast chromosome II (R. Stucka and H. Feldmann, unpublished).

3. Cosmid vectors and procedures used in cloning of yeast DNA

3.1 Cosmid vectors

A great variety of cosmid vectors are available that could be used to construct yeast genomic libraries. In connection with the yeast genome project (3), mainly three types of vectors have been employed, two of which are depicted in *Figure 1*. In all cases, size-fractionated fragments obtained by partial digestion with *Sau*3A of high molecular weight yeast DNA can be cloned into the unique *Bam*HI site of the respective vector.

3.1.1 pWE15

pWE15 (and pWE16), based on pJB8 (7), are cosmid vectors that have been designed for genomic walking and rapid restriction mapping (8). They contain bacteriophage T3 and T7 promoters flanking a unique *Bam*HI cloning site. By using the cosmid DNA containing a genomic insert as a template for either T3 or T7 polymerase, directional 'walking' probes can be synthesized and used to screen genomic cosmid libraries (or sublibraries). These vectors contain additional genes (*SV2-neo* or *SV2-dhfr*) which allow the expression, amplification, and rescue of cosmids in mammalian cells. *Not*I recognition sites have been placed near the *Bam*HI cloning site which allow the insert to be removed as a single large fragment. The presence of bacteriophage promoters flanking the insert DNA facilitates restriction mapping of the recombinant cosmid. pWE15 was the major cloning vehicle used to construct a yeast chromosome XI-specific cosmid library (5).

3.1.2 pOU61cos

pOU61cos, an amplifiable, low copy number, cosmid vector that was originally designed to generate *E. coli* cosmid libraries (9), has also been used in cloning yeast chromosome XI DNA (5).

3.1.3 Shuttle vector pYc3030

pYc3030 is a cosmid vector that most conveniently allows DNA to be shuttled between *E. coli* and yeast cells (see Section 5). It was generated from pHC79 (10) by adding the yeast 2 μm plasmid origin of replication (on a 2.2 kb *Eco*RI fragment) and the yeast *HIS3* marker (on a 1.56 kb *Bgl*II fragment) (11). pYc3030 contains a unique *Bam*HI cloning site.

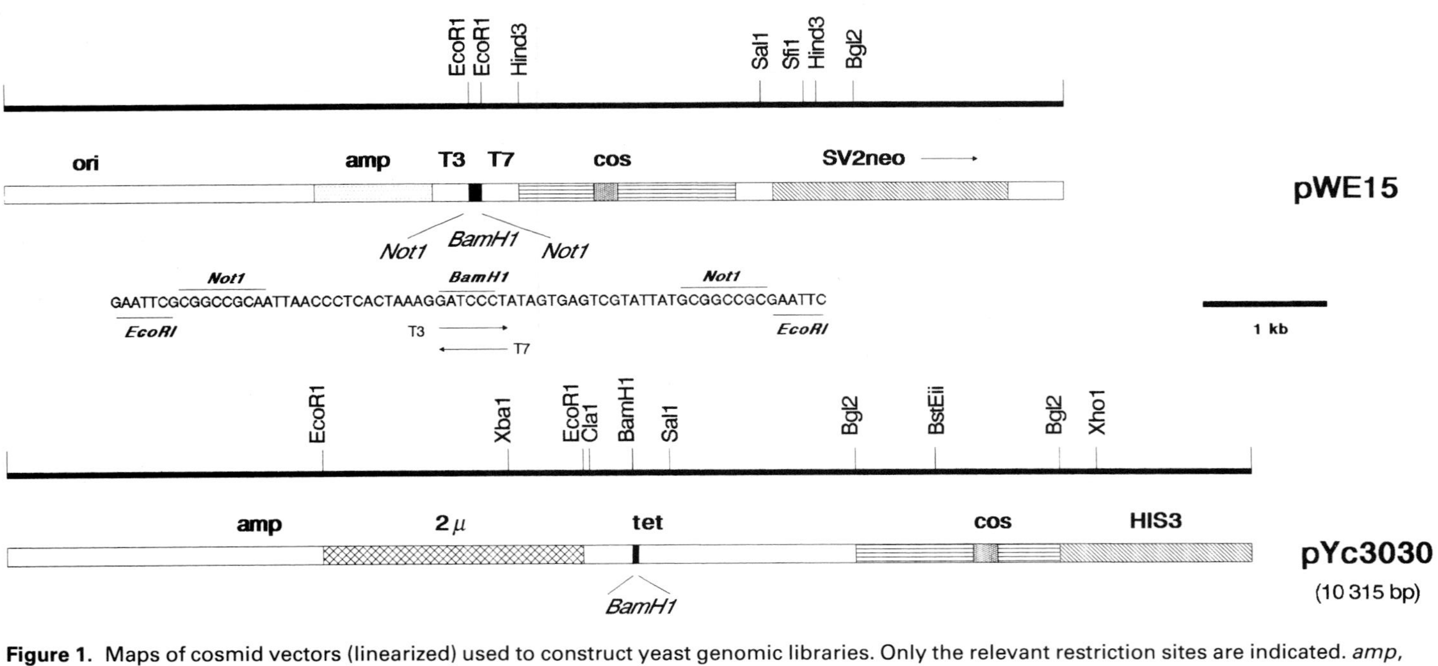

Figure 1. Maps of cosmid vectors (linearized) used to construct yeast genomic libraries. Only the relevant restriction sites are indicated. *amp*, ampicillin resistance gene; *tet*, tetracycline resistance gene; *ori*, bacterial origin of replication; *cos*, bacteriophage λ cohesive ends; *SV2neo*, neomycin resistance gene with SV40 expression control elements; *HIS3*, yeast imidazoleglycerolphosphate dehydratase gene as an auxotrophic marker; T3 and T7, bacteriophage promoters.

3.2 Cloning procedures

Various procedures have been described for the construction of genomic cosmid libraries (e.g. ref. 12). The following points, which have been found to be of general relevance in establishing cosmid libraries, are considered in the protocols given below:

(a) High molecular weight genomic DNA (> 150–200 kb) has to be used as starting material. Several procedures for the preparation of high molecular weight yeast DNA are available (12–14) (see also Chapters 1 and 5). The size of the DNA can be ascertained in 0.3% agarose gels run at a low-voltage gradient (see *Protocol 2*) or in pulsed field gel electrophoresis (12).

(b) Large, randomly distributed, genomic DNA fragments (30–45 kb) are generated by partial digestion of high molecular weight DNA with the appropriate restriction enzyme (*Sau*3A in this case) followed by size fractionation. This is one means of preventing the formation of cosmid clones containing fragments that are not contiguous in the genome. Several procedures employing size fractionation in gradients or gels have been described (2, 15); a procedure using NaCl gradient centrifugation is given below (see *Protocol 5*). Usually size fractionation will lead to some loss of material, but this is tolerable for yeast DNA, which can be obtained in sufficient amounts. As a further precaution to prevent ligation artefacts, DNA fragments are dephosphorylated prior to combination with the vector arms (e.g. ref. 16).

(c) Separate vector arms are prepared. The vector DNA is digested with restriction enzymes that cleave at either side of the *cos* site. The single-stranded protruding termini of the resulting molecules are rendered

Table 2. Composition of buffers

Buffer	**Composition**
Loening buffer	36 mM Tris, 30 mM Na_2HPO_4, 1 mM EDTA, pH 7.7
SCE	1.0 M sorbitol, 0.1 M sodium citrate, 60 mM EDTA (pH 7.0)
STE	0.8 M NaCl, 20 mM Tris–HCl, 10 mM EDTA, pH 8.0
SM phage buffer	100 mM NaCl, 8 mM Mg_2SO_4, 50 mM Tris, pH 7.5, 2% gelatin
Restriction buffers	buffers for the various enzymes are recommended and/or supplied by the manufacturers; keep 10 × restriction buffers on stock
Lysis buffer	3% sodium *N*-lauroyl sarcosine, 0.5 M Tris–HCl, pH 9.0, 0.2 M EDTA
Phosphatase buffer (10 × concentration)	500 mM Tris, 1 mM EDTA, pH 8.5; keep 10 × concentrated phosphatase buffer on stock
Ligation buffer	66 mM Tris–HCl, 5 mM $MgCl_2$, 1 mM DTE, 1 mM ATP, pH 7.5; keep 10 × concentrated ligation buffer on stock
Phenol–chloroform	phenol (saturated with TE): chloroform (1:1) (v/v)

incapable of ligation by dephosphorylation. In this way, the formation of tandem vectors during ligation is suppressed (7).

Note: it is important to note that polycosmid or multiple-insert-containing clones have never been observed.

Media and solutions used in the procedures described below can be found in *Table 2* and Appendix 1.

Protocol 1. Preparation of yeast high molecular weight DNA (14)

1. Prepare a 1 litre flask with 500 ml of YEP medium (see Appendix 1) and inoculate with 1/500 of an overnight culture of the yeast strain of your choice.
2. Grow the culture overnight (16 h) at 28°C to late-log phase. Pellet the cells by centrifugation at 1200 *g* and 4°C and wash the pellet with water once. With most strains, this growth regime produces 6–8 g of cells (wet weight).
3. The scale of the preparation is dictated by the capacity of an SW27 bucket used during the sucrose gradient (step **8**). The following steps are given for 2 g of cells used for one bucket. Run several preparations in parallel, if you want to scale up the preparation.
4. Resuspend 2 g of cells in 4 ml of SCE. Add 30 μl of β-mercaptoethanol and 2 mg of Zymolase 5000 (Sigma). Mix gently.
5. Incubate for 2–3 h at 37°C with occasional gentle agitation.
6. During step 5, prepare a sucrose gradient in a Beckman SW27 polyallomer tube containing 3 ml 50% sucrose in STE and a 26 ml 10–20% sucrose (w/v) gradient in STE.
7. Add the sphaeroplast suspension (from step **5**) slowly to 7 ml of lysis buffer with continuous swirling and incubate the lysate for 15 min at 65°C.
8. Layer the sample from step 7 immediately on top of the sucrose gradient and centrifuge at 26 000 r.p.m. for 3 h at 20°C.
9. Collect the DNA (which is invariably highly viscous) from the bottom of the tube as a single 6 ml fraction. Use a wide needle to perforate the tube at the bottom.
10. Dialyse the sample against TE buffer (see Appendix 1).
11. Extract the DNA solution with an equal volume of phenol–chloroform twice, and then with an equal volume of chloroform (see *Protocol 2*).
12. Dialyse the DNA solution against TE (changing the outer phase four times).
13. Determine the DNA concentration by UV measurement (260/280 nm) in an aliquot. A typical yield at this stage is about 80–100 μg DNA per SW27 gradient.

Protocol 2. Partial digestion of yeast DNA with *Sau*3A

A. *Analytical reactions*

To test the conditions for preparative digests, an analytical time course of the reaction is carried out. The samples are analysed on a 0.3% agarose gel with reference to appropriate DNA molecular weight markers (λ DNA; λ DNA digested with *Hin*dIII).

1. Mix 2–4 μg of DNA with 6 μl of restriction buffer (10 × concentration)[a] and water to 60 μl.
2. Take a 10 μl aliquot *before* adding 0.1–0.2 units of *Sau*3A enzyme and then 10 μl aliquots at 5, 10, 20, 40, and 60 min of incubation at 37°C. Stop the reactions by adding 6 μl of 0.2 M EDTA to each sample, then heat for 10 min at 68°C.
3. Apply the samples to a 0.3% agarose gel in Loening buffer (containing ethidium bromide at 1 μg/ml). Include samples of DNA molecular weight markers. Run the gel at 0.7–1 V/cm overnight. Visualize the DNA under UV light.
4. Should the size distribution not be satisfactory, repeat the analytical reactions using different amounts of *Sau*3A enzyme (0.5, 0.05, or 0.01 units).

B. *Preparative digests*

1. Select an enzyme concentration and three time intervals for incubation that resulted in DNA fragments of the following sizes:
   - $>$ 40 kb (short incubation time)
   - 35–45 kb (optimal incubation time)
   - $<$ 30 kb (longer incubation time)
2. Use 50–100 μg of chromosomal DNA per preparative digest and scale up the amounts of the other ingredients proportionally to the pilot reactions.
3. Stop the reactions by adding 0.1 vol. of 0.2 M EDTA to each sample, then heat for 10 min at 68°C.
4. Remove a 5 μl aliquot from each sample and apply it to a 0.3% agarose gel to ensure that the appropriate size distribution has been attained.
5. Combine the rest of the samples.
6. Extract the DNA with an equal volume of phenol–chloroform. Mix gently for 10 min and centrifuge in a microfuge for 15 min.
7. Remove the upper aqueous layer and repeat the extraction procedure.
8. Extract the aqueous layer with an equal volume of chloroform to remove residual phenol.

Protocol 2. *Continued*

9. Adjust the DNA solution to 0.3 M sodium acetate (pH 6) and precipitate the DNA with 2 volumes of ethanol. Leave in dry ice for 10 min.
10. Centrifuge the sample for 15 min in a microfuge, remove the supernatant as completely as possible, and wash the pellet once with 70% ethanol and once with absolute ethanol.
11. Dry the DNA pellet under vacuum for 5 min and dissolve the DNA in TE buffer at 1 μg/μl.

[a] Use the appropriate restriction buffer recommended or supplied by the manufacturer.

Protocol 3. Phosphatase treatment of insert DNA fragments

1. Mix 200 μl of *Sau*3 A-digested DNA (at 1 μg/μl) with 700 μl of water and 100 μl of phosphatase buffer (10 × concentration). Add calf intestine alkaline phosphatase (CIP, Boehringer Mannheim; 0.05 units/μg DNA) and incubate for 30 min at 37°C.
2. Stop the reaction by adjusting to 20 mM EDTA; heat for 10 min at 68°C.
3. Extract the sample with phenol–chloroform and ethanol precipitate as described in *Protocol 2*.
4. Dissolve the DNA pellet in TE buffer 1 μg/μl.

Since it is important to dephosphorylate insert DNA completely, this step should be checked by self-ligation of phosphatase-treated and untreated DNA and comparisons of both samples with unligated DNA (see *Protocol 4*).

Protocol 4. Test ligation of dephosphorylated DNA fragments

1. Mix 1 μl of DNA at 0.25 μg/μl (either treated or untreated with phosphatase) with 1 μl of 10 × ligation buffer, 2 μl of 1 mM ATP, 5 μl of water, and 1 μl of T4 DNA ligase (80 units/μl).
2. Incubate at 15°C overnight.
3. Run the samples on a 0.3% agarose gel (see *Protocol 2*, step **3**) together with control samples of DNA fragments (treated or untreated with phosphatase).

None of the phosphatase-treated DNA samples (whether or not ligated) should reveal any bands larger than those corresponding to the size of the untreated DNA fragments. If they do, repeat the phosphatase treatment to completion.

Protocol 5. Size fractionation by NaCl gradient centrifugation

1. Dissolve DNA fragments from *Protocol 3* (100–200 μg) in 500 of μl TE buffer.
2. Prepare a linear NaCl gradient (1.25–5 M NaCl in TE buffer) in a Beckman SW41 polyallomer tube (13.2 ml). Layer the sample on top and centrifuge at 39 000 r.p.m. for 3 h at 18°C.
3. Collect 0.5 ml fractions from the bottom of the tube and analyse 20 μl aliquots on a 0.3% agarose gel as described in *Protocol 2*.
4. Pool fractions containing fragments in the size range 35–45 kb and ethanol precipitate as in *Protocol 2*.
5. Dissolve the pelleted DNA in TE buffer at 1 μg/μl.

Protocol 6. Preparation of vector arms

1. Digest 20 μg of closed circular pYc3030 vector DNA with 40 units of *Bst*EII, and another 20 μg with 40 units of *Cla*I, in 400 μl of the appropriate restriction buffer[a]) for 60 min at 37°C, in two separate reactions.
2. Ethanol precipitate the DNA samples as in *Protocol 2*.
3. Redissolve each pellet from step **2** in 200 μl of phosphatase buffer, add 1 unit of calf intestine alkaline phosphatase (CIP), and incubate for 30 min at 37 °C.
4. Extract each sample with phenol–chloroform and ethanol precipitate as in *Protocol 2*.
5. To test the effectiveness of the phosphatase treatment, a series of test ligations similar to those described in *Protocol 4* may be carried out.
6. If the results of the test ligations are satisfactory (i.e. if no ligation of the vector is detectable), redissolve each pellet from step **4** in 400 μl of the appropriate restriction buffer[a], add 40 units of *Bam*HI and incubate for 60 min at 37°C.
7. Ethanol precipitate the DNA samples as in *Protocol 2*.
8. Dissolve each pellet in approximately 100 μl of TE buffer and run on a preparative 0.8% agarose gel in 1 × TBE in separate lanes. Use appropriate DNA molecular weight markers for calibration.
9. Visualize the fragments (*Cla*I–*Bam*HI = 10 kb; *Bst*EII–*Bam*HI = 7.7 kb) under UV light and cut out the bands from the gel.

Protocol 6. *Continued*

10. Electro-elute the DNA from the respective gel slices using an appropriate device (ISCO Inc., sample concentrator cups) in 0.5 × TBE at 100 V for 1 h.

11. Ethanol precipitate the DNA from the eluates and redissolve the pellets in TE buffer at 1 μg/μl.

[a] Use the appropriate restriction buffer recommended or supplied by the manufacturer.

Protocol 7. Ligation and *in vitro* packaging

The total concentration of DNA in the ligation reaction should be high enough (> 200 μg/ml) to favour the formation of mixed concatemers between the arms of the cosmid vector and the insert DNA. Because of the inability of the vector to self-ligate, the reaction can be driven by an excess of vector DNA.

The packaging extracts can be prepared from *E. coli* lysogens BHB2688 and BHB2690 as described in refs 17 and 18. More conveniently, packaging extracts are commercially available from Boehringer Mannheim GmbH, Promega Biotec, or Stratagene (Gigapack II Gold).

1. Ligate 2 μg of size-fractionated DNA and 1 μg each of vector arms in 20 μl of ligation buffer for 15 h at 15°C, using 1 unit of T4 DNA ligase. A 1 μl aliquot of the reaction can be tested for efficient ligation by electrophoresis on a 0.3% agarose gel (see *Protocol 2*).

2. Place 10 μl of the freeze–thaw lysate (FTL) and 15 μl of sonicated extracts (SE) on ice. FTL will thaw first.

3. Add the FTL to the still frozen SE and then immediately add 5 μl of the ligation reaction (from step **1**). Mix gently with a fine glass rod; try to avoid introducing air bubbles.

4. Incubate for 60 min at 22°C.

5. Dilute the reaction mixture with 0.5–1 ml of SM phage buffer, add one drop of chloroform and store the sample at 4°C.

Protocol 8. Transduction and titring the cosmid library

1. Inoculate a single colony of *E. coli* 490A (r_k, m_k, *met*$^-$, *thr*$^-$, *leu*$^-$, *recA*$^-$) in 25 ml of LB medium containing 0.2% maltose (in a 250 ml flask) and incubate overnight at 37°C with vigorous shaking.

2. Pellet the cells by centrifugation at 4000 *g* for 20 min at 4°C.

3. Resuspend the cells in 12.5 ml of mM Mg_2SO_4 (19) and store at 4°C until needed. Although the cells are stable for some days, a freshly prepared culture is preferable.
4. Mix 200 μl of bacterial suspension with 10 μl of phage sample (from *Protocol 7*, step **5**) and incubate for 20 min at 37°C.
5. Add 800 μl of fresh LB medium and continue incubation for 45 min at 37°C.
6. Prepare 10-fold and 100-fold dilutions in LB medium (containing 100 μg/ml ampicillin) from the suspension of step **5**.
7. Spread aliquots of 100 μl of each of these suspensions on LB agar plates containing ampicillin at 100 μg/ml.
8. Incubate the plates at 37°C until colonies are big enough to count (usually 18 h). Titres usually range from 10^5 to 8×10^5 transformants per μg of size-fractionated chromosomal DNA.
9. At this point the remainder of the phage sample may be packaged and spread on agar plates as described above using appropriate dilutions.

For accurate counting, the number of colonies per plate should be in the following range: 10 cm diameter, 200–400 colonies; 15 cm diameter, 500–1000 colonies; 20 cm × 20 cm (Nunc plates), 2000–5000 colonies. Otherwise, and also for initial screening purposes, 10-times higher titres are tolerated. Note, however, that it may then be necessary to isolate individual colonies in a second round of plating (see Section 3.3.1).

3.3 Handling of yeast cosmid libraries

3.3.1 Storage, screening, and colony purification

i. Replica plating

Replica plating is one of the common procedures used for the storage and screening of cosmid libraries (e.g. ref. 2) and has been successfully applied to yeast cosmid libraries. We use a simple procedure outlined in *Protocol 9*.

Protocol 9. Replica plating of cosmid colonies

Several types of filter may be used:

- nitrocellulose filters (Millipore)—these have a low background in hybridizations;
- Colony/Plaque Screen (New England Nuclear Corp.) (13.7 mm diameter filters)
- Gene Screen (New England Nuclear Corp.)—robust filters for repeated use

Protocol 9. *Continued*

- Biodyne B Transfer Membrane (Pall)—large sheets that can be cut to appropriate sizes

1. Plate an appropriate amount of cells (*Protocol 8*) on LB agar plates containing 100 μg/ml ampicillin and incubate at 37 °C until the colonies are about 0.2 mm in diameter (this usually takes 8–12 h).
2. Keep the plates at 4 °C for 30 min prior to replica plating.
3. Place a blank filter on each plate. Mark the filter in several places (e.g. using a needle) so that it can be oriented with the master plate.
4. Remove the filter and use it for the screening procedure (see Section 3.3.1 *ii* below).
5. Incubate the master plate at 37 °C until the colonies reappear.
6. Steps **2–5** may be repeated two or three times to prepare further replica filters.
7. Seal the master plates with Parafilm and store at 4 °C in an inverted position. Colonies will stay intact for several months.

For long-term storage, replica filters can be kept on agar plates (containing 5% glycerol) at −20 °C.

ii. Screening

Replica filters can be screened with labelled probes by following one of the standard protocols for colony hybridization (12). If desired, the filters can be reused in several rounds of screening.

Once appropriate clones have been identified, it is possible to prepare the DNA from 5 ml cultures of these clones by the alkaline lysis procedure ('mini-preparation'; see Section 3.3.2) and to fix it on membrane filters by using semi-automated devices (for example, Schleicher & Schüll slot-blot apparatus). This approach gives much lower backgrounds in hybridization experiments.

iii. Colony purification

Positive colonies are surrounded by and frequently in contact with negative colonies. After an area on a replica filter has been identified, the positive colony is picked from the master plate or a wet replica filter (see *Protocol 9*) and suspended in 1 ml of LB medium containing 100 μg/ml ampicillin and 15% glycerol. (The sample may be stored at −70 °C.) Diluted aliquots of this bacterial stock are then plated on to fresh LB agar plates at a low density (100–1000 colonies per plate), replica filters are prepared, and screening is repeated.

iv. Storage

If desired, all colonies can be eluted from a master plate with an appropriate volume of LB medium containing 100 μg/ml ampicillin and 15% glycerol, and stored at −70°C. Aliquots of this bacterial stock (diluted with LB medium containing 100 μg/ml ampicillin) can be used for plating on new LB agar plates and subsequent preparation of replica filters.

Likewise, a number of single colonies sufficient to represent the yeast genome at a given probability (see *Table 1*) can be picked and grown separately in 1–5 ml cultures (in LB medium containing 100 μg/ml ampicillin). After adding 15% glycerol, these cultures can be stored at −70°C. Samples stored in this way are stable for many years. Aliquots from such bacterial stocks can be spotted on to filters in defined arrays at a high density by using automated devices. These filters may then be used in (automated) screening procedures.

3.3.2 Isolation and purification of DNA from cosmids

For analytical purposes, cosmid DNA can be prepared by one of the 'miniprep' procedures (alkaline lysis procedure or boiling procedure; 12) of cells grown from a single colony in a 2–5 ml overnight culture (in LB medium containing 100 μg/ml ampicillin inoculated with an isolated colony). These procedures will render semi-purified DNA, e.g. suitable for restriction analysis. When more demanding procedures are to follow, such as automated sequencing (20), the DNA can be very conveniently purified by the use of Quiagen tips. Detailed protocols are supplied by the manufacturer. On a preparative scale, cosmid DNA can be obtained from larger cultures by the same procedure using Quiagen 'maxi' instead of 'mini' kits for purification.

A rather simple and rapid 'miniprep' procedure has been devised by Del Sal *et al.* (21): it uses the cationic detergent cetyltrimethylammonium bromide (CTAB) to recover the extrachromosomal DNA. We have successfully applied this approach to the isolation of cosmid DNA of sufficient purity for screening procedures.

4. Cosmid mapping strategies

4.1 Chromosomal walking

We have used conventional chromosomal walking to construct a physical map of yeast chromosome II (R. Stucka and H. Feldmann, unpublished). Taking advantage of known markers, corresponding cosmids were sorted out by hybridization. Unique 'terminal' probes of these were then used to isolate extending cosmids. Standard techniques (12) were applied to the analysis of the single cosmids. Restriction maps were deduced from double and partial digests with the four enzymes, *Bam*HI, *Sal*I, *Xba*I, and *Xho*I, aided by the fast mapping procedure of Rackwitz *et al.* (22). Comparisons between digestion profiles of cosmids from neighbouring intervals were made to construct a high resolution (2 kb) physical map of chromosome II; hybridizations between these cosmids were used to confirm the map.

4.2 Chromosome fragmentation

4.2.1 Nested chromosome fragmentation using meganuclease I-*Sce*I

A rapid and efficient 'top down' mapping strategy has been developed recently (5). This generally applicable technique is based on chromosome fragmentation by the meganuclease I-*Sce*I, the first available member of a new class of endonucleases with very long recognition sequences. This enzyme allows complete cleavage at a single artificially inserted site in an entire genome. Sites can be introduced by homologous recombination using specific cassettes containing selectable markers. This strategy has been applied to the yeast chromosome XI as a first example to demonstrate the feasibility of this method: a set of transgenic yeast strains carrying the I-*Sce*I sites at various locations along chromosome XI defined physical intervals against which the cosmid clones were mapped by simple hybridizations. In the same way new genes or DNA fragments can easily be mapped.

4.2.2 Positional mapping with chromosome fragmentation vectors

Originally, this method was developed to characterize chromosomal positions in terms of physical distance between a gene and each of the telomeric ends of a linear chromosomal DNA molecule. It involves breaking the chromosome at this particular site and measuring the lengths of two chromosome fragments (proximal and distal to the gene) on pulsed field gels (23, 24). Specific vectors have been designed for this purpose, which—in addition to sequences required for propagation in *E. coli*—carry *SUP11* as a colour marker for recognition of copy number, the yeast *URA3* gene as a selectable marker, a yeast telomere sequence, and a polylinker region for cloning.

On this basis, a series of improved vectors, pMACS (Mapping Artificial Chromosome Segments), suitable for cosmid mapping has been developed (Hamberg, K., Fischer, D., and Philippsen, P., in preparation). In addition to the elements described above, these vectors contain the *lacZ-I* region plus a polylinker for convenient subcloning of cosmid fragments. Taking advantage of these vectors, mapping of the yeast chromosome XIV has been achieved (Hamberg, K., Fischer, D., and Philippsen, P., in preparation). When a linear molecule that carries a particular chromosomal site at one end, a centromere, and a telomere sequence at the opposite end, is introduced into the cell, a chromosome-copying mechanism is induced, which results in an additional stable chromosome segment in this transformant.

5. Complementation analysis in yeast using cosmids

One advantage of using pYc3030 as a vector is that cosmids from a yeast genomic library can be shuttled between *E. coli* and yeast cells. Yeast cells are transformed with closed circular cosmids by the usual procedure (25) and are easily selected by spreading on selective medium.

Whenever a gene has been located to an approximate position on the genetic map, a set of nested cosmids covering this region may be used in complementation experiments. For example, we have identified the genes *TYR1* (26) and *CIF1* (27) on yeast chromosome II by this procedure. When the chromosomal location of a gene is unknown, complementation experiments can be carried out using cosmid DNA from a whole cosmid library. In this case, sufficient DNA (0.5–1 mg) has to be used for yeast transformation to guarantee the presence of each single cosmid in the preparation.

Complementation of gene disruptions in haploids or diploids by a cosmid carrying the wild-type allele has also been achieved. From our experience, it appears that pYc3030 recombinant cosmids are maintained in yeast cells in only a few copies, despite the fact that they contain the 2 μm plasmid origin of replication.

References

1. Collins, F. and Hohn, B. (1979). *Proc. Natl Acad. Sci. USA*, **75**, 4242.
2. DiLella, A. G. and Woo, S. L. C. (1987). In *Methods in enzymology*, Vol. 152 (ed. S. L. Berger and A. R. Kimmel), pp. 199–212. Academic Press, London.
3. Vassarotti, A. and Goffeau, A. (1992). *Trends Biotechnol*, **10**, 15.
4. Clarke, L. and Carbon, J. (1976). *Cell*, **9**, 91.
5. Thierry, A. and Dujon, B. (1992). *Nucleic Acids Res.*, **20**, 5625.
6. Hauber, J., Stucka, R., Krieg, R., and Feldmann, H. (1988). *Nucleic Acids Res.*, **16**, 10623.
7. Ish-Horowicz, D. and Burke, J. F. (1981). *Nucleic Acids Res.*, **9**, 2989.
8. Evans, G. A. and Wahl, G. M. (1987). In *Methods in enzymology*, Vol. 152 (ed. S. L. Berger and A. R. Kimmel), pp. 604–10. Academic Press, London.
9. Knott, V., Rees, D. J. C., Cheng, Z., and Brownlee, G. G. (1988). *Nucleic Acids Res.*, **16**, 2601.
10. Hohn, B. and Collins, J. (1980). *Gene*, **11**, 291.
11. Hohn, B. and Hinnen, A. (1980). In *Genetic engineering* (ed. J. K. Setlow and A. Hollaender), pp. 169–83. Plenum Press, New York.
12. Sambrook, J., Fritsch, E. F., and Maniatis, T. (1989). *Molecular cloning. A laboratory manual*, 2nd edn. Cold Spring Harbor Laboratory Press, Cold Spring Harbor, NY.
13. Johnston, J. R. (1988). In *Yeast. A practical approach* (ed. I. Campbell and J. H. Duffus), pp. 107–23. IRL Press, Oxford.
14. Olson, M. V., Loughney, K., and Hall, B. D. (1979). *J. Mol. Biol.*, **132**, 387.
15. Herrmann, B. G. and Frischauf, A.-M. (1987). In *Methods in enzymology*, Vol 152 (ed. S. L. Berger and A. R. Kimmel), pp. 180–3. Academic Press, London.
16. Frischauf, A.-M. (1987). In *Methods in enzymology*, Vol. 152 (ed. S. L. Berger and A. R. Kimmel), pp. 183–9. Academic Press, London.
17. Hohn, B. and Murray, K. (1977). *Proc. Natl Acad. Sci. USA*, **74**, 3259.

18. Scalenghe, F., Turco, E., Edström, J. E., Pirrotta, V., and Melli, M. (1981). *Chromosoma*, **82**, 205.
19. Leder, P., Tiemeier, D., and Enquist, L. (1977). *Science*, **196**, 175.
20. Voss, H., Zimmermann, J., Schwager, C., Erfle, H., Stegemann, J., Stucky, K., and Ansorge, W. (1990). *Nucleic Acids Res.*, **18**, 5314.
21. Del Sal, G., Manfioletti, G., and Schneider, C. (1988). *Nucleic Acids Res.*, **16**, 9878.
22. Rackwitz, H. R., Zehetner, G., Frischauf, A.-M., and Lehrach, H. (1984). *Gene*, **30**, 195.
23. Vollrath, D., Davis, R. W., Connelly, C., and Hieter, P. (1988). *Proc. Natl Acad. Sci. USA*, **85**, 6027.
24. Gerring, S. L., Connelly, C., and Hieter, P. (1991). In *Methods in enzymology*, Vol 194 (ed. C. Guthrie and G. R. Fink), pp. 57–77. Academic Press, San Diego.
25. Itoh, H., Fukuda, Y., Murata, K., and Kimura, A. (1983). *J. Bacteriol.*, **153**, 163.
26. Mannhaupt, G., Stucka, R., Pilz, U., and Feldmann, H. (1989). *Gene*, **85**, 303.
27. Gonzalez, M. I., Stucka, R., Blazquez, M. A., Feldmann, H., and Gancedo, C. (1992). *Yeast*, **8**, 183.

4

The construction and use of cDNA libraries for genetic selections

JOHN T. MULLIGAN and STEPHEN J. ELLEDGE

1. Introduction

The popularity of the yeast *Saccharomyces cerevisiae* as a research system is due in large part to its facility as a genetic organism and its usefulness as a model eukaryote. It combines the straightforward collection and analysis of mutations with the ability to isolate the corresponding genes by transformation and complementation. All of the genetic elements necessary for promoting chromosomal replication and segregation reside on small fragments which have been incorporated, along with selectable markers, into small cloning vectors capable of replicating in *Escherichia coli* or yeast.

This paper describes the construction and use of cDNA and genomic DNA libraries in a new class of regulated expression vectors, the λ YES vectors (1). These vectors allow the regulated expression of inserts in yeast and in *E. coli* by fusing the coding regions to the strong regulated *GAL1* promoter in *S. cerevisiae*, and the *lacZ* promoter in *E. coli*. They have several advantages over the standard vectors. First, since the *GAL1* promoter is regulated, it can be manipulated to express cDNAs at low, medium, or high levels to achieve optimal effects. Secondly, the use of cDNA or small genomic DNA fragments circumvents the potential problem of linkage to toxic genes. Thirdly, once a clone with the desired properties has been isolated, the location of the gene is known, as it must be adjacent to the promoter. Often with 2 μm genomic libraries, one must pare down the 10 or 15 kb of genomic DNA present in the clone to identify the position of the gene of interest. Using the λ YES vectors, the sequence of the 5′ and 3′ ends of the isolated gene can be ascertained immediately to determine whether it is a novel or previously identified gene. Finally they can also be used for antibody screening as with λgt11.

2. cDNA cloning vectors

The class of λ YES (yeast–*E. coli* shuttle) vectors (1) were designed for the construction of large cDNA libraries with a high percentage of inserts, regulated expression in yeast or *E. coli*, and simple conversion from a phage to a plasmid to facilitate both recovery of inserts and introduction of libraries into yeast. λ YES-R (RNA) is a 42.6 kb phage with a 8.4 kb insert capacity. The plasmid portion of λ YES is flanked by direct repeats of *lox* sites; *cre*-mediated site-specific recombination between these sites can be used to convert the phage form of the vector into a plasmid. Infection of a *cre*-producing strain (e.g. BNN132) with λ YES results in the production of ampicillin-resistant colonies (by circularization of the plasmid portion of the vector and the loss of the phage λ portion of the vector) with an efficiency of about 50% relative to plaque forming units.

The plasmid version of the vector (see *Figure 1*) contains the ColE1 origin, *lac* promoter, and *bla* gene for replication, expression, and selection in *E. coli*, and the *URA3, ARS1, CEN4*, and *GAL1* promoter sequences for replication, expression, and selection in *S. cerevisiae*. The *GAL* and *lac* promoters are placed in convergent orientations on opposite sides of the *Xho*I cloning site. Since the typical cDNA cloning strategy is non-directional, half of the inserts are in the proper orientation for expression in yeast and half in that for expression in *E. coli*. Directional libraries can be made in λ YES using *Eco*RI or *Xho*I and *Xba*I although the *E. coli* translational start is removed. The *lac* promoter contains a ribosome binding sequence optimally spaced from an AUG start codon and is capable of making protein fusions. The *GAL1* promoter used lacks a translational start. Adjacent to the *lac* promoter is a termination sequence derived from *HIS3* in the same orientation as *GAL1*-initiated transcripts. A second vector λ YES-P (protein), contains the *GAL1* promoter and the first 29 amino acids of the *GAL1* protein adjacent to the cloning site, and is capable of making protein fusions in yeast. Such fusion proteins expressed in yeast may have unique genetic properties which are described below.

One disadvantage of cDNA libraries is that a given cDNA appears in a library with a frequency in direct proportion to the abundance of the mRNA. If the abundance of the mRNA of the gene of interest is very low, then many more clones will have to be screened or placed under selection to identify it. Furthermore, if there are many different genes that can produce the desired phenotype, one will have to sort through many examples of the abundant classes before isolating rare clones. In this respect a genomic library is advantageous because the abundance of each gene is equal.

The advantages of regulated expression cDNA libraries and of equal representation genomic libraries have recently been combined in a new type of library, a random-shear genomic library under the control of the *GAL1* promoter. A library of this type, constructed by Ramer *et al.* (2), was made in λ YES-R and has 5×10^7 total recombinants, representing a break at every nucleotide in the yeast genome on the average. Thus, most yeast genes are present at approximately equal frequency under the control of the *GAL1* promoter. If complementation is under *GAL* control, the position of the gene of interest is defined. This type of library is even more useful for genetic screens other than complementation, as described below. Many specialized λ YES derivatives exist including λ ACT which is described below.

3. Genetic uses for regulated expression libraries

3.1 Dosage-dependent suppressors

2 μm libraries and regulated cDNA and genomic expression libraries have proven very useful for the isolation of dosage-dependent suppressors of mutations. Apparently, the properties of many genetic systems are such that overproduction of a protein in the pathway can compensate for certain types of mutations in the pathway. This property allows isolation of additional genes that can function in the pathway of interest. Often the new proteins physically associate with the mutant protein of interest. For example, the *CLN1* and *CLN2* genes were isolated as dosage-dependent suppressors of a *cdc28* mutation (3). These genes encode regulatory subunits of the Cdc28 protein kinase. Other examples of this type of suppression are too numerous to list here. It should be noted that simply increasing the dosage of a gene does not guarantee overproduction of the protein encoded by the gene. By placing a gene under the control of a regulated promoter, the potential for overproduction or temporally inappropriate (heterochronic) expression is increased, and with it the chances of observing a phenotype distinct from a 2 μm genomic library.

3.2 Dominant phenotypes

Gene isolation is not limited to complementation or suppression of mutations. Genes can also be isolated based on their ability to produce phenotypes in a wild-type background when overproduced. For example, overproduction of genes encoding target proteins of drugs can provide resistance to those drugs and a convenient method of cloning the gene, as demonstrated by Rine for HMG-CoA reductase (4). Other examples of dominant phenotypes exploited for gene isolation are α-factor resistance (5) and overproduction lethality (2). In the latter example, regulated expression libraries are essential for identification of this class of genes.

Expression of mutant proteins may also produce dominant phenotypes. It is possible to mutagenize a library and screen for dominant phenotypes in wild-type cells or as suppressors. This approach has been used in bacteria, but has not yet been exploited in yeast. λ YES-P libraries can be constructed that overexpress only mutant proteins (i.e. fusions of truncated open reading frames to the first 29 amino acids of the *GAL1* protein). In many cases the additional amino acids will not alter the function of the protein. However, expression of truncated proteins that lack certain functional domains will occasionally result in dominant effects. The truncated proteins may have lost

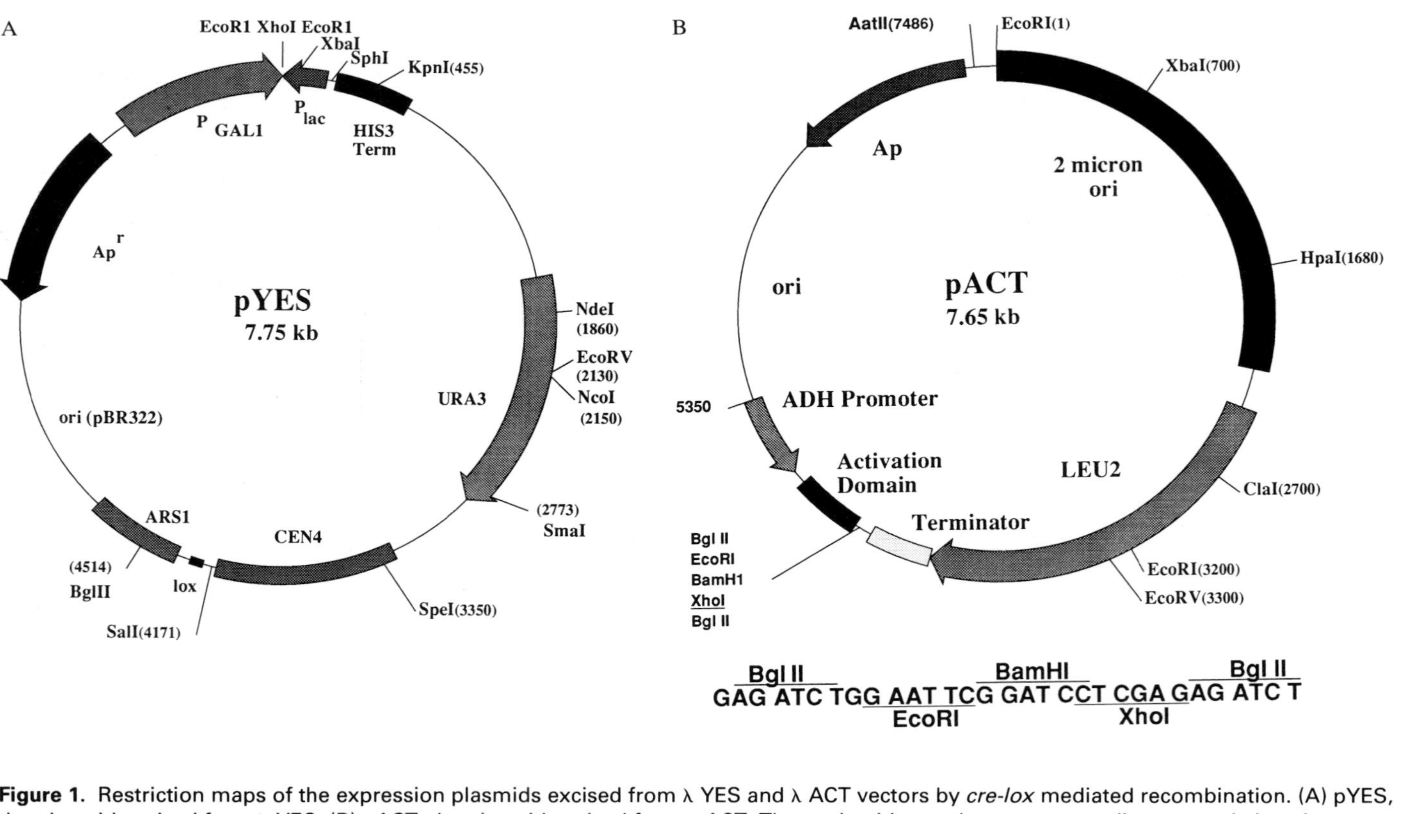

Figure 1. Restriction maps of the expression plasmids excised from λ YES and λ ACT vectors by *cre-lox* mediated recombination. (A) pYES, the plasmid excised from λ YES. (B) pACT, the plasmid excised from λ ACT. The nucleotide numbers corresponding to restriction sites were generated from published sequences and are only approximate due to uncertainties in the sequence of junction fragments. The sequence of the polylinker in pACT is shown below in-frame coding triplets. cDNAs are inserted at *XhoI* sites for both vectors.

a domain which regulates their activity; alternatively, a truncated protein which is inactive may interfere with the function of the wild-type protein by ineffective participation in complexes or by competing for substrates. The λ YES-P libraries express mutant proteins that have properties distinct from mutants produced by chemical mutagenesis. Protein fusion libraries can over-produce proteins lacking amino-terminal amino acids. Chemical mutagenesis will occasionally inactivate a domain, but will rarely do this and overproduce the mutant protein at the same time. Thus, λ YES-P libraries provide a new genetic tool for the isolation of genes.

3.3 Gene isolation from other organisms by selection in yeast

Most higher eukaryotic systems lack the facile genetics and ease of gene isolation that yeast enjoys. However, a surprising number of proteins from other organisms retain their function when expressed in yeast. This property has been exploited to isolate genes from other organisms using cDNA expression libraries in yeast coupled with genetic screens. The most notable examples of cross-species complementation have been in the isolation of the human *CDC2Hs* and *CDK2* genes by complementation of *cdc2* in *Schizosaccharomyces pombe* (6) and *cdc28* in *S. cerevisiae* (7), respectively. The protein kinases encoded by these genes play important roles in control of cell cycle progression and thus have been highly conserved throughout evolution.

Gene isolation is not limited to complementation. The genetic interactions described above apply for heterologous genes as well as yeast genes. Heterologous genes can be isolated as suppressors of yeast mutants or by dominant phenotypes as described above. Heterologous genes may even have a slight advantage over homologous genes for producing dominant phenotypes. Many of these have diverged significantly and may no longer retain the same specificity of protein–protein interaction. Thus, they may behave as mutant proteins and produce novel phenotypes that the yeast homologues would not.

Of course, the yeast genome is much smaller than the genomes of higher eukaryotes and cannot have homologues to all their genes. Therefore, cloning by complementation and suppression is limited primarily to the isolation of genes essential to basic cellular processes. Recently, more general cloning schemes have been devised that broaden the menu of genes that can be isolated using expression in yeast. In particular, the two-hybrid system has been devised which allows genes to be cloned that physically associate with a target protein *in vivo* (8). This cloning scheme is based on the fact that transcription factors have separable DNA binding and transcriptional activation domains. These need not be encoded in the same protein to activate transcription, only to associate physically *in vivo* so as to link DNA binding and transcriptional activation. To carry out this approach, two sets of fusion proteins are constructed. One generates an in-frame hybrid between sequences encoding the DNA binding domain of the yeast transcription factor Gal4 (amino acids 1–147), and sequences encoding a portion of the target protein. A second expression plasmid contains sequences encoding the Gal4 activation domain II (amino acids 768–881) fused to a cDNA library generated from the organism of interest. If the two proteins expressed in yeast are able to interact, the resulting complex will regain the ability to activate transcription from promoters containing Gal4 binding sites, UAS_G. Recently, yeast strains have been constructed in which activation of gene expression through UAS_G results in expression of a selectable marker so that interacting clones can be genetically selected (9). Furthermore, a new lambda based cloning vector, λ ACT (activation domain), has been designed that allows the simple construction of activation domain-tagged cDNA libraries (10) (see *Figure 1*). Many new genes have been isolated using the two-hybrid system, including genes interacting with the retinoblastoma protein (9), serum response factor (10), and a protein kinase, *SNF1* (11). Artificial selections, such as those designed to isolate DNA binding proteins by having their binding sites control expression of selectable markers in yeast, are also being explored.

3.4 Isolation of genes by complementation of *E. coli* mutations using λ YES

Like yeast, many mutations are available in *E. coli* which affect many of the basic processes of cellular metabolism. With the λ YES vectors one can easily attempt the complementation of these mutations with cDNA or genomic libraries. In addition, artificial selections have been devised that allow selection of activities not normally present in *E. coli*. For example, a genetic selection for sequence-specific DNA binding proteins exists that can be used with cDNA expression libraries (12).

4. Construction of cDNA libraries in the λ YES vectors

4.1 General considerations

The construction of a good cDNA library represents a few weeks of work by one or two people; the isolation of genes from a library and the analysis of these genes can involve many person-years of effort. It will pay to spend the time necessary to make the best possible library and ensure that the subsequent effort of analysing this library is well rewarded. Controls on each critical step will help to ensure that many or most of the cDNAs are full-length, that few clones in the library lack inserts, and that the total number of recombinants is large enough that each member of the mRNA population has a high chance of being represented in the cDNA library.

Since much of the effort of constructing a cDNA library is in assembling and testing all of the necessary components, making several libraries at once involves only slightly more effort than making one library. If you are going to make one, you might as well make several from related organisms, other strains, different tissue types, or other developmental stages.

4.2 cDNA synthesis

Protocol 1 for cDNA synthesis is a combination of the first strand synthesis protocol from Krug and Berger (13) and the second strand synthesis protocol of D'Alessio and Gerard (14). We strongly recommend reading these two papers for background information on cDNA synthesis. Three controls are recommended for this protocol. First, probe a Northern blot of the poly(A)-purified mRNA with, for example, tubulin to verify that the starting template is mostly full-length. The first and second strand synthesis reactions should be tested separately on a control RNA to ensure that these steps are efficient. We use a 7.5 kb poly(A)-tailed RNA from BRL (catalogue no. 5621SA) as a control template, add radioactively labelled dCTP to either the first strand reaction or the second strand reaction, and monitor the efficiency of the synthesis by alkaline agarose gel separation of the synthetic products. More than 50% of the total incorporation should be in a 7.5 kb band.

Dilution of the AMV reverse transcriptase is reported to increase the activity of the enzyme, as measured by the production of full-length transcripts of control RNAs (13). Incorporation of radioactive dCTP in the second strand reaction is optional, as 5 μg of mRNA should yield enough cDNA to detect on an ethidium bromide stained agarose gel; however, radioactivity will help keep track of the pellet when using the spermine precipitation in step **6**. We use only trace amounts of radioactivity such that very few (<1%) of all cDNAs contain radioactivity. cDNA stored at −70°C will then remain fully functional for future libraries.

Protocol 1. cDNA synthesis

Equipment and reagents

- 16°C, 37°C, and 42°C heat blocks
- poly (A)-purified RNA (oligo (dT)-purified twice to remove rRNA)
- AMV reverse transcriptase (Sekagaku); DNA polymerase I (Pharmacia 27–0926); RNase H (Pharmacia 27–0894); *E. coli* DNA ligase (Pharmacia 27–0872)
- AMV dilution buffer (0.2% Triton, 2 mM DTT)
- 10 × AMV reverse transcriptase buffer (500 mM Tris–HCl, pH 8.3, 500 mM KCl, 100 mM $MgCl_2$, 10 mM dithiothreitol (DTT), 10 mM EDTA, 100 μg/ml acetylated bovine serum albumin (BSA) (RNase-free))
- dNTP mix (25 mM each of dATP, dCTP, dGTP, and dTTP) (Pharmacia)
- 2 × second strand buffer (50 mM Tris–HCl, pH 8.3, 200 mM KCl, 10 mM $MgCl_2$, 10 mM DTT)
- 10 × T4 DNA polymerase buffer (500 mM NaCl, 100 mM Tris–HCl, pH 7.9, 100 mM $MgCl_2$, 10 mM DTT)
- 10 × ligation buffer (500 mM Tris–HCl, pH 7.6, 100 mM $MgCl_2$, 100 mM DTT, 500 μg/ml acetylated BSA)
- RNAsin (40 units/μl) (Promega)
- other solutions and reagents: 10 mM spermidine; 15 mM β-NAD (Pharmacia); 80 mM sodium pyrophosphate; 1 mg/ml oligo (dT)
- [α-^{32}P]dCTP (3000 Ci/mmol; 10 mCi/ml) (Amersham)

Protocol 1. *Continued*

Method

1. Dilute AMV reverse transcriptase 1:4 in AMV dilution buffer and store on ice for 30 min (12 units per cDNA synthesis).
2. Resuspend 5 μg of poly $(A)^+$ mRNA in 25 μl of water, heat to 65°C for 3 min, and cool on ice.
3. Mix on ice:

• diluted AMV reverse transcriptase	12 units
• 10 × AMV reverse transcriptase buffer	5.0 μl
• 1 mg/ml oligo (dT)	2.5 μl
• 25 mM dNTP mix	2.0 μl
• 10 mM spermidine (**not spermine**)	2.5 μl
• RNAsin (40 units/μl)	3.0 μl
• 80 mM sodium pyrophosphate	2.5 μl
• water to a total volume of 25 μl	

4. Mix RNA into enzyme solution from step **3** and incubate at 42°C for 2 h, then transfer to ice.
5. (a) Mix on ice just before the first strand reaction from step **4** is finished:

• 2 × second strand buffer	200 μl
• dNTP mix (25 mM each)	2 μl
• 15 mM β-NAD	4 μl
• DNA polymerase I	100 units
• RNase H	4 units
• *E. coli* DNA ligase	4 units
• [α-^{32}P]dCTP	2 μl

   (b) Add water to 350 μl, and add to the finished first strand reaction from step **4.** Incubate for 2 at 16°C.
6. Spermine precipitate (see *Protocol 2*) *or* add 40 μl of 250 mM EDTA, extract once with 400 μl phenol and once with 400 μl of chloroform, and precipitate by addition of 24 μl 5 M NaCl and 800 μl of ethanol.
7. (a) Resuspend the cDNA in 40 μl of TE buffer and then add, in this order,

• 10 × T4 DNA polymerase buffer	5 μl
• dNTP mix (25 mM each)	1 μl
• T4 DNA polymerase (1 unit/μl)	4 μl

   (b) Incubate at 37°C for 30 min.

8. Stop the reaction with 5 μl of 250 mM EDTA and 3 μl of 5 M NaCl. Extract once with 50 μl of phenol, once with 50 μl of chloroform, and precipitate with 125 μl of ethanol. Rinse with 70% ethanol and dry briefly.

Protocol 2. Spermine precipitation

Reagents

- 100 mM spermine-HCl
- spermine wash buffer (70% ethanol, 10 mM $MgCl_2$, 0.3 M $Na(OAc)_2$, pH 7)
- 70% ethanol
- 1 M KCl

Method

1. Add 100 mM spermine to a final concentration of 5 mM for cDNA and mix well. The presence of 100 mM KCl will inhibit the precipitation of oligonucleotides shorter than 50 bp and it should be added when removing adaptors. Spermine precipitation of λ DNA requires less spermine (0.5–1 mM final concentration).
2. Incubate on ice for 30 min. Collect the precipitate by microfuge centrifugation for 15 min. For λ DNA only a very brief (2 sec) spin is needed; prolonged pelleting makes λ DNA extremely difficult to resuspend later.
3. Carefully remove the supernatant with a Pipetman. Add 1 ml of spermine wash buffer and incubate on ice for 1 h. Remove supernatant taking care not to disturb the pellet, add 1 ml of wash buffer, and incubate for an additional hour. Rinse with 70% ethanol to remove buffer and dry briefly.

4.3 cDNA cloning

Adaptors on the inserts and a partial fill-in strategy for the vector are employed to select for inserts (1, 13). The partially filled-in vector can only ligate to the cDNA, and vice versa. This protocol routinely results in libraries with 95–99% inserts and approximately 10^8–10^9 clones per μg of starting mRNA (1). The adaptors are added to the insert at a molar ratio of more than 10:1 to reduce the chance of producing chimeric clones by the ligation of two cDNAs. Using two adaptors reduces the chance that secondary structure in the resulting mRNA will interfere with expression of the cDNA gene product. With one adaptor, all of the inserts would be flanked by 19 bp inverted repeats. Use of two adaptors ensures that half of the cDNAs will be flanked by adaptors of differing sequence. Adaptors should be designed to recreate *XhoI* sites and to lack start and stop codons.

Removal of residual adaptor dimers and size selection of the cDNA are accomplished by spermine precipitation coupled with agarose gel purification. The gel is run in order to size-select the fragments, which are subsequently purified. Our experience has been that the isolation of DNA fragments from agarose gels is more efficient when the concentration of the DNA is high and the volume of the gel slice is low. One way to optimize this is to run the gel and remove the desired size class in one piece, then insert that slice into a new low melting point gel at a lower position and then run the gel in reverse for the same amount of time to concentrate the cDNAs before purification out of the agarose gel.

Protocol 3. Adaptor addition and size selection of the cDNA

Equipment and reagents

- 16°C incubator
- gel box
- low melting point agarose (FMC Bioproducts)
- 1 × TAE buffer
- ethidium bromide (10 mg/ml)
- bromophenol blue (gel-loading buffer) (16)
- long wavelength (360 nm) UV illuminator
- 10 × ligation buffer
- 10 mM ATP
- T4 DNA ligase (New England Biolabs, 400 units ml)
- 100 μM adaptor mix (100 μM each of 5′-P-CGAGAGTTCAC-3′, 5′-P-GTGAACTC-3′, 5′-P-CGAGATTTACC-3′, and 5′-P-GGTAAATC-3′), all synthesized either with the 5′ phosphate on or kinased, the first two and the second two should be mixed to allow them to anneal before the two pairs are mixed together

Method

1. Resuspend the cDNA from *Protocol 1* in 12 μl of TE buffer.
2. (a) Add on ice:
   - 10 × ligation buffer — 2 μl
   - 10 mM ATP — 1 μl
   - 100 μM phosphorylated adaptor mix — 3 μl
   - T4 DNA ligase — 2 μl

   (b) Incubate overnight at 12°C.
3. Add 160 μl of TE buffer, 20 μl of 1 M KCl and spermine precipitate (see *Protocol 2*). This should remove most adaptors and is **highly recommended.** Resuspend in 20 μl of TE buffer.
4. While washing the spermine pellet, prepare a 1% low melting point agarose gel with 1 × TAE buffer and 1 μg/ml ethidium bromide. Chill to 4°C for 20–60 min. Add 1 × TAE buffer to the height of the top of the gel: **do not** cover the gel with buffer.
5. Add 1/10 volume bromophenol blue to the cDNA and load on to gel.

Run the gel forward at constant voltage until the bromophenol blue is about halfway or less to the end of the gel.

6. Remove and discard the buffer and move the gel to a long-wavelength UV illuminator (360 nm, **NOT** 254 or 300 nm). Using a clean scalpel blade, slice across the sample lanes at 600 and at 9000 bp. Cut out a patch of agarose that is bounded by these cuts and contains all of the samples. Remove and discard the agarose containing the low molecular weight DNA (<600 bp) which contains the adaptor dimers. You might want to save a second slice from 600 to 400 bp, since occasionally some mRNAs—and so full-length cDNAs—are small, although most molecules of this size class are truncated cDNAs of little value.
7. If the cDNA is not to be concentrated by a second reverse electrophoresis in a new low melting point gel, then trim the gel slice carefully to remove agarose containing no DNA and determine the Cerenkov counts.
8. Purify the size-selected cDNA using a low melting point phenol extraction or glass bead protocol, according to the instructions of the manufacturer, or your preferred isolation technique. Resuspend the cDNA in 5 μl of TE buffer. Estimate the yield of cDNA based on Cerenkov counting (if a radioactive label was incorporated into the cDNA), by fluorescence (14), or by gel electrophoresis.

The most reliable way to prepare the partially filled-in vector is to digest 200–300 μg of vector DNA with *Xho*I, spermine precipitate, and carry out *Protocol 5* on several lots of vector in parallel with varying amounts of *Taq* polymerase or varying lengths of incubation time. We have found that brief filling-in reactions give the largest libraries, but average about 5–10% empty vectors in the final libraries. Longer incubations result in 10- to 100-fold smaller libraries with a lower percentage of empty vectors. Prepare an adapted restriction fragment, about 1 kb in length, as a positive control for packaging.

Note: the Packagene packaging extract from Promega is very useful for controls on vector preparation and testing of the ligation mixes. It is less expensive than the Stratagene packaging extract, and each 50 μl aliquot can be split into five aliquots to reduce further the cost of these controls. The final packaging of successful libraries should be done with a high efficiency packaging extract like Stratagene GigaPack Gold II to ensure the largest possible libraries.

Amplification of the libraries is carried out on a *lacI^q^* strain of *E. coli* to reduce the expression of the *lacZ* promoter and reduce the selection against any cDNA that has a deleterious effect if expressed in *E. coli*. Any subsequent selection for functional expression in *E. coli* must be carried out on a non-catabolite-repressing carbon source such as LB, mannitol, or lactose because this promoter is subject to catabolite repression.

The libraries should be stable indefinitely at −70°C in 7% DMSO. They are aliquoted into small tubes for working stocks, and to simplify distribution of the libraries. We routinely send libraries at room temperature by the Postal Service in the same tubes in which they were frozen at the time of initial amplification. Lambda stocks are stable for weeks under these conditions and will arrive safely unless exposed to extreme temperatures. The remainder of the lysate can be stored in sterile 50 ml Falcon tubes at −70°C.

Protocol 4. Library construction, initial characterization, and storage

Equipment and reagents

- screw-capped 1–2 ml tubes for storage at −70°C
- λ dilution buffer (10 mM Tris–HCl pH 7.5, 10 mM $MgSO_4$)
- 10 × ligation buffer
- 10 mM ATP
- T4 DNA ligase (New England Biolabs, 400 units μl)
- prepared λ YES-R vector arms (see *Protocol 5*)
- Packagene packaging extract (Promega)
- Gigapack II Gold packaging extract (Stratagene)
- lambda plating bacteria
- BNN132 lambda plasmid excision bacteria
- LB top agar (0.8% agar)
- 30 150 mm LB plates
- fresh bottle of DMSO

Method

1. Mix on ice:
- 10 × ligation buffer — 2.5 μl
- 10 mM ATP — 2.5 μl
- T4 DNA ligase — 2.5 μl
- water — 2.5 μl

2. (a) Mix on ice:
- T-filled λ YES-R vector arms (1 μg/μl) (see *Protocol 5*) — 2 μl
- ligation mix from step **1** — 2 μl
- cDNA (100 ng/μl) — 1 μl

(b) Prepare a control reaction with TE buffer in place of the cDNA. Incubate at 4°C overnight.

3. Package 1 μl of each of the ligations with 10 μl of Packagene packaging extract (Promega) according to the manufacturer's instructions.

4. Prepare lambda plating cells as described in Sambrook *et al.* (16), and titre the packaged libraries and controls. The cDNA ligations should give a total of 10^5–10^6 plaques (10- to 100-fold more than the no-insert controls). If so, proceed to step **5**.

5. Package the remaining 4 μl of the cDNA ligations with one Gigapack Gold packaging extract (Stratagene) according to the manufacturer's instructions, and titre as in step **4**. You should see a total of 10^7–10^8 p.f.u./ml from this packaging.
6. Test the library for inserts by performing *cre-lox* automatic subcloning.
   (a) Mix 0.1 μl of the packaged library with 100 μl of logarithmically growing BNN132 bacteria in LB + 10 mM $MgCl_2$.
   (b) Incubate at 37 °C for 30 min.
   (c) Add 1 ml of LB and shake at 37 °C for 30 min.
   (d) Spread 1 μl, 10 μl, and 100 μl on LB/ampicillin plates, and incubate overnight at 37 °C. The number of colonies should be 0.2–0.8 times the number of plaques expected from the same amount of packaging mix.
   (e) Prepare DNA from 24 colonies and restriction map with *XhoI* to determine the proportion of the clones with inserts and the size distribution of the inserts.

If 85–95% of the clones in the library have inserts of the appropriate size, proceed to step **7**.

7. Amplify and store the library:
   (a) Mix 1 × 10^7 p.f.u. of the library with 0.5 ml of a fresh overnight culture of *$lacI^q$* lambda plating bacteria (**not** BNN132) grown in LB + 10 mM $MgCl_2$. For λ ACT, LE392 is the host of choice because cDNAs are not expressed in *E. coli*.
   (b) Incubate for 15 min at 37 °C.
   (c) Add 6.5 ml of molten (47 °C) LB top agar to each tube and spread on a fresh, prewarmed at 37 °C, 150 mm LB plate. Allow to harden at 25 °C.
   (d) Incubate for 6–8 h at 37 °C, until the plaques meet. Do not incubate for longer than 8 h because the titre will be reduced.
   (e) Add one drop of chloroform, incubate for 5 min, add 15 ml of λ dilution buffer to each plate, and at 4 °C rotate or rock gently overnight if possible.
   (f) Pool the phage in λ dilution buffer, extract once with chloroform, centrifuge, and collect the supernatant, taking care to avoid the bacterial debris on the chloroform/aqueous interface.
   (g) Add DMSO (7% by volume) to the supernatant.
   (h) Aliquot 1–2 ml into each of 100 sterile, screw-capped tubes and freeze in liquid nitrogen. Store at −70 °C. Some of the library can be kept at 4 °C.
   (i) Titre the amplified library. It should be about 10^{10} p.f.u./ml.

Protocol 5. Preparation of 'T-filled' λ YES-R arms

Reagents

- *Xho*I restriction enzyme
- 10 mM spermine
- *Taq* DNA polymerase (Perkin Elmer)
- *Taq* DNA polymerase buffer (supplied by manufacturer)

Method

1. Prepare λ YES-R plasmid DNA:
   (a) Prepare λ plating bacteria from an *E. coli* strain (**not** BNN132).
   (b) Mix 1 μl of λ YES-R phage with the plating bacteria.
   (c) Incubate for 20 min at 28°C
   (d) Add 1 ml of LB and incubate for 30 min at 28°C with shaking.
   (e) Spread on LB/ampicillin plates and incubate overnight at 28°C.
   (f) Prepare CsCl-purified plasmid DNA from the ampicillin-resistant lysogens as you would for any plasmid, but grow at room temperature in 3 or 6 litres of LB with 50 μg/ml ampicillin. For highest purity, do two CsCl gradients. This should result in 2 mg or more of vector.
2. Digest 100 μg of λ YES-R plasmid DNA with *Xho*I. Less can be used if needed.
3. Precipitate with ethanol, dry briefly (but **not to completion**) and resuspend in 190 μl of TE buffer.
4. Add 10 μl of 10 mM spermine to the side of the tube. Mix into the DNA solution by rapid inversion. An immediate and obvious precipitate should form; spin for 2 sec in a microfuge, and wash as described in step **3** of *Protocol 2*. Resuspend in 90 μl of TE buffer.
5. Add 10 μl of 10 × *Taq* DNA polymerase buffer and 1 μl of 2.5 mM dTTP.
6. Incubate at 72°C for 5 min, add 1 unit of *Taq* DNA polymerase, and incubate at 72°C for 1 or 2 min.
7. Add 90 μl of water, spermine precipitate by addition of 5 μl of 10 mM spermine, and wash as in step **4**. Resuspend in 100 μl of TE buffer and use for ligations with adapted cDNA.

Protocol 6. Conversion of a library in λ YES-R into a plasmid library

Reagents

- logarithmically growing BNN132 bacteria
- 10 fresh 150 mm LB plates containing ampicillin (50 μg/ml) and 0.2% glucose
- 3 litres of 'terrific broth' (16)

Method

1. Mix 10^8 phage from the amplified library with 2 ml of logarithmically growing BNN132 bacteria (3×10^8 cells/ml) + 10 mM $MgCl_2$ and incubate at 30 °C without shaking for 30 min. The number of phage and cells used can be scaled up for better representation.
2. Add 2 ml of LB and shake at 30 °C for 1 h; cells can be concentrated or plated directly.
3. Spread 200 μl of the cells on each of 10 fresh, 150 mm LB plates containing 50 μg/ml ampicillin and 0.2% glucose, and incubate overnight at 37 °C. You should see a lawn of cells. Occasionally small plaques are seen in the lawn. These can be safely ignored. Dilutions of the original infection should be plated to determine excision efficiency. Poor efficiency is usually due to using stationary phase BNN132 cells, using too much ampicillin in the plates, or adding no $MgCl_2$ during absorption.
4. Add 10 ml of LB to each plate and resuspend the ampicillin-resistant colonies. Pool the bacteria from all 10 or more plates and use to inoculate 3 litres of 'terrific broth' (16) with 50 μg/ml ampicillin. Grow to stationary phase and make plasmid DNA by CsCl gradients.

5. Complementation of mutants with cDNA libraries

Cloning by complementation depends on a good selection scheme, and on an adequate representation of the complementing activity in the library. The representation of an activity in the libraries may be much lower than the representation of the corresponding mRNA in the original RNA population. The libraries are non-directional and expression may in some cases require a particular circumstance in the cloning event (e.g. fusion in the correct reading frame or deletion of part of the untranslated region of the message).

The ease of screening the products of a selection with these vectors makes it feasible to isolate complementing clones even when the selection is not tight. *E. coli* complementing clones can be isolated and re-transformed into the mutant strain; in yeast, plasmid loss can be selected with 5-fluoro-orotic acid (5-FOA) to test for reversal of phenotype which allows one to distinguish genuine complementation from revertants.

Complementation in yeast is limited by transformation efficiency. The best transformation protocol is by Schiestl and Gietz (17) (see Chapter 8) and uses sonicated single-stranded DNA or total RNA as a carrier. We have found RNA to be the most reliable carrier. Transformation efficiency varies from strain to strain, so you might want to start with a strain known to have a reasonable transformation efficiency, $10^4/\mu g$ or better (frequencies of $>10^6/\mu g$ have been observed). With λ YES, transformants must eventually be plated on media containing galactose. However, yeast grown on galactose transform very poorly and glucose-grown cells show poor efficiency when plated directly on to galactose plates. It has also been observed that the *GAL1* promoter takes about one generation to induce fully when switched from glucose to galactose. This can be overcome by first establishing transformants on glucose liquid medium (or on plates), then transferring to galactose medium for about one generation before plating under selective conditions on 2% galactose plates. The expression level of the *GAL1* promoter can be varied by growth on different carbon sources with the following order of strength (2% galactose > 1% galactose + 1% glucose > 2% glycerol > 2% glucose) and different conditions should be explored.

Protocol 7. Complementation of *S. cerevisiae* mutants: transformation protocol

Equipment and reagents

- 42°C waterbath
- YPD broth (see Appendix 1)
- LiTE (100 mM LiAc, 10 mM Tris–HCl, pH 8, 1 mM EDTA)
- LiSorb (LiTE + 1 M sorbitol)
- plasmid DNA from a λ YES-R cDNA library (see *Protocol 6*)
- SC-uracil (SC-Ura) medium (liquid and plates) (18) (see Appendix 1)
- 10 mg/ml yeast total RNA
- LiPEG (LiTE + 40% PEG 3350 (w/v))

Method

1. Grow the recipient strain of yeast to mid-log phase (10^7 cells/ml) in 1 litre of YPD.
2. Pellet the cells and resuspend in 500 ml of LiTE.
3. Pellet the cells and resuspend in 25 ml of LiSorb.
4. Incubate for 20 min at 30°C with shaking.
5. Pellet the cells and resuspend in 2.6 ml of LiSorb.
6. Reserve 100 μl of cells for a negative control and add 50 μg of cDNA plasmid DNA and 500 μl of 10 mg/ml yeast total RNA.
7. Mix well and then incubate for 10 min at 30°C without shaking.
8. Add 22.5 ml of LiPEG and mix well. Transfer to a 125 ml flask.
9. Incubate the flask in a 42°C waterbath for 12 min.
10. Spread 5 μl on an SC-Ura plate to check the transformation efficiency. This should give 1000 or more transformants with a strain which transforms well.

11. Add the remainder of the transformation mix to 500 ml of SC-Ura liquid medium and incubate with shaking at 30°C for 4 h.
12. Pellet the cells. At this point the transformants have been established. They can be resuspended in galactose medium to induce *GAL1*, plated directly under selective conditions, or frozen by addition of DMSO to 9% (v/v) final concentration and stored at −70°C for later use. After thawing, the stored transformants can be plated directly under selective conditions.

(Also see Chapter 8.)

References

1. Elledge, S. J., Mulligan, J., Ramer, S., Spottswood, M., and Davis, R. W. (1991). *Proc. Natl Acad. Sci. USA*, **88**, 1731.
2. Ramer, S. W., Elledge, S. J., and Davis, R. W. (1992). *Proc. Natl Acad. Sci. USA*, **89**, 11589.
3. Hadwiger, J. A., Wittenberg, C., Richardson, H. E., Lopes, M., and Reed, S. I. (1989). *Proc. Natl Acad. Sci. USA*, **86**, 6255.
4. Rine, J., Hansen, W., Hardman, E., and Davis, R. W. (1983). *Proc. Natl Acad. Sci. USA*, **80**, 6750.
5. Courchesne, W., Kunisawa, R., and Thorner, J. (1989). *Cell*, **58**, 1107.
6. Lee, M. G. and Nurse, P. (1987). *Nature*, **327**, 31.
7. Elledge, S. J. and Spottswood, M. (1991). *EMBO J.*, **10**, 2653.
8. Fields, S. and Song, O. K. (1989). *Nature*, **340**, 245.
9. Durfee, T., Becherer, K., Chen, P., Yeh, S., Yang, Y., Kilburn, A., Lee, W., and Elledge, S. J. (1993). *Genes Dev.*, **7**, 555.
10. Dalton, S. and Triesman, R. (1992). *Cell*, **68**, 597.
11. Yang, X., Hubbard, E. J. A., and Carlson, M. (1992). *Science*, **257**, 680.
12. Elledge, S. J., Sugiono, P., Guarente, L., and Davis, R. W. (1989). *Proc. Natl Acad. Sci. USA*, **86**, 3689.
13. Krug, M. and Berger, S. (1987). In *Methods in enzymology*, Vol. 152 (ed. S. L. Berger and A. Kimmel), pp. 316–25. Academic Press, London.
14. D'Alessio, J. and Gerard, G. (1988). *Nucleic Acids Res.*, **16**, 1999.
15. Yang, Y. C., Ciarletta, A., Temple, P., Chung, M., Kovacic, S., Witeck-Giannotti, J., Leary, A., Kriz, R., Donahue, R., Wong, G., and Clark, S. (1986). *Cell*, **47**, 3.
16. Sambrook, J., Fritsch, E., and Maniatis, T. (1989). *Molecular cloning. A laboratory manual*, 2nd edn. Cold Spring Harbor Laboratory Press, Cold Spring Harbor, NY.
17. Schiestl, R. H. and Gietz, R. D. (1989). *Curr. Genet.*, **16**, 339.
18. Guthrie, C. and Fink, G. R. (ed.) (1991). *Methods in enzymology*, Vol. 194, pp. 3–21. Academic Press, London.

5

Pulsed field gel electrophoresis

JOHN R. JOHNSTON

1. Introduction

Pulsed field (gradient) gel electrophoresis (here abbreviated to PFGE) (1, 2) allows separation of large DNA molecules, typically ranging in size from 50 to 10 000 kb (10 Mb). Pulsed electric fields in different directions are applied to agarose gels. Since the DNA molecules of the 16 (haploid) chromosomes of *Saccharomyces cerevisiae* range in size from approximately 200 kb to 3 Mb, they are eminently suitable for separation by this technique. This provides definition of 'electrophoretic karyotypes' (3) of strains by sizing chromosomes according to how far they migrate in the gel, after staining of DNA bands with ethidium bromide (EthBr).

Not only do laboratory-bred strains possess different karyotypes, because of chromosome length polymorphisms (CLP) and chromosomal rearrangements (4), but so do industrial strains of *S. cerevisiae* to a greater extent because of their greater and varied number of chromosomes. A large number of commercial brewing, baking, wine, and distilling yeasts has now been classified by this method (for example, refs 5–11).

Although outside the scope of this book, PFGE has also been applied to an extensive range of yeasts of genera other than *Saccharomyces*, such as, notably, *Schizosaccharomyces pombe* and *Candida* species (for example, refs 5, 12–15). Perhaps the most important application of PFGE in the last few years has been the isolation of long insert sequences of large genomes incorporated into yeast artificial chromosome (YAC) vectors (see Section 3). The physical mapping of these genomes, and also of yeast itself, has been assisted by production of large chromosomal fragments by infrequently-cutting restriction enzymes (see 16), including the recently described meganuclease I-*Sce*I (17, also see Chapter 3).

2. Equipment

The original types of apparatus (1, 2) have since been modified and developed in various ways, particularly in the design and geometry of electrodes and the electric fields produced. Since most systems have been reviewed recently in detail (18–20), only a summary is presented here. All systems depend upon the fact that periodic changes in the direction of electric field retard the migration through the matrix of agarose gels of larger DNA molecules and this retardation is dependent on the size of the DNA molecules. Although there are several other variables involved in PFGE, such as field strength and agarose gel and salt concentrations, those of electric field geometry and switching intervals of pulses are paramount.

2.1 Orthogonal field alternation (OFAGE)

This early system (2) features two pairs of electrodes, placed to produce electric fields diagonally across the gel and at an angle greater than 90° to each other. Because these fields are inhomogeneous, DNA in wells increasingly distal from the centre of the gel migrates in increasingly curved tracks. Sharp DNA bands are, however, obtained. This apparatus is now largely redundant.

2.2 Field inversion (FIGE)

A single pair of electrodes produces a field, the polarity of which is switched through 180° per cycle. A longer pulse time in the forward direction, often twice that in the opposite direction, results in net DNA migration down the gel. The relationship between rate of migration and DNA size is, however, complex (21) and programmed ramping, increasing the switching cycle during the run, is usually used (for example, ref. 22). DNA bands are also wider and more diffuse. A conventional horizontal gel box can, however, be used. The FIGE system has been relatively infrequently exploited, although it has been recently used to analyse chromosomes V and VIII (23). In addition, it has been used, without ramping, to separate the larger chromosomes carried by yeasts of several genera other than *Saccharomyces* (13). Commercial models include 'DNASTAR PULSE' (DNAStar Inc.) AND 'GENETIC' (Biocent).

2.3 Contour-clamped homogeneous electric fields (CHEF)

Typically, this system features a hexagonal array of 24 small electrodes connected by a series of resistors (24). The resulting electric fields, angled at 120°, are highly uniform, due to the clamping of electrode potentials, and therefore produce straight tracks. The CHEF system has been reviewed in detail (18) and recent modifications described (25, 26). Commercial models include the DRII Megabase System and Mapper XA (Bio-Rad) and Hex-a-field Apparatus (BRL). The former allows up to 30 DNA samples to be run simultaneously.

2.4 Rotating gel (RGE)

In this apparatus rotation of the gel through more than 90° at each switching cycle in a single uniform electric field produces homogeneous crossed fields (27–29). At least two models, RAGE (Stratagene) and Waltzer (Tribiotics) are commercially available.

2.5 Transverse alternating field (TAFE)

In this case, because the electric fields are transverse to the gel, the pulsed directions of the DNA molecules are across the thickness of the gel (30). The fields are homogeneous across the depth of the gel, resulting in straight lanes. However, field strength and orientation angles vary along the length of the gel from the top to bottom surfaces. Due to the gel's vertical position, handling is more difficult and gel size is restricted. A system, GeneLine (Beckman), is commercially available.

2.6 Pulsed homogeneous orthogonal field (PHOGE)

This is a versatile system, allowing two degrees of resolution in different size ranges (31). Lower resolution is produced over a wide range of sizes (50 kb–1 Mb), using ramped pulsed times (5–90 sec), while maximal resolution is achieved for two-fold size differences. The system gives straight tracks and uses a large gel area, allowing a large number of samples to be run. It has been proposed that the molecular mechanism of separation in OFAGE may be the superimposition of the mechanisms produced under PHOGE and FIGE. A recent modification of the apparatus has been described (32).

2.7 Programmable autonomously controlled electrode (PACE)

This term has been used to describe a modification of a CHEF-type apparatus, which actually does clamp electric fields (33). Because of its precise and variable control of all parameters, it can be used to study the relationships between, on the one hand, voltage, field angles, switch time, agarose concentration, and temperature and, on the other, DNA mobility.

2.8 Other modifications

It should be appreciated that the above terminology contains inconsistencies and, in some cases, is misleading. In this respect, Olson's 'glossary of pulsed-field acronyms' (18) is worth consulting. Modifications of apparatus, such as simultaneous electrophoresis in a double-decker gel arrangement (34), at least at the laboratory level, are reported fairly regularly. Some of these are aimed at reducing the length of the electrophoresis time required to separate very large DNA molecules, in the 3–12 Mb range, which need lower voltage gradients.

3. Preparation of high molecular weight DNA

The essential requirement in keeping the fragile large DNA molecules intact upon their release from cells (or sphaeroplasts) is to embed them in agarose. Further processing is by the diffusion of reagents into the agarose plugs. A procedure for preparation which has much in common with others from various laboratories, based upon reference (35) and here referred to as 'standard', is given in *Protocol 1*.

Protocol 1. Standard DNA plug preparation (without 2-mercaptoethanol)

Equipment and reagents

- disposable 50 ml screw-capped centrifuge tubes
- Eppendorf tubes
- bench centrifuge
- microcentrifuge
- waterbath or temperature block
- plastic insert moulds (Pharmacia–LKB, cat. no. 80–1102–5 or Bio-Rad, cat. no. 170–3713)
- plastic 6-well culture dishes, with lids (e.g. Corning Cell Wells, cat. no. 25810)
- 1 M sorbitol
- 0.5 M EDTA, pH 7.8, as stock solution
- SCE buffer (1 M sorbitol, 0.1 M sodium citrate, 10 mM EDTA, pH 7.5)
- SE buffer (75 mM NaCl, 25 mM EDTA, pH 7.5)
- ES buffer (0.5 M EDTA, pH 9.5, 1% *N*-lauroylsarcosine (Sarkosyl))
- TE buffer (see Appendix 1)
- lyticase (Sigma)[a]: 10 u/μl in 50% (v/v) glycerol–TE buffer (store at −20°C).
- proteinase K (Sigma)[b]: 10 mg/ml in 50% (v/v) glycerol–TE buffer (store at −20°C)
- low melting point agarose (LMPA) (e.g. BRL Ultra pure)
- phenylmethylsulphonyl fluoride (PMSF) (100 mM in isopropanol) (store at −20°C)

Method

1. Grow overnight cultures in 5–10 ml of YPD. (The more usual flasks may be replaced by 50 ml centrifuge tubes, vigorously shaken.)
2. Count cells with a haemocytometer or estimate cell densities using a spectrophotometer[c].
3. Transfer volumes containing approximately 3–4 $\times$ 10^8 cells to Eppendorf tubes. This quantity of cells will make nine or 10 plugs, each of 100 μl volume.
4. Spin briefly in a microcentrifuge.
5. Wash with 0.125 M EDTA, pH 7.8.
6. Resuspend cells in 1 ml of SCE buffer and add 200 units of lyticase[d].
7. Incubate, without shaking, for 30 min at 30°C.
8. Centrifuge sphaeroplasts at 300 *g* for 5 min; remove supernatant using a Pasteur pipette.

9. Process pellets into agarose plugs, strain by strain.
   (a) Add 430 μl of 1 M sorbitol, thus bringing the total volume to approximately 500 μl.
   (b) Resuspend sphaeroplasts well by gently pipetting up and down or, briefly, gently vortexing.
   (c) Add, and quickly mix by inversion or vortexing, 500 μl of 1.5% LMPA, dissolved in SE buffer, from a volume in a tube placed at 45°C in a waterbath or temperature block.
   (d) Quickly pipette 100 μl aliquots into wells of a plastic insert mould, its back sealed by tape. Place in a refrigerator or on ice to accelerate gelling.

10. Remove plugs from the mould (by removing tape from the back and pressing gently with a rubber teat) into 5 ml of 1 mg/ml proteinase K, dissolved in SE buffer, in a culture dish well.

11. Very gently shake plugs overnight on a platform shaker (approximately 90 r.p.m) at 30°C (or warmer room-temperature)[e]. Plugs should look clear after proteinase K treatment.

12. On the next morning, rinse plugs with TE buffer. Change the TE buffer three times, diluting out most of the Sarkosyl. Return plugs to the shaker and change the TE buffer every 1.5 h or so.

13. Plugs can now be used. Store at 4°C in 0.5 M EDTA (or, for shorter-term storage, in TE buffer).

14. If, however, DNA is to be restricted, wash plugs with 0.1 mM PMSF in TE buffer for 4 h after the initial three rinses (step **12**). Shake plugs gently overnight and the next day with occasional changes of TE buffer (to eliminate Sarkosyl)[f].

[a] Alternatively, use Yeast Lytic Enzyme (ICN, cat. no. 152270). It is very stable in a 10 mg/ml solution of 50% glycerol and 50 mM potassium phosphate (mono- and dibasic mixed to pH 7.8). Store at −20°C (keeps for 1 year).

[b] Alternatively, use proteinase K from Boehringer-Mannheim (cat. no. 1373 196) (100 mg supplied in 5 ml of solution), which keeps well.

[c] Alternatively, measure pellet size. A 50 μl pellet in a microcentrifuge tube will yield 1 ml of plugs.

[d] An alternative to sphaeroplasting cells before embedding them in plugs is to embed cells in plugs first. Then place them into SCE buffer containing 1 mg/ml Zymolyase (ICN) for approximately 2 h, shaking gently at either 30°C or warm room-temperature. Pipette off the SCE buffer, add ES buffer with 1 mg/ml proteinase K, and shake overnight at room temperature.

[e] Incubation at 50°C is preferred by some workers for proteinase K.

[f] Dissolve PMSF in isopropanol in fume cupboard. It is best used fresh as it deteriorates even when stored at −20°C. A recent non-toxic alternative to PMSF is 'PEFA-Bloc-SC' (Boehringer-Mannheim, cat. no. 1429–868), which is soluble in water.

There is a number of variations along this general theme (for example, ref. 36). A simpler, more rapid version is listed in *Protocol 2*. The last few years have seen the widespread use of YAC vectors (37) in the cloning of higher eukaryote genomes, such as human and mouse (38–40), and in transgenic animals (41), so methodology for preparation of YAC DNA is presented in *Protocol 3*.

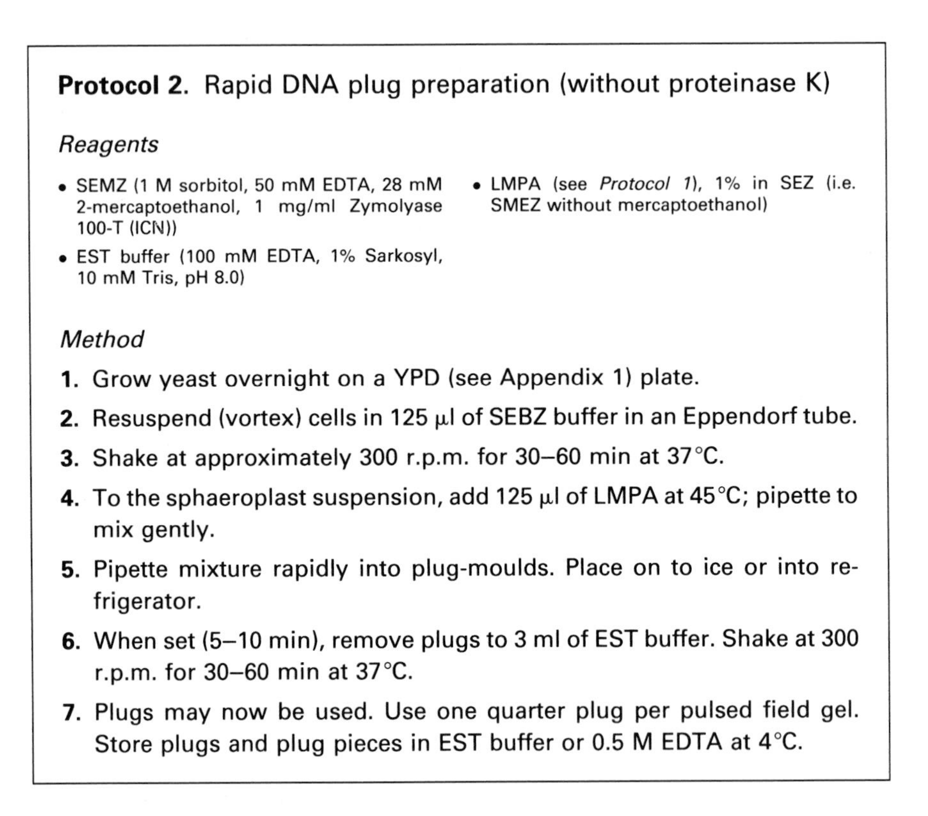

Protocol 2. Rapid DNA plug preparation (without proteinase K)

Reagents

- SEMZ (1 M sorbitol, 50 mM EDTA, 28 mM 2-mercaptoethanol, 1 mg/ml Zymolyase 100-T (ICN))
- EST buffer (100 mM EDTA, 1% Sarkosyl, 10 mM Tris, pH 8.0)
- LMPA (see *Protocol 1*), 1% in SEZ (i.e. SMEZ without mercaptoethanol)

Method

1. Grow yeast overnight on a YPD (see Appendix 1) plate.
2. Resuspend (vortex) cells in 125 μl of SEBZ buffer in an Eppendorf tube.
3. Shake at approximately 300 r.p.m. for 30–60 min at 37°C.
4. To the sphaeroplast suspension, add 125 μl of LMPA at 45°C; pipette to mix gently.
5. Pipette mixture rapidly into plug-moulds. Place on to ice or into refrigerator.
6. When set (5–10 min), remove plugs to 3 ml of EST buffer. Shake at 300 r.p.m. for 30–60 min at 37°C.
7. Plugs may now be used. Use one quarter plug per pulsed field gel. Store plugs and plug pieces in EST buffer or 0.5 M EDTA at 4°C.

Protocol 3[a]**.** Preparation of YAC DNA (without proteinase K)

Equipment and reagents

- microcentrifuge
- CHEF system
- filter unit (Ultrafree-MC, 30 000 NMWL, Millipore, cat. no. UFC3 TTK00)
- plastic moulds (2 mm × 26 mm × 8 mm)
- AHC medium (see Appendix 1)
- 0.5 M EDTA, pH 8.0 (stock solution)
- 1 M and 1.2 M sorbitol
- 10 mM Tris–HCl, pH 7.5 and pH 8.0
- TE buffer, pH 7.5 (see Appendix 1)
- 2-mercaptoethanol
- 1 mg/ml Zymolyase-20T (ICN)
- 1% agarose
- 2% low gelling temperature agarose (LGTA) (Seaplaque, FMC Bioproducts)
- agarase buffer (10 mM Bis-Tris–HCl, pH 6.5, 1 mM EDTA, 100 mM NaCl)
- 1 unit/μl agarase (New England Biolabs)
- 1% lithium dodecyl sulphate
- ethidium bromide (EthBr) (10 mg/ml stock)

Method

1. Grow cells from one colony in 25 ml of AHC medium (without uracil and tryptophan) overnight.
2. Wash cells in 50 mM EDTA, pH 8.0.
3. Resuspend cells in 150 μl of 1 M sorbitol, 20 mM EDTA, pH 8.0, 14 mM 2-mercaptoethanol, 1 mg/ml Zymolyase.
4. Add an equal volume of 2% LGTA, dissolved in the resuspension solution (step **3**) but lacking Zymolyase.
5. Pour this mixture into plastic moulds and allow to gel into plugs.
6. Remove the plugs into a solution of 1.2 M sorbitol, 20 mM EDTA, 14 mM 2-mercaptoethanol, 10 mM Tris–HCl, pH 7.5, 1 mg/ml Zymolyase 20T.
7. Incubate at 37°C for 2 h.
8. Lyse cells in 1% lithium dodecyl sulphate, 100 mM EDTA, 10 mM Tris–HCl, pH 8.0 at 37°C for 1 h.
9. Replace with fresh solution and incubate overnight at 37°C.
10. Wash plugs twice for 30 min in 700 μl TE buffer, pH 7.5 at room temperature.
11. Load plugs into a preparative gel of 1% GTG agarose (Seaplaque, FMC) (see *Protocol 4*).
12. Run the gel in a CHEF system in 0.5 × TBE buffer at 5 V/cm with switching time appropriate for the particular YAC.
13. Cut strips 2 cm wide from either side of the preparative gel.
14. Stain with EthBr to locate the position of the YAC in the gel.
15. Excise a slice containing the YAC from the unstained central section of the gel and cut into 1 cm pieces.
16. Equilibrate the pieces in agarase buffer for two periods of 30 min.
17. Place a piece (weighing ~0.4 g) into a 1.5 ml polypropylene tube and spin for a few seconds in a microcentrifuge to bring it to the bottom.
18. Melt the piece of gel by placing the tube at 68°C in a waterbath or temperature block.
19. Equilibrate to 40°C for 5 min, carefully add agarase (1 unit per 100 mg agarose), and mix by pipetting once, using a cut-off tip.
20. Digest at 40°C for 2 h.
21. Remove undigested agarose and small particles by centrifuging at 12 000 *g* for 15 min at room temperature.
22. Transfer to a new tube using a cut-off tip and store at 4°C.

Protocol 3. *Continued*

23. To concentrate the DNA:
   (a) load, using a cut-off tip, 400 μl of the agarase-treated solution into a filter unit. Centrifuge the filter at 3000 *g* in a microcentrifuge for 5 min.
   (b) measure the volume of liquid which has passed through the filter and continue centrifugation until approximately 320 μl has passed.
   (c) allow the filter containing approximately 80 μl of DNA solution to sit for at least 1 h at room temperature. Carefully resuspend for transfer to a new tube by pipetting three times, using a cut-off tip.

24. Check the quality of the DNA by adding 5 × gel-loading buffer to an aliquot of the YAC DNA preparation.

25. Carefully mix the sample, using a cut-off pipette tip, and load into the well of a pulsed field gel.

26. Estimate DNA concentrations on a 0.8% agarose minigel, using DNA as a standard.

[a] Adapted from Gnirke *et al.* (42), Huxley *et al.* (43), and Southern *et al.* (27)

4. Gel preparation and runs

Procedures for preparation of a pulsed field gel and a PFGE run are described in *Protocol 4*. The conditions used for a run depend upon the size range of bands required to be resolved and the equipment in use. For example, in one analysis of human genomic DNA, using the YAC cloning system (44), DNA fragments of sizes 25 kb–1 Mb were run for 22–24 h on a CHEF system. The conditions used were a 1% agarose gel, TBE buffer, 200 V, 8°C, and pulse times of either ramping from 50 sec to 90 sec, or 360 sec. Larger fragments (1–3 Mb), however, were separated on an RGE apparatus, using a 1% gel, TAE buffer, 70 V, 8°C, and a pulse time of 6 min over 54 h. In another example, on fractionation of yeast chromosomes by VDE endonuclease (45), a pulse programme of 100 sec for 36 h, 600 sec for 36 h, and 300 sec for 24 h (with field strength of 5.5 V/cm) was used. For maximal resolution of the smallest yeast chromosomes, particularly chromosome I, a 1% gel was run on a CHEF apparatus with TBE buffer at 160 V and 12°C, with a pulse time of 18 sec (52). Following PFGE runs, many gels are, of course, further analysed by hybridization of probes to bands. For experimental details of this process, see ref. 35.

Protocol 4. Separation of high molecular weight DNA by PFGE

Equipment and reagents

- PFGE apparatus (see Section 2)
- PFGE glass gel plate[a], comb, and gel box
- UV transilluminator
- Polaroid camera (for UV exposures)
- agarose[b] (high purity, Molecular Biology grade)
- 10 × TBE buffer (see Appendix 1) (stock solution)
- EthBr (10 mg/ml as stock solution)

Method **(use disposable gloves throughout)**

1. Pour hot agarose[c] on to *level* glass plate in a suitably sized gel box, with appropriate slotted (or open) comb. Allow gel to set.
2. Load plugs, usually either one-half or one-quarter size pieces, either into slots or against front and bottom surfaces if total comb area is open. If the latter, then overlay inserts with melted agarose (use of a wider-bore disposable pipette or a warmed glass pipette prevents blocking by gelling agarose). For sizing DNA bands[d], load small pieces of λ ladders, such as wild-type (48.5 kb increments) or deletion mutants, e.g. Δ 39 (39 kb increments).
3. Place the glass plate holding the gel into the PF system, ensuring its orientation is correct (i.e. with the well end closest to the negative electrodes). Ensure that the gel is *level* and covered by pumped, cooled 0.5 × TBE buffer[e] to a depth of a few millimetres.
4. Check all tubing and connectors for leaks.
5. Check all safety features are functioning correctly before switching on power-pack.
6. Adjust voltage/current, timer, and ramping programme, if used. Check buffer temperature[f].
7. During and/or at end of the run (timed), check the variables, i.e. voltage/current, pulse time, and buffer temperature.
8. Switch off power-pack. Follow safety procedures before removing glass plate and gel from apparatus.
9. Using a razor-blade, cut between the gel and the Velcro, then slide gel from glass plate into a box or tray containing a solution of a few drops of EthBr in either distilled water or 0.5 × TBE buffer.
10. Stain (usually for 10–30 min).
11. Rinse gel in distilled water, place on transilluminator, view with UV

Protocol 4. *Continued*

(protective glasses or visor **MUST** be used) and, if required, photograph.

12. Destain in distilled water overnight or for 1–2 days to improve contrast for photography of bands.

[a] For example, a CHEF gel plate of 5 × 5 inches × 3 mm. Pieces of Velcro glued to two corners of the plate prevent sliding of the gel during the run.
[b] Suppliers such as BRL and Bio-Rad, etc. Agarose specifically for PFGE is now available, e.g. Bio-Rad Chromosomal and Beckman LE grades.
[c] An agarose solution can be prepared rapidly in a microwave oven. Although the concentration is sometimes varied, for example to 0.8% (48), a typical solution is 1% (w/v). The volume depends on the system used (100 ml for the plate mentioned in footnote *a*).
[d] Too high a concentration of DNA in plugs may slow migration and lead to overestimation of DNA size from bands (46).
[e] Cooled distilled water can be used to dilute from 10 × TBE buffer to save time cooling buffer in apparatus. TBE buffer may be replaced by 0.5 × TAE buffer (see Appendix 1) if gels are to be blotted on to nitrocellulose filters.
[f] Variable, but typically 8–14°C.

5. Concluding remarks

PFGE has formed an essential basis for projects aiming at total sequencing of eukaryote genomes notably human and *S. cerevisiae*. With the sequencing of yeast chromosome III (47), sequence inhomogeneities (48) and regional variations in base composition (49) are being investigated. Long chromosome length polymorphisms (CLP), some involving sequences around 20 kb, are also the subject of investigation. Several have been defined in chromosome I (and in chromosome III) in strains which are descendants of the 'parental' haploid laboratory strain, S288C (X2180–1A and X2180–1B) (50, 4), or in the natural diploid ancestral strain, EM93 (51, 52) (see *Figure 1*).

In addition to using PFGE to detect and analyse chromosomal rearrangements, such as translocations (4), the mitotic stability of both haploid and diploid karyotypes has recently been studied (53). The karyotype of this diploid strain, a wine yeast, underwent relatively frequent alterations, including rearrangements between chromosomes I and VI. PFGE analysis is constantly being extended to industrial strains of *Saccharomyces* and to different species of yeast, such as *S. kluyveri* (5, 54).

Acknowledgements

It is a pleasure to record my appreciation to Maren Bell, Rebecca Contopoulou, and Robert Mortimer for their involvement and assistance in various aspects of the work presented in this chapter. Travel grants from NATO and SERC for work undertaken in the University of California, Berkeley are also gratefully acknowledged.

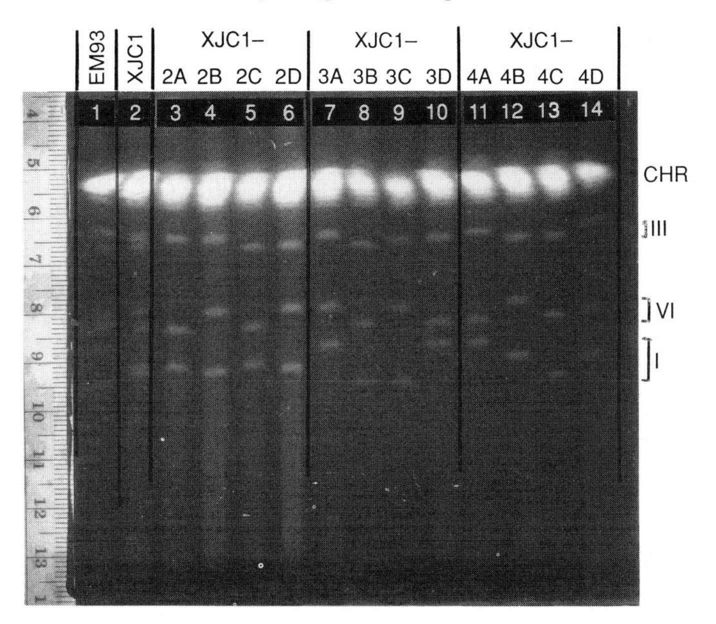

Figure 1. Pulsed field (CHEF) gel for resolution of chromosome I, in particular, but also chromosomes VI and III. Run conditions were: 160 V, 160mA, 12°C, for 23 h with pulse time of 18 sec. Lane 1 is of ancestral diploid strain EM93 (51, 52), showing resolution of longer (*L*) and shorter (*S*) homologues of each of chromosomes I, VI and III. Tetrads of EM93 produce segregation ratios for CHR I of 2*L* : 2*S*, 1*L* : 1*S* : 2*I*ntermediate lengths and 4*I* : 0, suggesting that *L* and *S* homologues CHRI differ by two distinct CLP. This proposal is confirmed by tetrad analysis of a cross (diploid strain XJC1) between EM93 segregants 29C and 29D from a tetrad with a ratio of 1*L* : 1*S* : 2*I*, where 29C and 29D carry *I* lengths of CHRI. Lane 2 shows this cross, with a doublet band for CHRI. Lanes 3–14 are three tetrads of strain XJC1. Tetrad 2 shows a ratio of 4*I* : 0, tetrad 3 a ratio of 2^{L} : 2*S*, and tetrad 4 a ratio of 1^{L} : 1*S* : 2*I*. These are, respectively, PD, NPD and T tetrads for the two distinct CLP (*clp x* and *clp y*) or CHRI originating in strain WM93. Tetrads 2, 3, and 4 also show segregations of *L* and *S* homologues of CHR VI and CHR III. (The assistance of Maren Bell is gratefully acknowledged.)

References

1. Schwartz, D. C. and Cantor, C. R. (1984). *Cell*, **37**.
2. Carle, G. F. and Olson, M. V. (1984). *Nucleic Acids Res.*, **12**, 5647.
3. Carle, G. F. and Olson, M. V. (1985). *Proc. Natl Acad. Sci. USA*, **82**, 3756.
4. Mortimer, R. K., Game, J. C., Bell, M., and Contopoulou, C. R. (1990). In *Methods: a companion to methods in enzymology*, Vol. 1, pp. 169–79. Academic Press.
5. Johnston, J. R. and Mortimer, R. K. (1986). *Int. Syst. Bacteriol.*, **36**, 569.
6. DeJonge, P., Dejonge, F. C. M., Meijers, R., Steensma, H. Y., and Scheffers, W. A. (1986). *Yeast*, **2**, 193.
7. Sheehan, C. A., Weiss, A. S., Newsom, I. A., Flint, V., and O'Donnell, D. C. (1991). *J. Inst. Brew.*, **97**, 163.

8. Johnston, J. R. (1990). In *Yeast technology* (ed. J. F. T. Spencer and D. M. Spencer), pp. 55–104. Springer-Verlag, Heidelberg.
9. Casey, G. P., Pringle, A. T., and Erdmann, P. A. (1990). *J. Amer. Soc. Brew. Chem.*, **48**, 100.
10. Vezinhet, F., Hallet, J. N., Valade, M., and Poulard, A. (1992). *Amer. J. Enol. Vitic.*, **43**, 83.
11. De Zoysa, P. (1992). Ph.D. Thesis, University of Strathclyde, Glasgow.
12. Vollrath, D. and Davis, R. W. (1987). *Nucleic Acids Res.*, **15**, 7865.
13. Johnston, J. R., Contopoulou, C. R., and Mortimer, R. K. (1988). *Yeast*, **4**, 191.
14. Fan, J.-B., Chikashige, Y., Smith, C. L., Niwa, O., Yamagida, M., and Cantor, C. R. (1989). *Nucleic Acids Res.*, **17**, 2801.
15. Doi, M., Homma, M., Chindamporn, A., and Tanaka, K. (1992). *J. Gen. Microbiol.*, **138**, 2243.
16. Sherman, F. and Wakem, P. (1991). In *Methods in enzymology*, Vol. 194 (ed. C. Guthrie and G. R. Fink), pp. 38–57. Academic Press, London.
17. Thierry, A. and Dujon, B. (1992). *Nucleic Acids Res.*, **20**, 5625.
18. Olson, M. V. (1989). In *Genetic engineering* (ed. J. K. Setlow), Vol. 11, pp. 183–227. Plenum Press, New York.
19. Anand, R. and Southern, E. M. (1990). In *Gel electrophoresis of nucleic acids: a practical approach*, 2nd edn (ed. D. Rickwood and B. D. Hames), pp. 101–23. IRL Press, Oxford.
20. Eby, M. J. (1990). *Biotechnology*, **8,** 243.
21. Carle, G. F., Frank, M., and Olson, M. V. (1986). *Science*, **232,** 65.
22. Heller, C. and Pohl, F. M. (1990). *Nucleic Acids Res.*, **18,** 6299.
23. Tanaka, S. and Isono, K. (1992). *Nucleic Acids Res.*, **20,** 3011.
24. Chu, G., Vollrath, D., and Davis, R. W. (1986). *Science*, **234,** 1582.
25. Meese, E. and Meltzer, P. S. (1990). *Technique*, **2,** 26.
26. Chu, G. and Gunderson, K. (1991). *Anal. Biochem.*, **194,** 439.
27. Southern, E. M., Anand, R., Brown, W. R., and Fletcher, D. S. (1987). *Nucleic Acids Res.*, **15,** 5925.
28. Serwer, P. (1987). *Electrophoresis*, **8,** 301.
29. Sutherland, J. C., Emrick, A. B., and Trunk, J. (1989). *Electrophoresis*, **10,** 315.
30. Gardiner, K. and Patterson, D. (1989). *Electrophoresis*, **10,** 296.
31. Bancroft, I. and Wolk, C. P. (1988). *Nucleic Acids Res.*, **16,** 7405.
32. Bancroft, I., Wesphal, L., Schmidt, R., and Dean, C. (1992). *Nucleic Acids Res.*, **20,** 6201.
33. Birren, B. W., Hood, L., and Lai, E. (1989). *Electrophoresis*, **10,** 302.
34. Nagy, A. and Choo, K. H. (1991). *Nucleic Acids Res.*, **18,** 5317.
35. Van Ommen, G. J. B. and Verkerk, J. M. H. (1986). In *Human genetic disease: a practical approach* (ed. K. E. Davies), pp. 113–33. IRL Press, Oxford.
36. Gerring, S. L., Connelly, C., and Hieter, P. (1991). *Methods in enzymology*, Vol. 194 (ed. C. Guthrie and G. Fink), pp. 57–77. Academic Press, London.
37. Burke, D. T. and Olson, M. V. (1991). In *Methods in enzymology*, Vol. 194 (ed. C. Guthrie and G. R. Fink), pp. 251–70. Academic Press, London.
38. Sears, D. D., Hegemann, J. H., and Hieter, P. (1992). *Proc. Natl Acad. Sci. USA*, **89,** 5296.
39. Lairmore, T. C., Dou, S., Howe, J. R., Chi, D., Carlson, K., Veile, R., Mishra,

S. K., Wells, S. A., and Donis-Keller, H. (1993). *Proc. Natl Acad. Sci. USA*, **90,** 492.
40. Riley, J. H., Morten, J. E. N., and Anand, R. (1992). *Nucleic Acids Res.*, **20,** 2971.
41. Schedl, A., Beeran, F., Thies, E., Montoliu, L., Kelsey, G., and Schutz, G. (1992). *Nucleic Acids Res.*, **20,** 3073.
42. Gnirke, A., Huxley, C., Peterson, K., and Olson, M. V. (1993). *Genomics*, **15,** 659.
43. Huxley, C., Hagino, Y., Schlessinger, D., and Olson, M. V. (1991). *Genomics*, **9,** 742.
44. Vetrie, D., Bobrow, M., and Harris, A. (1993). *Genomics*, **15,** 631.
45. Bremer, M. C. D., Gimble, F. S., Thorner, J., and Smith, C. L. (1992). *Nucleic Acids Res.*, **20,** 5484.
46. Doggett, N. A., Smith, C. L., and Cantor, C. R. (1992). *Nucleic Acids Res.*, **20,** 859.
47. Oliver, S. G. *et al.* (146 co-authors) (1992). *Nature*, **357,** 38.
48. Karlin, S., Blairsdell, B. E., Sapolsky, R. J., Cardon, L., and Burge, C. (1993). *Nucleic Acids Res.*, **21,** 703.
49. Sharp, P. M. and Lloyd, A. T. (1993). *Nucleic Acids Res.*, **21,** 179.
50. Ono, B. and Ishino-Arao, Y. (1988). *Curr. Genet.*, **14,** 413.
51. Mortimer, R. K. and Johnston, J. R. (1986). *Genetics*, **113,** 35.
52. Johnston, J. R., Curran, L., and Mortimer, R. (1992). *Yeast*, **8,** S642. (Abstract.)
53. Longo, E. and Vezinhet, F. (1993). *Appl. Environ. Microbiol.*, **59,** 322.
54. Weinstock, K. G. and Strathern, J. N. (1993). *Yeast*, **9,** 351.

Note added in proof: a method for the preparation of intact chromosomes for PFGE without the use of lytic enzymes has recently been published (Gardner, D. C. J., Heale, S. M., Stateva, L. I., and Oliver, S. G. (1993). *Yeast*, **9**, 1053).

6

Plasmid shuffling and mutant isolation

ROBERT S. SIKORSKI, JILL B. KEENEY, and JEF D. BOEKE

1. Introduction

Genetic analysis of mutants in *Saccharomyces cerevisiae* offers a unique approach to the study of biological processes. To initiate genetic studies in this organism one can mutagenize a population and screen or select for those mutants which affect the process of interest. Alternatively, one can use techniques selectively to mutagenize a single gene already known to play an important role in this process. The generation of mutant yeast strains starting from a cloned non-essential yeast gene is relatively straightforward. To remove the wild-type gene product, an essential step in analysing recessive alleles, DNA at the wild-type non-essential locus can be deleted entirely from the genome. A collection of mutant alleles can be made *in vitro* and introduced into the deleted host cell via episomal cloning vectors. The generation of mutant yeast strains from an essential yeast gene is a more complicated task since removal of the wild-type gene in one simple step yields an inviable genotype.

This chapter describes a plasmid shuffling technique which can be used to produce mutant alleles of any chosen yeast gene. The principles of plasmid shuffling are first presented followed by a detailed description of the strains and plasmids which will be required. The four steps of plasmid shuffling are then reviewed, and a step-by-step protocol for efficient *in vitro* mutagenesis is provided.

2. Plasmid shuffling: principles

Plasmid shuffling is a method for constructing mutant yeast strains which uses replicating yeast episomes as a means of exchanging a selected wild-type gene for mutated alleles (1–3). Targeted disruption by DNA-mediated transformation is first performed to inactivate a single copy of the selected yeast gene (termed here gene *ABC*) in a diploid strain (see Section 4.1). After meiosis, viability of the resulting haploid mutant strain is maintained by a wild-type

ABC gene carried on an episome. Mutagenized copies of *ABC* (see Section 4.2) are then introduced into this special haploid mutant carried on a second episome and exchanged or shuffled for the wild-type copy of *ABC* (see Section 4.3). The phenotypes of any mutants are then examined in the background of a null allele of gene *ABC* (see Section 4.4).

3. Plasmid shuffling: materials

3.1 Yeast–*E. coli* shuttle vectors

Two plasmid constructs and consequently two distinct shuttle vectors are needed for plasmid shuffling. One construct consists of a vector carrying the wild-type gene and the other consists of a vector carrying a 'library' of mutant alleles. The vector requirements for each construct are different. The mutant-allele construct can be derived from any pre-existing YCp (centromeric), YEp (2 μm plasmid-derived), or YRp (*ARS* plasmid) shuttle vectors. The only stipulation is that the selectable marker on the vector be compatible with the genotype of the *Escherichia coli* (to check level of mutagenesis) and yeast (to select transformants) host strains. The pRS series of shuttle vectors (4) is particularly desirable for *in vitro* mutagenesis because these vectors are small and yield large amounts of plasmid DNA in *E. coli*. The backbone of these plasmids presents a minimal target for the mutagen and the high yield of DNA in bacteria allows easy amplification of a mutant library (see Section 4.2). As for the wild-type allele construct, the vector chosen must contain both an appropriate selectable marker and a counterselectable one. We have made a series of pRS shuttle vector derivatives specifically for the purpose of plasmid shuffling (5). In addition to the *URA3*-based YCp vector pRS316 (4), we have constructed plasmids containing the *CYH2* (pRS318), *CAN1* (pRS319), and *LYS2* (pRS317) counterselectable genes. Care should be taken when using pRS319 since the truncated *CAN1* gene of this construct does not produce wild-type levels of canavanine sensitivity in some yeast strain backgrounds; pRS318 will be the vector of choice in most instances, as the *LYS2* gene, being very large, contains many restriction sites, limiting the usefulness of pRS317.

3.2 Yeast strain

Any yeast strain can be readily converted into a strain suitable for plasmid shuffling. The four shuffling strategies outlined above all require the presence of a counterselectable element within the plasmid carrying the wild-type version of the manipulated gene. Some of these counterselectable markers (*CAN1* and *CYH2*) do not lend themselves to positive growth selection, a feature necessary for the initial maintenance of the plasmid in yeast. *CAN1* can be selected for, but only in Arg^- *can1* mutants, which grow poorly (6).

Some counterselectable genes (*URA3* and *LYS2*) readily confer both positive and negative selection (7, 8).

For most applications, *URA3* is probably the most convenient marker to use, since it is a small, readily manipulable gene and many yeast strains in use will already contain a *ura3* mutation. The *URA3* gene encodes orotidine-5′-phosphate decarboxylase, an enzyme required for uracil biosynthesis. Selection of *ura3* cells (e.g. cells which have lost a *URA3* plasmid) is accomplished by plating the cells on media containing 5-fluoro-orotic acid (5-FOA). This compound is apparently converted to a toxic product, 5-fluorouracil, by the action of the decarboxylase, preventing growth of *URA3* cells; *ura3* cells are 5-FOA-resistant. The 5-FOA negative selection procedure is very efficient and selective, and under appropriate conditions only one in several hundred 5-FOA resistant cells will be Ura^+ (7). Quite conveniently, one can also positively select for *URA3* cells (and transformants) simply by excluding uracil from the medium.

LYS2 encodes α-aminoadipate reductase, an enzyme required for lysine biosynthesis. Yeast cells with wild-type *LYS2* activity will not grow on media containing α-aminoadipate (α-AA) as a primary nitrogen source (8). High levels of α-AA are thought to cause the accumulation of a toxic intermediate, and *lys2* (as well as rarer *lys5*) mutants block the formation of this intermediate (31). Although *lys5* mutants are α-AA-resistant, they arise at a frequency much lower than that of loss of a centromere plasmid, and, therefore, present no problem in plasmid shuffling. *LYS2* can be selected in a positive fashion using lysine-free medium. In a recent study in which 379 α-AA-resistant mutants were selected in strain F763 (*MATα trp1Δ1 ura3–52*), 25 were Lys^+, 338 were in *LYS2*, 12 in *LYS5*, and (unexpectedly) three were in *LYS14*, a gene not previously reported to mutate to α-AA resistance.

CAN1 encodes an arginine permease (32) which is the route of entry into yeast cells for the arginine analogue canavanine. Canavanine is incorporated into proteins with lethal consequences. *CAN1* ($permease^+$) cells are sensitive to canavanine whereas *can1* ($permease^-$) cells are resistant. The sensitivity is dominant so that *CAN1/can1* cells also fail to grow on canavanine. *CAN1* cells can be selected directly by transformation of the *CAN1* gene, but the procedure is complicated by the fact that the host cell must be both *can1* and Arg^- for this to work. For routine use of *CAN1* in a negative selection scheme it is more convenient to include another selectable marker on the plasmid which can be used in selecting transformants.

CYH2 encodes the L29 protein of the ribosome (33). Cycloheximide blocks polypeptide elongation during translation and prevents the growth of cells that are *CYH2*. Cycloheximide resistance results from an amino acid substitution in the *CYH2* protein (33). The lethality of the drug is dominant and cells containing both the sensitive (wild-type) and the resistant (mutant) *CYH2* alleles fail to grow on media containing cycloheximide. The loss of a *CYH2*-containing plasmid can be selected directly if the host carries a resistant allele at the chromosomal locus. No positive selection exists for *CYH2* and another selectable marker must be used in combination for the introduction of a *CYH2* vector.

3.2.1 Building *cyh2* strains

Cycloheximide counterselection for plasmid shuffling purposes requires the presence of appropriate auxotrophies in the strain to allow positive selection of the plasmids in later steps. *cyh2* mutations can be selected as follows.

(a) grow yeast cells in 10 ml of YPD broth (see Appendix 1) to saturation.
(b) plate ten 0.3 ml aliquots on to YPD-agar plates supplemented with 10 μg/ml cycloheximide. (Store cycloheximide at −20°C and add to medium as a 10 mg/ml solution in ethanol.)
(c) after 2–4 days, a few colonies will appear, often on only a subset of the plates.
(d) pick resistant colonies from a plate giving the smallest number of resistant colonies. Under these conditions, only *cyh2* mutants will grow.

Note: use of lower cycloheximide concentrations can result in a very wide spectrum of other cycloheximide resistance-conferring genotypes.

3.2.2 Building *can1* strains

As for the cycloheximide counterselection (see Section 3.2.1), canavanine counterselection must be used in conjunction with an appropriate auxotrophy in the strain to allow selection of the plasmid in later steps. This is done as follows.

(a) wash a 10 ml YPD (see Appendix 1) overnight culture of the strain of interest and resuspend in 10 ml water.
(b) plate 0.1 ml of the washed culture and 0.1 ml of 10-fold and 100-fold dilutions (in water) on SC-Arg agar plates (see Appendix 1) containing 60 μg/ml canavanine.
(c) autoclave and store canavanine at room temperature as an aqueous solution of 60 mg/ml.
(d) pick resistant colonies from a plate giving the smallest number of resistant colonies.
(e) only *can1* mutants will form colonies under these conditions.

Note: canavanine SD plates (supplemented with other required compounds) can also be used for this selection, although growth of the resistant colonies will be slower than on SC-Arg plates. Arginine in the plates will allow growth of canavanine-sensitive cells and so must be omitted from the medium.

3.2.3 Building *lys2* strains

Follow the directions above for *can1* selection (see Section 3.2.1), except plate the cells on α-aminoadipic acid medium (see *Table 1*).

Table 1. Growth media for common counterselection schemes[a]

5-fluoro-orotic acid (URA3)		**D,L-α-aminoadipic acid (LYS2)**		**Canavanine (CAN1)**		**Cycloheximide (CYH2)**	
A. YNB (+NH_4SO_4)[b]	7 g	YNB (−NH_4SO_4)[b]	2 g	YNB (+NH_4SO_4)[b]	7 g	yeast extract	10 g
H_2O	375 ml	H_2O	360 ml	H_2O	400 ml	peptone	20 g
						H_2O	400 ml
B. 20% glucose	100 ml	20% glucose	100 ml	20% glucose	100 ml	20% glucose	100 ml
C. 4% agar	500 ml	4% agar	500 ml	4% agar	500 ml	4% agar	500 ml
D. uracil (2 mg/ml)	25 ml	lysine (4 mg/ml)	7.5 ml	—		—	
E. 5-fluoro-orotic acid	1 g	D,L-α-aminoadipate[c]	2 g	canavanine (60 mg/ml)	1 ml	cycloheximide (10 mg/ml)	1 ml
		H_2O	30 ml				
		adjust pH to 6.0 with 1 M KOH (requires ~12 ml)					

[a] Autoclave solutions A, B, and C separately. Cool these to about 50°C and combine with filter-sterilized D and E. For 5-FOA medium, mix the 5-FOA powder with A, B, and D and filter sterilize. The powder will take some time to go into solution; heating to 50°C will help.

[b] Note that YNB (Difco; see Appendix 1) is available with (+) and without (−) added NH_4SO_4, a distinction which is important for *LYS2* counterselection.

[c] A solution of α-AA is made and pH adjusted separately (final volume ~42 ml).

3.2.4 Building *ura3* strains

In most cases, if a useful counterselectable marker does not already exist in the strain of interest, introducing a *ura3* mutation is desirable. The best markers for this purpose are probably the *ura3–52* mutation (a non-reverting Ty insertion in the *ura3* gene (9) or one of several constructed *ura3* deletion mutations (1). Joachim Li (personal communication) recently constructed a plasmid, pJL162, bearing a complete deletion of the *URA3 Hin*dIII fragment and this is probably the most desirable allele for most purposes. pJL162 is an integrating plasmid that also bears *LEU2*. Thus the deletion is readily introduced into cells as follows.

(a) Transform strain of interest to Leu^+ with pJL162.
(b) After colony purification, grow the Leu^+ colonies on non-selective (YPD) medium to allow plasmid loss by recombination, and then replica plate to 5-FOA plates.
(c) Check the 5-FOA-resistant colonies obtained for uracil auxotrophy, non-revertibility of Ura^- phenotype and leucine prototrophy.
(d) Confirm that the desired removal of the *URA3* gene has been effected by Southern blotting with a *URA3* gene probe.

3.3 Media

Positive selection for the presence of a shuttle vector in yeast can be accomplished using standard SD or SC dropout media (12). Counterselection against the presence of a shuttle vector will require the use of special media. Recipes for preparing media for use in conjunction with the *URA3, LYS2, CAN1*, and *CYH2* genes are given in *Table 1*. The required compounds can be purchased from the following vendors: 5-FOA from SCM Specialty Chemicals; D,L-α-aminoadipic acid (α-AA), canavanine, and cycloheximide from Sigma Chemicals. 5-FOA is quite expensive if purchased in small amounts, but it can be purchased via the Genetics Society of America at a greatly reduced price by members of the Society. A relatively easy synthesis procedure (13) has also been used to make 5-FOA suitable for use in yeast genetics at a cost of about $20/gm (R. L. Keil, personal communication).

3.4 Bacterial strains

The overall level of mutations produced by *in vitro* mutagenesis (see Section 4.2) of a plasmid-borne yeast gene can be roughly estimated by measuring the level of induced null mutations in a shuttle vector's selectable marker. This testing is easily performed in *E. coli*, taking advantage of the fact that many yeast biosynthetic enzymes (including the products of the *URA3, HIS3, TRP1*, and *LEU2* genes (14) are expressed in bacteria directly from a genomic yeast DNA fragment. The expression levels are sufficient to complement the growth requirements of certain *E. coli* auxotrophs, and null alleles can be readily identified in bacteria by replica plating on to the appropriate minimal medium. A very useful *E. coli* strain in this regard is *MH1066* (*ΔlacX74 hsr⁻ rpsl pyrF*: Tn5 *leuB600 trpC9830 galE galK*) (15), in which one can select for the function of the yeast *URA3, TRP1*, or *LEU2* genes on minimal M9 medium (Appendix 1) containing the appropriate supplements (16).

4. Plasmid shuffling: method

4.1 Building an appropriate yeast host strain

The basis of any plasmid shuffling scheme involves the construction of a special yeast strain, the recipient strain, in which only one functional copy of the essential gene of interest (we will call this gene *ABC*) is present. *ABC* is strategically positioned within an episomal plasmid vector. The strain should also bear markers, e.g. *cyh2, can1, ura3, leu2, trp1*, and/or *his3*; at least one of these must be counterselectable. This strain is usually made by first inactivating one chromosomal copy of *ABC* in a suitably marked homozygous **a**/α diploid with a deletion or deletion–insertion mutation, using one-step or two-step gene replacement strategies (17, 18). Sporulation of the resultant heterozygote (19) should result in two viable and two inviable spores, the latter containing only a non-functional copy of *ABC*. By introducing (through transformation) a replicating plasmid such as a YCp vector containing the wild-type *ABC* gene (YCp-*ABC*) into the heterozygote before sporulation, some of the *abc*Δ spores will be 'rescued' for viability by the *ABC* gene within the episome. The vector chosen to carry the wild-type gene must be compatible with one of the negative selection schemes described in Section 4.3. This new haploid strain with the genotype *abc*Δ, YCp-*ABC* can now serve as the recipient for plasmids containing mutations in *ABC*.

4.2 Mutagenesis

The method chosen for mutagenesis will depend on the structure or function of the gene under study. If the gene and its function are well characterized, then local, directed mutagenesis of chosen residues or domains may be preferred, although one is likely only to confirm one's biases with this method, and may miss new and unexpected information. Several excellent protocols for local, *in vitro* mutagenesis using mutant oligonucleotides (20, 21), gapped-duplex target DNA (22), and nucleotide misincorporation (23) have been developed (24). Many of these methods work best with a single-stranded DNA template to serve as a target. Thus, yeast vectors which have a single-stranded DNA form (by virtue of an f1 phage origin of DNA replication) are most useful.

Often, however, the biochemical function of the selected gene (and/or its nucleotide sequence) may not be known and a method for generating an assortment of mutations randomly throughout the coding sequence may be more informative. In this regard, mutagenesis using nitrous acid, bisulphite, or hydroxylamine may be used. Nitrous acid produces a relatively wide variety of base modifications and consequently of point mutation types (25) and so may be the best mutagen if the aim is to obtain as wide a variety of mutant alleles as is possible. However, its use for mutagenizing yeast plasmids has not to our knowledge been reported. UV light also produces a very wide spectrum of mutant types (26) but may be less desirable due to the fact that frameshift mutations are quite common. Bisulphite reacts with DNA *in vitro* causing deamination of cytosine residues and C to T transition mutations. Because bisulphite reacts preferentially with unpaired bases, the target DNA should be in single-stranded form. This limits its usefulness in generalized mutagenesis because a relatively narrow spectrum of mutation types is obtained if only one strand is mutagenized. Hydroxylamine reacts with a double-stranded target DNA to create N^4-hydroxycytosine, which can base pair with adenosine, and results in both C to T and G to A transition mutations when double-stranded plasmid DNA is mutagenized. A variety of other mutations have been reported to occur with hydroxylamine under extended treatment conditions (27).

One can omit the amplification step in the protocol opposite and transform hydroxylamine-mutagenized DNA directly into yeast (28). All yeast transformants should be independent if they are plated directly on to solid medium without an outgrowth period. Care should be taken in the analysis of the resulting mutants when using this shortened version, however, since the mutagenic effect of any lingering hydroxylamine on yeast cells is not clear.

We have found that while the hydroxylamine mutagenesis procedure works well with relatively small plasmids, the transformation efficiency of large plasmids decreases very rapidly with increased incubation time in hydroxylamine. We examined these plasmids by agarose gel electrophoresis and discovered that the hydroxylamine treatment resulted in significant plasmid linearization (see *Figure 1*). This is of little consequence for transforming DNA into *E. coli*, as linearized plasmids will not give rise to transformants, but could confound analysis of yeast directly transformed with the mutagenized DNA, because the linearized DNA could either integrate via homology or recover the chromosomal mutant sequence by gene conversion.

An alternative to chemical mutagenesis is '*in vivo*' mutagenesis using a mutator strain of *E. coli* such as *mutD5* (29). In this strain, spontaneous mutations involving all possible base pair changes are elevated by 10^3- to 10^4-fold, and the frequency of base-pair insertions is increased (30). DNA changes can be introduced into a plasmid by simple passage through a round of amplification in a *mutD5* strain.

4.3 Removal of wild-type gene by counterselection

To remove the wild-type gene, *ABC*, from the wild-type/mutant cotransformant, we take advantage of two aspects of yeast biology. First, even relatively

Protocol 1. Hydroxylamine mutagenesis[a]

1. Make several millilitres of fresh hydroxylamine solution: 1 M hydroxylamine (Sigma Chemicals), 50 mM sodium pyrophosphate (pH 7.0), 100 mM sodium chloride, and 2 mM EDTA.
2. Add 10 μg of the target plasmid to 500 μl of hydroxylamine solution.
3. Allow the reaction to proceed at 75°C. Titre the degree of mutagenesis by removing 100 μl aliquots at several different time points of incubation (0, 30, 60, 90, and 120 min) and stop the reaction by placing it on ice.
4. Remove the excess hydroxylamine by float-filter dialysis (type VS 0.025 μm filters; Millipore) or conventional dialysis in Micro-collodion bags (Sartorius SM132–02) against TE buffer at 4°C. This step is *essential* as hydroxylamine is inhibitory to bacterial transformation.
5. Transform 1–5 μl of each time point into bacteria to assess the degree of mutagenesis. Estimate the level of mutagenesis by scoring null mutations in a non-essential plasmid-encoded function, usually the yeast selectable marker gene. The strain *MH1066* is particularly useful here because, as mentioned previously, functional yeast *URA3, TRP1,* and *LEU2* genes complement the analogous bacterial mutations. For scoring, simply replica plate the bacterial transformants from rich medium to minimal medium that lacks the compound in question.
6. Amplify the original mutagenized DNA in *E. coli* in order to obtain enough DNA for transforming yeast. Collect and pool about 10–20 000 bacterial colonies from the original mutagenized DNA by scraping the transformation plates into L-broth or water. Isolate plasmid DNA from this pool. Remember that the base of this 'library' will determine the maximum number of individual yeast mutants that one can obtain.
7. The DNA is now ready to transform into the recipient yeast strain. Try to obtain 200–500 colonies per 9 cm diameter Petri dish, with 300 being optimal. Any more colonies than this will make replica plate analysis difficult. Since in our hands the competency of the yeast cells can vary greatly from batch to batch, we have found it important to titre the competency before transforming and plating large amounts of the mutagenized DNA. Titred batches of cells made competent by the lithium acetate procedure[b] can be frozen or left in 0.1 M LiOAc at 4°C for up to 2 weeks without a significant change in the transformation frequency.

[a] Adapted from Busby *et al.* (27).
[b] Ito *et al.* (11).

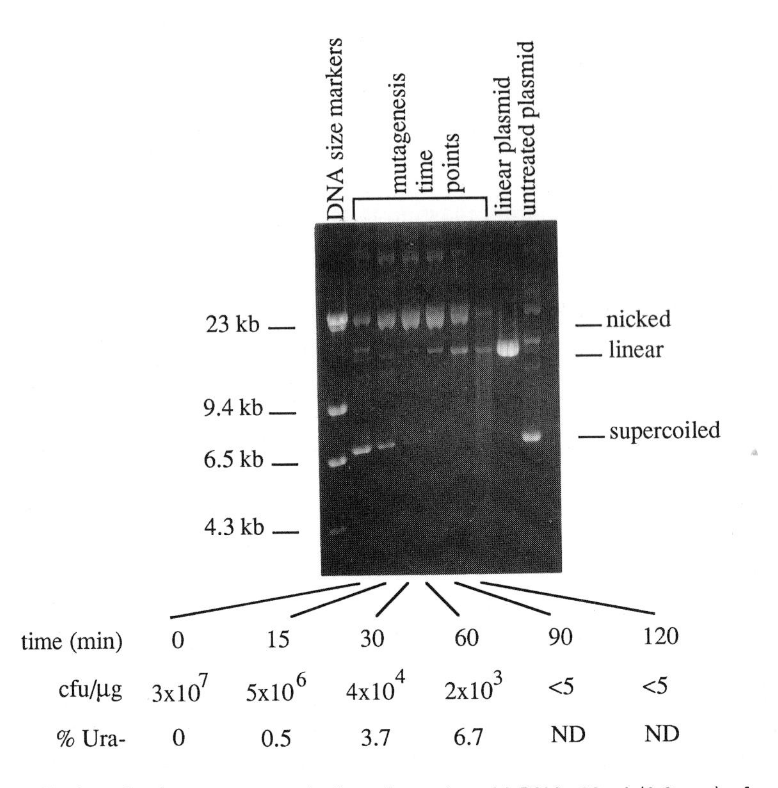

Figure 1. Hydroxylamine mutagenesis linearizes plasmid DNA. 10 μl (0.2 mg) of each timepoint of a mutagenesis reaction of a 14.5 kb plasmid were run on a 0.35% agarose gel. The number of colony forming units (c.f.u.) per mg of input DNA into *E. coli* and the percent of Ura− colonies are shown for each timepoint. Lanes containing linearized plasmid and untreated plasmid are shown on the right-hand side of the figure. Note the transition from super-coiled plasmid to nicked plasmid, and eventually linear plasmid. After 60 min of mutagenesis, as linear plasmid DNA dominates, the transformation efficiency drops off dramatically.

stable YCp plasmids (such as YCp-*ABC*, in our example) are lost from a cell by mis-segregation or misreplication at a high rate. Moreover, drugs are available that prevent the growth of yeast cells expressing particular genes. By including such counterselectable markers on the same plasmid containing the wild-type *ABC* gene one can select directly for the growth of plasmid-free cells. Several negative selection schemes have been described, but only four make use of easily handled compounds that are readily available commercially. These four, *URA3*, *LYS2*, *CAN1*, and *CYH2* have already been described (see Section 3) and the same media are used to select for plasmid-free cells (Section 3.2).

Why choose one counterselection scheme over the other? The *URA3*/5-FOA system is probably the best in most circumstances; only one gene is required for selection and counterselection, and the *URA3* gene is small and has been completely sequenced. Furthermore, a complete deletion allele (see Section 3.2.4) is now available and the extent of this deletion corresponds exactly to the *URA3* fragment used in most yeast shuttle plasmids. *LYS2*/α-AA is the next best choice but the *LYS2* gene itself is rather large with many restriction enzyme recognition sites that hinder cloning. In *URA3*$^+$ and *LYS2*$^+$ strains, counterselection with a *CAN1*- or *CYH2*-containing plasmid is useful since the resistant alleles can be easily selected for in these strains.

4.4 Identifying interesting mutant alleles

For the essential gene, *ABC*, in the strain derived in Section 4.1, the plasmid containing the counterselectable marker can be lost only if a second *ABC*-bearing plasmid is present to provide sufficient *ABC* function to allow growth. This feature provides the basis for identifying mutations in *ABC*. Mutants that cannot provide sufficient *ABC* function will not survive loss of the YCp-*ABC* plasmid and will not grow on media containing the compound used for negative selection of this plasmid. Mutants that provide *ABC* function conditionally do so only under appropriate conditions (selected by the investigator). Conditions such as low or high temperature, altered osmotic conditions (34), growth on medium containing heavy-water (35), etc. have been used successfully.

Conditional alleles of an essential gene can be identified by replica plating to the appropriate counterselective medium under chosen conditions and screening for strains which cannot grow. Lack of growth can at times (especially in dense platings) be difficult to score, and we have found that the visible red colour produced by an *ade2* mutation can make the identification much easier. Actively growing yeast cells with an *ade2* mutation form red colonies on solid medium containing limiting amounts of adenine (see ref. 36 for details). Non-growing cells, even in dense patches, are distinctly white on such medium. Therefore, one can transfer many cells from a fully adenine-supplemented plate, on which all colonies are white, to counterselection media with limiting adenine. Non-growing colonies (i.e. mutants) will remain white while growing colonies (i.e. wild-type, but still *ade2*) will turn red.

Recently, several groups have used a twist of the above strategy to develop plasmid shuffling systems and related screens that rely on a visual phenotype rather than a counterselectable marker (37–40). This makes use of the fact that the red pigment accumulation can be genetically reversed to the normal white colour. This white colour can result either from suppression of the *ade2* mutation or from having an additional mutation in a gene earlier in the adenine biosynthetic pathway than *ade2*, such as *ade3*, that blocks the accumulation of red pigment but not the adenine auxotrophy. An *abc*Δ strain that has mutations in *ade2* and *ade3* is kept alive with a YCp plasmid bearing *ADE3* and *ABC*; this strain makes red colonies due to the *ade2* mutation. Following transformation with a second plasmid bearing a different selectable marker, e.g. a YCp *LEU2 ABC* plasmid, the cells are able to lose the *ADE3* plasmid if it contains an active *ABC* gene. Thus, these colonies throw off white sectors which correspond to plasmid-free (*ade3*) sectors. A dissecting microscope may be necessary to see the small sectors. Mutant versions of the YCp plasmid which bear an inactive copy of ABC are unable to throw off sectors. A practical problem with this approach is that many other events (petite formation, etc.) can give rise to white sectors or affect pigment development. The red colour develops best if the plates are incubated at 30°C overnight and then incubated at room temperature for several days.

Acknowledgements

We thank P. Hieter, T. Davis, and D. Jenness for helpful discussions. This work was supported in part by postdoctoral fellowship PF 3503 from the American Cancer Society (J.B.K.), grants from the NIH, and an American Cancer Society Faculty Research Award (J.D.B.).

References

1. Boeke, J., Truehart, J., Natsoulis, G., and Fink, G. R. (1988). In *Methods in enzymology*, Vol. 154 (ed. R. Wu and L. Grossman), pp. 164–75. Academic Press, London.
2. Mann, C., Buhler, J., Treich, I., and Sentenac, A. (1987). *Cell*, **48**, 627.
3. Budd, M. and Campbell, J. L. (1987). *Proc. Natl Acad. Sci. USA*, **84**, 2838.
4. Sikorski, R. S. and Hieter, P. (1989). *Genetics*, **122**, 19.
5. Sikorski, R. S. and Boeke, J. (1991). In *Methods in enzymology*, Vol. 194 (ed. C. Guthrie and G. R. Fink), pp. 302–18. Academic Press, London.
6. Broach, J. R., Strathern, J. N., and Hicks, J. B. (1979) *Gene*, **8**, 121.
7. Boeke, J., LaCroute, F., and Fink, G. R. (1984). *Mol. Gen. Genet.*, **197**, 345.
8. Chatoo, B. B., Sherman, F., Azubalis, D. A., Fjellstedt, T. A., Mehnert, D., and Ogur, M. (1979). *Genetics*, **93**, 51.
9. Rose, M. and Winston, F. (1984). *Mol. Gen. Genet.*, **193**, 557.
10. Sherman, F. (1991). In *Methods in enzymology*, Vol. 194 (ed. C. Guthrie and G. R. Fink), pp. 3–21. Academic Press, London.
11. Ito, H., Fukuda, Y., Murata, K., and Kimura, A. (1983). *J. Bacteriol.*, **153**, 163.
12. Sherman, F., Fink, G. R., and Hicks, J. B. (1979). In *Methods in yeast genetics*. Cold Spring Harbor Laboratory Press, Cold Spring Harbor, NY.
13. Alam, S. N., Shires, T. K., and Aboul-Enein, H. T. (1975). *Acta Pharm. Suec.*, **12**, 375.
14. Ratzkin, B. and Carbon, J. (1977). *Proc. Natl Acad. Sci. USA*, **74**, 487.
15. Hall, M. N., Hereford, L., and Herskowitz, I. (1984). *Cell*, **36**, 1057.
16. Maniatis, T., Fritsch, E. E., and Sambrook, J. (1982). *Molecular cloning. A laboratory manual*, 2nd edn. Cold Spring Harbor Laboratory Press, Cold Spring Harbor, NY.
17. Rothstein, R. J. (1983). In *Methods in enzymology*, Vol. 101 (ed. R. Wu, L. Grossman, and K. Moldave), pp. 202–11. Academic Press, London.

18. Winston, F., Chumley, F., and Fink, G. R. (1983). In *Methods in enzymology*, Vol. 101 (ed. R. Wu, L. Grossman, and K. Moldave), pp. 211–28. Academic Press, London.
19. Sherman, F. and Hicks, J. (1991). In *Methods in enzymology*, Vol. 194 (ed. C. Guthrie and G. R. Fink), pp. 21–37. Academic Press, London.
20. Kunkel, T. A. (1985). *Proc. Natl Acad. Sci. USA*, **82**, 488.
21. Zoller, M. J. and Smith, M. (1984). *DNA*, **3**, 4789.
22. Kramer, W., Drutsa, H. W., Jansen, V., Kramer, B., Pflugfelder, M., and Fritz, H. J. (1984). *Nucleic Acids Res.*, **12**, 9441.
23. Shortle, D., Grisafi, P., Benkovic, S. J., and Botstein, D. (1982). *Proc. Natl Acad. Sci. USA*, **79**, 1588.
24. Shortle, D. and Botstein, D. (1985). *Science*, **229**, 1193.
25. Meyers, R. M. and Maniatis, T. (1985). *Science*, **229**, 242.
26. Miller, J. H. (1985). *J. Mol. Biol.*, **182**, 45.
27. Busby, S., Irani, M., and De Crombrugghe, B. (1982). *J. Mol. Biol.*, **154**, 359.
28. Rose, M. D. and Fink, G. R. (1987). *Cell*, **48**, 1047.
29. Fowler, R., Degnen, G., and Cox, E. (1974). *Mol. Gen. Genet.*, **133**, 179.
30. Enquist, L. W. and Weisberg, R. A. (1977). *J. Mol. Biol.*, **111**, 97.
31. Zaret, K. S. and Sherman, F. (1985). *J. Bacteriol.*, **162**, 579.
32. Hoffman, W. (1985). *J. Biol. Chem.*, **260**, 11831.
33. Kaufer, A. F., Fried, H. M., Schwindinger, W. F., Jasin, M., and Warner, J. R. (1983). *Nucleic Acids Res.*, **11**, 3123.
34. Hawthorne, D. C. and Friis, J. (1964). *Genetics*, **50**, 829.
35. Bartel, B. and Varshavsky, A. (1988). *Cell*, **52**, 935.
36. Koshland, D. and Hieter, P. (1988). In *Methods in enzymology*, Vol. 155 (ed. R. Wu), pp. 351–72. Academic Press, London.
37. Blinder, D., Bouvier, S., and Jeness, D. D. (1989). *Cell*, **56**, 479.
38. Bender, A. and Pringle, J. R. (1991). *Mol. Cell. Biol.*, **11**, 1295.
39. Geiser, J. R., van Tuinen, D., Brockerhoff, S. E., Neff, M. M., and Davis, T. N. (1991). *Cell*, **65**, 949.
40. Davis, T. N. (1992). *J. Cell Biol.*, **118**, 607.

7

Ty insertional mutagenesis

J. N. STRATHERN, M. MASTRANGELO, L. A. RINCKEL, and D. J. GARFINKEL

1. Introduction

The use of transposable elements as insertional mutagens is an extremely powerful technology for genetic analysis. This approach was brought to a fine art in prokaryotic genetics and is proving equally applicable in fruit flies, mice, and higher plants.

Saccharomyces cerevisiae harbours a family of endogenous, retrovirus-like elements, Ty elements (Ty1, Ty2, Ty3, Ty4, and Ty5), that spontaneously transpose into new sites at rates of 10^{-9} to 10^{-5} per generation (for recent reviews of Ty biology see references 1 and 2). Three major developments combined to make Ty1 and Ty2 elements useful as insertional mutagens. The first was the demonstration that retroelement transposition could be stimulated by expressing Ty1 from an efficient promoter (3). Fusing a Ty1 or Ty2 element to the *GAL1* promoter results in galactose-inducible transposition of the element at levels approaching one event per cell. The second was the demonstration that Ty1 and Ty2 elements could accommodate selectable markers within their genomes (4, 5). For example, Ty1 elements carrying *TRP1, HIS3, ADE2*, or resistance to neomycin, have been transcribed from *URA3*-based *GAL1* expression plasmids and shown to transpose into chromosomal targets. Such transposition events were identified as cells that retain the selected Ty marker when the plasmid is lost (plasmid loss is conveniently selected because Ura^- cells are resistant to 5-fluoro-orotic acid (5-FOA) (6). This approach has already proven useful for isolating mutations caused by the insertion of marked Ty1 elements. Ty1 transposon tagging has identified genes involved in mating, silencing of *HML* and *HMR*, and heat shock response, as well as a variety of metabolic genes (4, 7, 8).

The most recent step in the evolution of Ty1 as an insertional mutagenesis tool was the development of selectable markers that only function after transposition; thus, cells that have had a transposition event can be directly selected. Curcio and Garfinkel (9) developed a *his3* marker gene that becomes *HIS3* only after transposition (see *Figure 1*). The key feature of this allele (*his3-AI*) is an artificial intron (AI) that has been inserted into the *his3* gene in

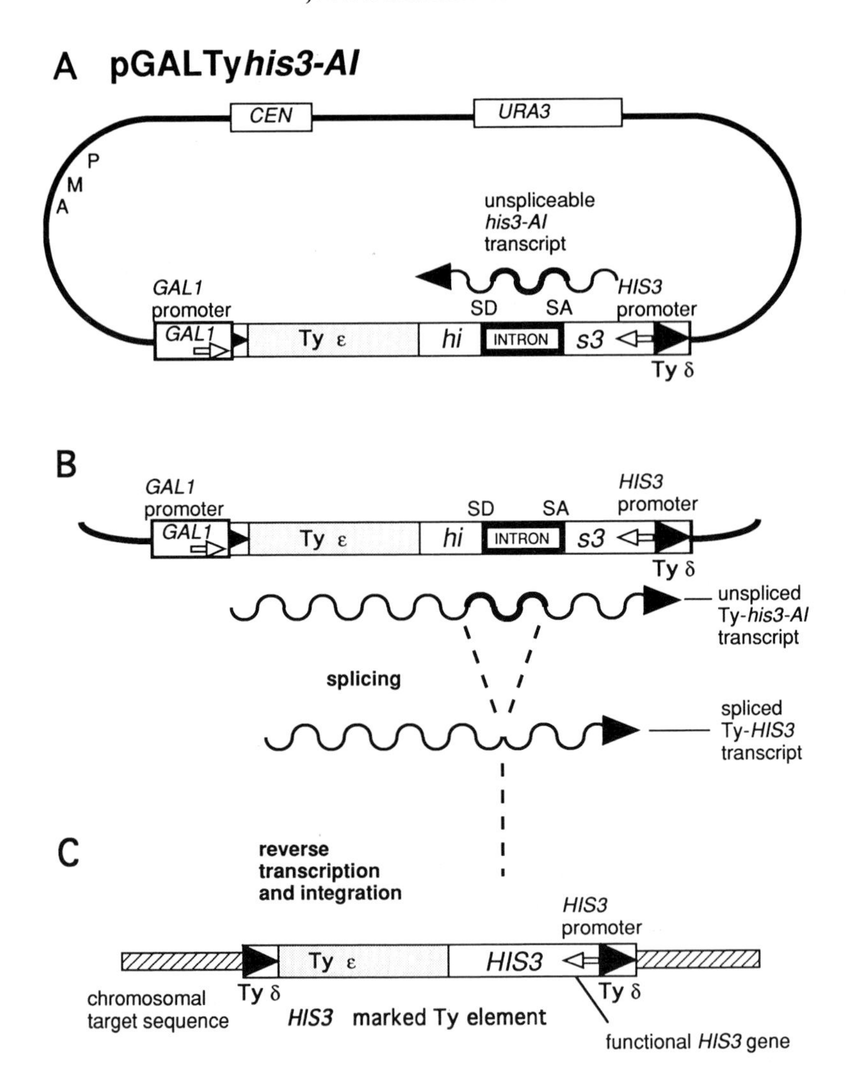

Figure 1. Transposon tagging with pGALTy*his3-AI*. (A) The plasmid pGALTy*his3-AI* carries the β-lactamase gene (AMP), a bacterial plasmid origin of replication, *URA3* gene, a centromere (CEN IV), a yeast origin of replication (ARSI), a TyI element under the transcriptional control of the *GAL1* promoter, and the *his3-AI* (*hi . . . . s3*) marker gene within the TyI. SD and SA are splice donor and splice acceptor sequences for the artificial intron. The long terminal repeat of the TyI element is called a Ty δ element. The transcription from the *HIS3* promoter is not spliceable. (B) The TyI-*his3-AI* transcript is induced upon growth on galactose as a carbon source. This transcript is spliceable, but is the antisense strand for the *HIS3* gene. (C) Reverse transcription results in a TyI element carrying a functional *HIS3* gene.

the wrong orientation for it to be spliced from the *his3* transcript. Cells carrying this allele are His$^-$. The *his3-AI* DNA was inserted into a *GAL*Ty so that the antisense strand of the *his3* gene is transcribed as a part of the Ty RNA. In this orientation, the artificial intron in the *his3* sequence can be spliced from the Ty1 transcript. Reverse transcription and transposition of the spliced transcript result in a Ty1 insertion carrying the *HIS3* allele (see *Figure 1C*). Thus, transposition events are directly selected. About 12% of the transpositions of the marked element are accompanied by splicing to produce a His$^+$ cell. With the combined abilities to induce high levels of Ty transposition and to select for cells that have had transposition events, marked Ty elements can be efficiently utilized as insertional mutagens and in transposon-tagging protocols.

The simplest use of this approach is to induce transposition of the pGALTy*his3-AI* element and then select for His$^+$ and another phenotype. Classical genetic crosses allow one to determine whether the mutation causing the phenotype for which one selected is linked to the *HIS3*-marked element. The Ty *HIS3* element provides not only a genetic tag, but also a physical one. The presence of the *HIS3* sequences can be used to determine which chromosome carries the insertion in Southern blots of chromosome separation gels. Further, the *HIS3* sequences can be used as probes for cloning sequences adjacent to the insertion site.

2. Successes

The pGALTy*HIS3* mutagenesis system has been used to isolate mutations causing a broad spectrum of phenotypes. Our most extensive analysis has been in isolating α-factor-resistant mutants. This selection is advantageous as a test system because at least 12 genes have been previously identified as essential for pheromone response. We were able to tag seven different genes (including a novel class of mutations) among fewer than 20 analysed mutants (7, 10). In a screen for genes essential for growth at high temperature (another screen with a number of potential targets), we identified not only three genes previously shown to have this phenotype, but also two new genes (8).

The pGALTy*his3-AI* system has been used for transposon tagging essential genes by inducing transposition of this element in **a**/α diploids. His$^+$ diploids were sporulated and screened for His$^+$ haploids. Haploids were selected as uncovering recessive, drug-resistant mutations carried in the heterozygous state in the starting diploid (*cyh2/CYH2 can1/CAN1*). About 0.5% of the His$^+$ diploids did not give rise to His$^+$ spores and yielded tetrad data consistent with the presence of a *HIS3*-marked insertion in an essential gene (M. Mastrangelo, unpublished data).

3. Problems

3.1 Insertions are not random

One limitation of using Ty as an insertional mutagen is that insertions are not random throughout the genome. This non-randomness can be observed both at the level of gene-sized targets and at the level of site preference within a gene. A simple illustration of the target preference is that selected gene disruptions are obtained at frequencies below that predicted from the target size. For example, the *CAN1* gene (arginine permease) is about 2.5 kb long, or over one part in 10^4 of the yeast genome. However, insertions into this gene (selected as resistant to the arginine analogue, canavanine) represent less than 10^{-5} of the His$^+$ colonies obtained from transpositions of pGALTy*his3-AI*. Thus, insertions into *CAN1* occur at less than 10% of the predicted frequency in cells that have had a Ty transposition. Similar calculations based on the frequency of insertions into the *LYS2* gene (selected as resistant to α-aminoadipate), Ty-induced α-factor-resistant mutants, and Ty-induced auxotrophs, all suggest that over 90% of the Ty transpositions occur into targets that do not cause mutagenesis. In practical terms, this means that one should deal with a large enough pool ($>10^7$) of Ty insertions to ensure that insertions into genes that are poor targets are included. It remains possible that some genes will be very poor targets for Ty insertions and difficult to mutagenize by this approach.

One possible explanation is that 90% of the events are not true insertions, but are instead recombination events between the marked Ty cDNA and an endogenous Ty. However, we have found no example of a Ty insertion consistent with a simple substitution event. Further, analysis of unselected Ty insertions into chromosome III showed that they were integrations into novel sites (J. D. Boeke, personal communication).

The analysis of large numbers of Ty insertions into the *URA3* (11) and *CAN1* loci (12; L. A. Rinckel, unpublished results) illustrate conspicuous site preferences. When His$^+$ and Can$^-$ (resistance to canavanine) were simultaneously selected, over 50% of the mutant *can1* alleles resulted from the insertion of a marked Ty element. The density of insertions is highest in the promoter region of *can1* and *ura3*. Whether this altered distribution represents a preference for regions of high A+T content found in these promoter regions, a reflection of some general aspect of chromatin structure, or a specific interaction with some component of the transcription machinery has not yet been determined. Clear evidence exists, however, of an interaction between the Ty3 integration machinery and the RNA polymerase III (pol3) transcription apparatus (13). Preference for particular regions within specific genes may be related to Ty insertions being underrepresented within genes. Regions of the genome may exist that are preferred integration sites and cause no phenotype when occupied by Ty elements. The apparent clustering of Ty element δ elements (Ty long terminal repeats) and the observation that many novel insertions are within or next to such δ elements, favours the view that Ty elements prefer certain chromosomal regions. On the other hand, we know of no example of a directly selectable mutation that cannot be obtained by a Ty insertion event.

3.2 Insertions can be complex

The majority of Ty insertions, including those carrying selectable marker genes, are simple insertions of the element into the target sequence accompanied by a 5 bp duplication of the target site. Insertions rarely involve multimeric or rearranged Ty elements, and some are accompanied by the deletion of target sequences. Multimeric Ty insertions were detected among events that caused the normally silenced *HMLα* locus to be expressed (7, 10). Ty monomers at the same positions did not result in this release from the silencing mechanism. Multimeric insertions were also found as a minor class of Ty insertions that activated a promoterless *his3* gene (7) and of insertions that activated *CAN1* (14). Such rare multimeric or rearranged Ty insertions can present problems for cloning strategies based on isolating sequences adjacent to the marker gene. Similarly, Ty insertions associated with a deletion can cause problems in identifying the gene responsible for the mutant phenotype. In a screen for Ty-induced mutations causing temperature sensitivity, two deletions of the *PET18* locus were identified that appear to be similar to the spontaneous deletions in this interval caused by recombination between endogenous Ty elements (8). A smaller deletion associated with a temperature-sensitive phenotype was obtained at a region with a cluster of endogenous δ elements. Deletions associated with Ty integration have also been reported for the *CYC1* region of chromosome X (15). In these cases, it was suggested that Ty integration can be accompanied by recombination with adjacent Ty elements resulting in deletions. These deletions emphasize the importance of comparing the structure of the mutant and parent strains at the insertion sites.

4. Future developments

An examination of the techniques applied to insertional mutagenesis in other organisms suggests a variety of modifications to Ty elements that would increase its usefulness as an insertional mutagen. Promoter tagging by insertion elements requires that a transposable element carries a detectable ORF situated so that it can be expressed from sequences adjacent to the insertion site. In contrast to prokaryotes, where genes are commonly found in multigenic operons, a promoter-tagging element in yeast would require that the insertion be in-frame with (or provide) the first AUG of the transcript.

Similarly, it may be possible to develop Ty elements that carry a conditional promoter that can be used to screen for phenotypes dependent upon the expression of genes adjacent to the integrated Ty. Natural Ty elements occasionally cause mutations that have phenotypes that are conditional upon cell type (16). The majority of such mutations are caused by Ty insertions into the promoter region of the affected gene and cause the gene's expression to be under the control of enhancer sequences present in the Ty element. These mutations allow the gene adjacent to the Ty element to be expressed in α- or **a**-mating type cells, but the affected gene is not expressed in **a**/α cells. These alleles are called ROAM mutations (regulated overproducing alleles under mating control).

5. Considerations in choosing a method

The protocols described below are designed for use with the pGALTy*his3-AI* plasmid shown in *Figure 1*. In this system, Ty insertions can be induced by growth in galactose, and the marked Ty insertions can be directly selected by growth in the absence of histidine. It is often advantageous to perform the mutagenesis on a series of parallel pools so that independently derived mutations can be obtained. Further, when using the Ty tagging method, it is preferable to avoid having multiply marked Ty transposition events in the mutagenized cells. The protocols described below are designed to give efficient transposition under conditions that limit the number of multiple insertions per cell. The key feature is to establish conditions that result in about 1% His$^+$ cells (spliced transposition events). Keep in mind that there may be additional Ty insertions carrying the unspliced *his3-AI* allele which could show up in the physical characterization of the strain. Keeping the level of His$^+$ insertions to approximately 1% helps minimize this problem.

There are three major strain requirements for using the pGALTy*his3-AI* system:

- the strain must be able to induce the *GAL1* promoter in response to growth in galactose
- it must be a *ura3* mutant
- it must be a *his3* mutant (we strongly recommend the *his3–200* deletion allele to preclude His$^+$ formation by recombination between the plasmid and chromosomal *his3* sequences)

We present two methods for using pGALTy*his3-AI* to make transposon-tagged mutants. The first is based upon inducing cells in liquid media and is particularly suited to producing large libraries of transposon-mutagenized cells. The second method involves inducing transposition on solid media and can be used to generate independent Ty transposition events in genes that can be directly selected.

Protocol 1. Galactose induction of pGALTy*his3-AI* in liquid

1. Streak a strain carrying the pGALTy*his3-AI* plasmid for single colonies on SC-URA (synthetic complete medium lacking uracil)[a].
2. Using single colonies as inocula, grow 10 ml overnight cultures in SC-URA liquid with shaking at 30°C. The cultures should grow to saturation at concentrations of about 5×10^7/ml.
3. Centrifuge cells and resuspend at 5×10^6/ml in SCGAL-URA (synthetic complete medium lacking uracil, with galactose as a carbon source[a]).
4. Incubate at 20°C, with shaking, for 24 h. (Ty transposition is temperature-sensitive.)
5. Centrifuge cells, resuspend in sterile H_2O, and spread on SC-HIS plates at 30°C. (In our strains, about 1% of the cells at this point are His^+. This number is sensitive to the efficiency and timing of galactose induction which can vary from strain to strain.)
6. In some cases, plate the mutagenized cultures under conditions that select for His^+ and the desired phenotype. In other cases, first plate the cultures for His^+ and screen the resulting colonies for the desired phenotype.
7. Remove the pGALTy*his3-AI* vector from the mutagenized cells before genetic or physical characterization. The *URA3*-based vector is readily lost during growth without selection. Direct selection for cells that have lost the plasmid can be accomplished by growth on FOA media[a].

[a] See *Protocol 3* for preparation of media.

Protocol 2. Galactose induction of pGALTy*his3-AI* on plates

1. Plate a strain carrying pGALTy*his3-AI* for single-cell colonies on SCGAL-URA. This can be done either by dilution or streaking for single cells.
2. Incubate at 20°C for 4 days to allow colonies to form.
3. Replica plate the colonies to plates selecting for His^+ and the condition required for the mutation of interest.
4. Pick papillae that result from cells that satisfy both the His^+ and the additional selection from separate colonies, ensuring that they represent independent events. (For example, in our strains, about 0.1% of the colonies give papillae that are His^+ and resistant to canavanine.) Streak to obtain single colonies.

Protocol 3. Preparation of media

1. Synthetic complete (SC) (1 litre):
   (a) Make up 'Mix I' by adding the following (autoclave to sterilize):
   - yeast nitrogen base (without amino acids) 6.7 g
   - adenine 10.0 mg
   - uracil (omit for SC-URA) 40.0 mg
   - tyrosine 50.0 mg
   - agar (for plates) 30.0 g
   - H_2O to 500 ml

   (b) Make up 'Mix II' (100× stock; filter sterilize and refrigerated):
   - arginine 200 mg
   - histidine (omit for SC-HIS) 200 mg
   - isoleucine 600 mg
   - leucine 600 mg
   - lysine 400 mg
   - methionine 100 mg
   - phenylalanine 600 mg
   - threonine 1000 mg
   - tryptophan 400 mg
   - H_2O to 100 ml

   (c) To Mix I, add 435 ml of sterile H_2O, 10 ml of Mix II, 50 ml of sterile 40% glucose solution, and 5 ml of 1 M Na_2HPO_4 (pH 7).
2. SC+FOA (1 litre):
   - SC Mix I (see above) 500 ml
   - SC Mix II (see above) 10 ml
   - glucose (40% solution) 50 ml
   - Na_2HPO_4 (1 M, pH 7) 5 ml
   - 5-fluoro-orotic acid (filter sterilize in 435 ml H_2O) 800 mg
3. SCGAL-URA, is the same as SC except that 50 ml of 40% galactose is substituted for the glucose, and the uracil is omitted.

Acknowledgements

The research performed in this laboratory was supported in part by the National Cancer Institute, DHHS, under contract no. N0-C0-74101 with ABL. The contents of this publication do not necessarily reflect the views or policies of the DHHS, nor does mention of trade names, commercial products, or organizations imply endorsement by the US Government. Additional support was obtained from National Science Foundation grant DMB 8812128.

References

1. Boeke, J. D. and Sandmeyer S. B. (1990). In *The molecular and cellular biology of the yeast Saccharomyces: genome dynamics, protein synthesis, and energetics*, Vol. 1 (ed. J. R. Broach, J. Pringle, and E. Jones), pp. 1–68. Cold Spring Harbor Laboratory Press, Cold Spring Harbor, NY.
2. Garfinkel, D. J. (1992). In *The retroviruses*, Vol. 1 (ed. J. A. Levy), pp. 107–58. Plenum Publishing Corp., New York.
3. Boeke, J. D., Garfinkel, D. J., Styles, C. A., and Fink, G. R. (1985). *Cell*, **40**, 491.
4. Garfinkel, D. J., Mastrangelo, M. F., Sanders, N. J., Shafer, B. K., and Strathern, J. N. (1988). *Genetics*, **120**, 95.
5. Boeke, J. D., Xu, H., and Fink, G. R. (1988). *Science*, **239**, 280.
6. Boeke, J. D., LaCroute, F., and Fink, G. R. (1984). *Mol. Gen. Genet.*, **197**, 345.
7. Weinstock, K. G., Mastrangelo, M. F., Burkett, T. J., Garfinkel, D. J., and Strathern, J. N. (1990). *Mol. Cell. Biol.*, **10**, 2882.
8. Kawakami, K., Shafer, B. K., Garfinkel, D. J., Strathern, J. N., and Nakamura, Y. (1992). *Genetics*, **131**, 821.
9. Curcio, M. J. and Garfinkel, D. J. (1991). *Proc. Natl Acad. Sci. USA*, **88**, 936.
10. Mastrangelo, M. F., Weinstock, K. G., Shafer, B. K., Hedge, A.-M., Garfinkel, D. J., and Strathern, J. N. (1992). *Genetics*, **131**, 519.
11. Natsoulis, G., Thomas, W., Roghmann, M.-C., Winston, F., and Boeke, J. D. (1989). *Genetics*, **123**, 269.
12. Wilke, C. M., Heidler, S. H., Brown, N., and Liebman, S. W. (1989). *Genetics*, **123**, 655.
13. Chalker D. L. and Sandmeyer, S. B. (1990). *Genetics*, **126**, 837.
14. Wilke, C. M. and Liebman, S. W. (1989). *Mol. Cell. Biol.*, **9**, 4096.
15. Sutton, P. R. and Liebman, S. W. (1992). *Genetics*, **131**, 833.
16. Errede, B., Cardillo, T. S., Sherman, F., Dubois, E., Deschamps, J., and Wiame, J. M. (1980). *Cell*, **22**, 427.

8

High efficiency transformation with lithium acetate

R. DANIEL GIETZ and ROBIN A. WOODS

1. Introduction

The transformation of lithium acetate (LiAc) treated yeast cells with plasmid DNA was first reported in 1983 (1). The procedure was less efficient than transformation of cells converted to sphaeroplasts by treatment with Zymolyase (2, 3), but was simpler and faster. The original procedure, and subsequent improvements reported by a number of workers, involved a pre-incubation of the cells in LiAc, followed by an incubation in LiAc plus plasmid DNA, then by the addition of polyethylene glycol (PEG), and finally by exposure to a heat shock. The yields of transformants with this procedure rarely exceeded $1 \times 10^4/\mu g$ plasmid DNA (4), compared with $2 \times 10^5/\mu g$ for the sphaeroplast procedure (5). In 1989 it was reported that inclusion of single-stranded carrier nucleic acids (DNA or RNA) in the transformation reaction increased the efficiency of the LiAc procedure dramatically; to 1×10^5 transformants/μg (6). Further modifications to this 'LiAc/ssDNA/PEG' method reduced the time required and increased the efficiency more than 20 fold (7, 8). The current version of the method, described here, is markedly simpler and has yielded up to 2.2×10^7 transformants/μg plasmid DNA.

Although the LiAc transformation method is widely used, little is known about the mechanism. Transformation is stimulated by agents that alter the porosity of the cell wall, for example Li^+, proteases and β-mercaptoethanol (9). We have found that LiAc is the most effective of these agents, and that combinations of LiAc with the others are neither additive nor synergistic. The presence of single-stranded carrier DNA in the transformation reaction is essential for high efficiency. Although the yield of transformants is directly proportional to the average length of the fragments of carrier DNA, sonication is necessary to reduce the length sufficiently to prevent the denatured carrier from forming a gel when cooled (6). The addition of PEG to the mixture of cells, carrier, and plasmid DNA is absolutely essential for the production of transformants; in its absence none are obtained. One of us has shown that PEG concentrates the carrier and plasmid DNA on to the cell surface, probably by molecular exclusion (R. D. Gietz, unpublished results). Also, we have found that the molecular weight of the PEG is important: PEG 2000 and PEG 3350 are most effective and the concentration of PEG in the transformation mixture is critical (8). A heat shock at 42°C for 15–20 min dramatically increases the level of transformation for most strains (6, 8).

1.1 Applications of high efficiency LiAc transformation

The procedure is readily applied to routine molecular manipulations of the yeast genome such as the construction of specific genotypes by gene disruption (10). The method has also been used to isolate genes from humans and *Arabidopsis thaliana* by transformation of cDNA expression libraries, amplified in *Escherichia coli*, into yeast strains carrying appropriate mutations for complementation (11). The high levels of transformation make additional molecular manipulations possible. Gene-pool library ligations can be transformed directly into yeast without prior amplification in *E. coli* (12). This allows the recovery of yeast genes, such as *RAD4* (13, 14), which are toxic or under-represented in libraries prepared in *E. coli*. Cells transformed using a normal plasmid by the high efficiency procedure show an increase in co-transformation by normal, integrative, and gene replacement plasmids (12). This facilitates the introduction of non-selectable gene deletion cassettes into specific strains. The procedure can also be used to transform specific gene deletions into strains with no auxotrophic markers (R. D. Gietz, unpublished results).

In recent years there has been an increasing emphasis on the use of yeast as a model system for the study of eukaryotic molecular biology. Functional homologies between structural genes and between the systems for the regulation of gene activity and expression in yeast and mammals are well documented (see, for example 15, 16). The efficiency of the LiAc method ($> 10^6$ transformants per μg of plasmid) is high enough to allow genomic and cDNA libraries prepared from higher eukaryotes to be transformed directly into yeast, without prior amplification in *E. coli*. In addition, the transformants obtained can be replica plated directly on to nylon transfer membranes for colony hybridization. It has also been noted (17) that the high transformation frequencies possible with the protocol will facilitate use of the Fields and Song 'two-hybrid system' (18) to identify cDNAs, carried on one plasmid, which code for proteins that interact with a target protein specified by a DNA sequence carried on another plasmid.

The transformation frequencies obtainable with the protocols described below, coupled with direct replica plating and colony hybridization, make the LiAc/ssDNA/PEG procedure more suitable for the analysis of eukaryotic molecular biology than the transformation of sphaeroplasts (5), and, in contrast to electroporation (19), the procedure does not necessitate the purchase of specific equipment.

2. Preparation of media and reagents

2.1 Complex medium (YPAD)

Cells for transformation can be grown to stationary phase in YPAD (20, also see Appendix 1), and then diluted and regrown for two divisions in the same medium. Alternatively, cells can be grown to mid-log phase in YPAD, to a density not higher than 2×10^7 cells/ml, and used directly. Adenine hemisulphate (100 mg/l) inhibits the accumulation of red pigment by strains carrying mutations in *ADE1* or *ADE2* and improves their growth. The pH of the medium should be adjusted to 6.0 with NaOH. For convenience we grow all our strains in YPAD.

2.2 Synthetic complete (SC) media

Supplements to minimal medium can be added as a powder mix (21) or in liquid form (20). Commonly used plasmids carry either the *URA3*, *TRP1*, or *LEU2* gene and are selected for on 'SC-minus' media (see Appendix 1). It is convenient to make three mixtures: SC minus uracil, SC minus tryptophan, and SC minus leucine. SC-minus media should be adjusted to pH 5.6–6.0 with NaOH after the addition of the SC-minus mix and agar. SC medium is light-sensitive, so poured plates should be allowed to dry overnight in the dark at room temperature, and then stored in airtight bags or containers in the dark in a cold room or refrigerator. Cells plated on media exposed to the light exhibit reduced plating efficiency.

2.3 LiAc stock solution (1.0 M)

Previous protocols for the LiAc/ssDNA/PEG method used a Tris–EDTA/LiAc mix in which the Tris–EDTA and LiAc stock solution were both adjusted to pH 7.5 (6–8, 10). We have found that transformation efficiency is increased if the LiAc is used without pH adjustment and if the Tris–EDTA is omitted. A 1.0 M solution of LiAc has a pH between 8.4 and 8.9. The following are required for preparation of the stock solution:

- lithium acetate dihydrate (Sigma cat. no. L-6883)
- 0.22 μm filter sterilization unit (disposable or glass) and suitable vacuum source
- water purified using a cartridge filtration system (e.g. Nanopure, Barnstead/Thermolyne Corporation)
- sterile, securely capped bottle for storage of stock solution

Dissolve 10.2 g of lithium acetate in 100 ml of water. Check the pH, which should be 8.4–8.9, and sterilize the solution by filtration. Decant into the storage bottle and store at room temperature.

2.4 PEG stock solution, 50% (w/v)

The stock solution should be stored in a tightly sealed container to prevent evaporation. If the solution is stored in a loosely capped bottle, the concentration of PEG steadily increases, causing a slow but noticeable decline in transformation efficiency with time.

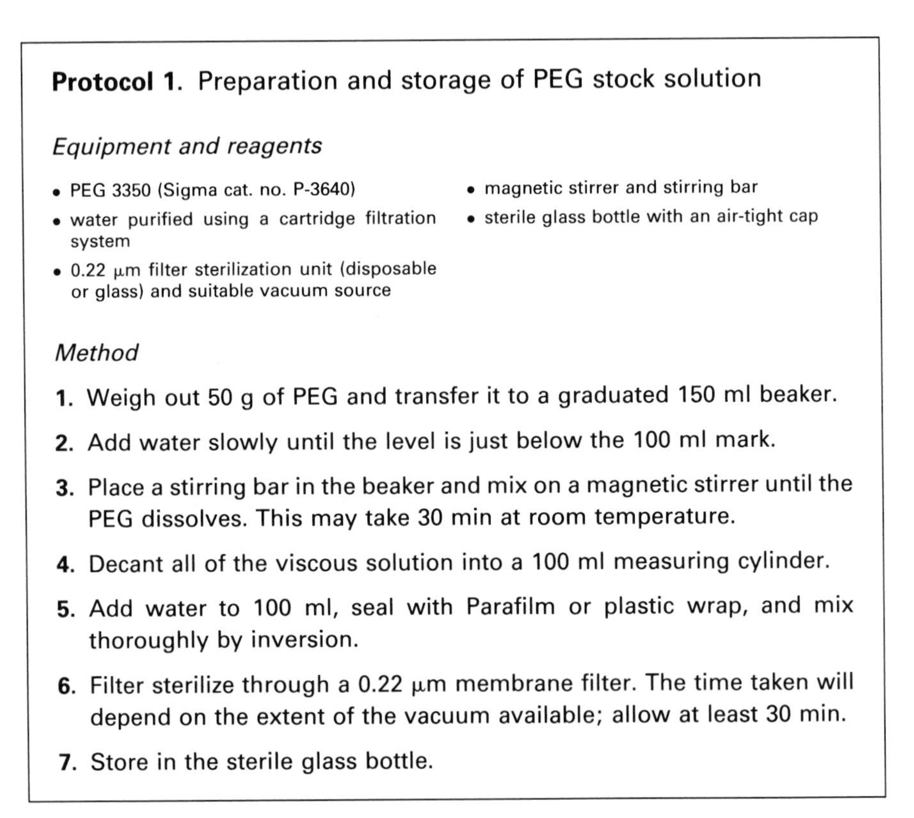

Protocol 1. Preparation and storage of PEG stock solution

Equipment and reagents

- PEG 3350 (Sigma cat. no. P-3640)
- water purified using a cartridge filtration system
- 0.22 μm filter sterilization unit (disposable or glass) and suitable vacuum source
- magnetic stirrer and stirring bar
- sterile glass bottle with an air-tight cap

Method

1. Weigh out 50 g of PEG and transfer it to a graduated 150 ml beaker.
2. Add water slowly until the level is just below the 100 ml mark.
3. Place a stirring bar in the beaker and mix on a magnetic stirrer until the PEG dissolves. This may take 30 min at room temperature.
4. Decant all of the viscous solution into a 100 ml measuring cylinder.
5. Add water to 100 ml, seal with Parafilm or plastic wrap, and mix thoroughly by inversion.
6. Filter sterilize through a 0.22 μm membrane filter. The time taken will depend on the extent of the vacuum available; allow at least 30 min.
7. Store in the sterile glass bottle.

2.5 Single-stranded carrier DNA

Careful preparation of the carrier DNA is important for successful transformation. If the fragments of DNA are too long, transformation efficiency is reduced because the DNA forms a gel and does not mix easily with the cells and plasmid DNA. If the fragments are too short the yield of transformants is also reduced. *Protocol 2* should give DNA of an appropriate length, 2–15 kb with a mean of 7 kb. If a preparation gives good transformation frequencies it does not matter if the size distribution differs from the above. It is important that the DNA, after boiling and quick cooling, is fluid enough to be dispensed with a micropipette.

Protocol 2. Preparation and storage of single-stranded carrier DNA

Equipment and reagents

- high molecular weight salmon sperm DNA (Sigma cat. no. D-1626)
- TE buffer pH 8.0 (see Appendix 1)
- magnetic stirrer and stirring bar
- sonicator with a large horn
- hot plate to boil water in a beaker
- Parafilm or plastic wrap
- buffer-saturated phenol/chloroform (1:1)
- chloroform
- centrifuge capable of taking 50 ml polypropylene tubes and of reaching 5000 *g*
- ice/water bath
- freezer (−20°C)
- sodium acetate solution, 3.0 M, pH 6.0
- ice-cold 95% ethanol
- sterile 15 ml polypropylene or glass centrifuge tubes with screw caps
- sterile 1.5 ml microcentrifuge tubes

Method

1. Dissolve 1.0 g of salmon sperm DNA in 100 ml of TE buffer, pH 8.0. Draw the DNA up and down in a 10 ml pipette to break up the large clumps. Add a stirring bar, cover with Parafilm or plastic wrap, and stir overnight on a magnetic stirrer at maximum speed in a cold room.
2. Sonicate, using a large horn, at 300–500 W for 30 sec. Move the horn around in the solution to ensure that the shearing is not localized.
3. Remove 100 μl of the DNA solution and pipette it into a 1.5 ml microcentrifuge tube.
4. Place the microcentrifuge tube in boiling water for 10 min.
5. Chill the boiled sample in ice water and test with a micropipette to see if it has gelled.
6. If it does *not* gel, the DNA has been sonicated sufficiently.
7. If it does gel, repeat steps **2–5**. Two or three 30 sec bursts at the same power setting should be sufficient.
8. Extract the sonicated DNA with an equal volume of phenol/chloroform[a]. Separate the organic and aqueous phases by centrifugation at 2000 *g* for 5 min and remove the aqueous phase to a clean centrifuge tube.
9. Extract the aqueous phase once with an equal volume of chloroform. Separate the organic and aqueous phases by centrifugation at 2000 *g* for 5 min and remove the aqueous phase to a clean centrifuge tube.
10. Precipitate the DNA by adding 0.1 volume of 3.0 M sodium acetate (pH 6.0) and 2.5 volumes of ice-cold 95% ethanol and mixing gently.
11. Recover the DNA by centrifugation at 5000 *g* for 10 min at 4°C, or by spooling it on to a Pasteur pipette with a hooked end.

Protocol 2. *Continued*

12. Wash the DNA pellet twice in 10 ml of room temperature 70% ethanol.
13. Blot the DNA pellet dry by gently squeezing with a paper towel and allow to air dry for about 20 min.
14. Dissolve the DNA in TE buffer to give a concentration of approximately 10 mg/ml. Break up the pellet with a pipette and mix thoroughly as in step **1**.
15. Dispense 10 ml aliquots into 15 ml polypropylene tubes.
16. Dispense 1.0 ml samples from one 10 ml aliquot into 1.5 ml microcentrifuge tubes in ice water.
17. Use the DNA solutions in two of these tubes to make 20 working aliquots by pipetting 100 μl samples into 1.5 ml microcentrifuge tubes. Each of these working aliquots is sufficient for 20 transformations.
18. Store all of the DNA samples at −20°C. Remember that the DNA must be boiled for 10 min and cooled in an ice water bath prior to use.

Note: *Protocol 2* will give sufficient single-stranded carrier DNA for 20 000 'standard' transformations (see *Protocol 4*), sufficient for many years!

[a] We have recently found that some batches of Sigma salmon sperm DNA (Sigma cat. no. D-1626) do not need to be extracted with phenol/chloroform, allowing the omission of steps **8–14** from *Protocol 2*. Test your batch before phenol/chloroform extraction.

3. Growth of yeast

The yeast strain to be transformed can either be grown overnight in YPAD to mid log phase ($1–2 \times 10^7$ cells/ml), or to stationary phase in a small volume, diluted into fresh YPAD to give 5×10^6 cells/ml, and regrown for two cell divisions to 2×10^7 cells/ml. Regrowth is necessary for maintenance of a plasmid in the strain to be transformed. The volume of the culture is determined by the number of transformations planned. Optimizing efficiency for a particular strain/plasmid combination (see *Protocol 4*) will require 50 or 100 ml of culture; a single transformation plus control requires only 10–15 ml.

Protocol 3. Growth and preparation of cells for transformation

Equipment and reagents

- YPAD or SC-minus medium (see Appendix 1)
- shaking/rotary incubator at 30°C
- microscope and haemocytometer, or spectrophotometer
- centrifuge capable of taking 50 ml polypropylene tubes and of reaching 1500 *g*
- sterile water purified using a cartridge filtration system
- adjustable micropipettes (1000 μl, 200 μl, and 20 μl) and sterile tips
- sterile 1.5 ml microcentrifuge tubes

Method

A. *Growth to mid-log phase*

1. Prepare a suspension of the yeast strain in sterile water and determine the cell titre either by measuring the optical density (OD) at 600 nm of a 10-fold dilution in sterile water, or by counting the cells in a 10-fold dilution using a haemocytometer[a]. Inoculate 50 or 100 ml of YPAD to give 2×10^4 cells/ml with a loopful of the yeast strain and grow at 30 °C in a shaking/rotary incubator until the culture contains $1–2 \times 10^7$ cells/ml[b].
2. Harvest the cells by centrifugation at 1500 *g* for 5 min.
3. Note the volume harvested and calculate the total number of cells obtained. Determine the volume required to give 2×10^9 cells/ml.
4. Wash the cells in 20 ml of sterile water by vortexing vigorously.
5. Centrifuge the cell suspension at 1500 *g* for 5 min.
6. Resuspend the cell pellet with 1.0 ml sterile water and transfer the suspension to a 1.5 ml microcentrifuge tube.

B. *Regrowth*

1. Inoculate 5 ml of YPAD with a loopful of the yeast strain and grow at 30 °C in a shaking/rotary incubator overnight. If maintenance of a plasmid is required grow the strain in 20 ml of SC-minus medium.
2. Leave sufficient YPAD at 30 °C overnight for the regrowth culture (step **4**).
3. Incubate overnight and determine the cell titre. After 16 h the culture should contain $\geq 1.5 \times 10^8$ cells/ml (YPAD) or $\geq 1 \times 10^7$/ml (SC-minus medium).
4. Pipette 2.5×10^8 cells into 50 ml of warm YPAD (step **2**) in a 250 ml culture flask, to give 5×10^6 cells/ml[c].
5. Incubate in a shaking incubator at 200 r.p.m. until the titre reaches 2×10^7 cells/ml.
6. Harvest and wash the cells as detailed for a log phase culture.

[a] It is necessary to determine the relationship between OD and viable cell number. This can be done by measuring the OD of samples taken at intervals throughout the growth of a culture, and determining cell number by plating appropriate dilutions on to YPAD. Make a standard curve for each strain and each spectrophotometer in common use.

[b] The time required to reach this cell density should be established for each strain in common use. Our strains have generation times ranging from 90 to 110 min; we adjust the starting inoculum accordingly and allow extra culture volume in case the strain does not grow as well as expected.

[c] With most strains this cell density is obtained by diluting 2.0 ml of an overnight culture in YPAD into 50 ml of fresh YPAD, or 10 ml of a culture grown in SC-minus medium into 40 ml of fresh YPAD. The volume of the regrowth culture depends on the number of transformations to be carried out; a 50 ml culture is sufficient for 10 standard transformations. The duration of regrowth is dependent on the generation time; allow 3–4 h.

4. Standard transformation procedure

This is the basic procedure of high efficiency transformation. For yeast strains AB1380 (22) and LP2752-4B (11) and the plasmids YCplac33 and YEplac195 (23) the highest efficiencies are achieved by treating 1×10^8 cells with 50 μg single-stranded carrier DNA and 100 ng plasmid DNA. The volumes of carrier DNA (5 μl), plasmid DNA (5 μl), cells (50 μl), and PEG/LiAc (300 μl), result in a final PEG concentration of 33.3%, which we have found to be optimal for most strains. The plasmid DNA can be purified by caesium chloride gradient centrifugation or by column chromatography. Miniprep DNA can also be used; it does not matter if miniprep DNA contains RNA—in fact this appears to enhance transformation. Transformation efficiencies with this protocol range from 1×10^5 to 2×10^7 transformants/μg plasmid DNA; this means between 1×10^4 and 2×10^6 per transformation tube. The protocol is also used to optimize and maximize transformation efficiency for a particular strain/plasmid combination (see *Protocols 5* and *6*). **Note**: it is important to include a control containing no plasmid DNA to check for possible reversion of the selectable marker in your yeast strain.

Protocol 4. Standard transformation procedure

Equipment and reagents

- 1.0 ml suspension of cells in sterile water (*Protocol 3*)
- LiAc stock solution, 1.0 M
- PEG stock solution, 50% (w/v)
- microcentrifuge
- sterile water
- single-stranded carrier DNA (100 μl working aliquot) (*Protocol 2*)
- plasmid DNA (20 ng/μl)
- adjustable micropipettes (1000 μl, 200 μl, and 20 μl) and sterile tips
- sterile 1.5 ml microcentrifuge tubes
- waterbaths at 30°C and 42°C (or an ice bucket or other suitable container with water at the appropriate temperature)
- plates of appropriate SC-minus medium
- incubator at 30°C

Method

1. Prepare a 100 mM solution of LiAc by diluting 500 μl of 1.0 M stock solution (see Section 2.3) into 4.5 ml of sterile water.
2. Prepare a PEG/LiAc solution (40% PEG, 100 mM LiAc) by mixing 500 μl of 1.0 M LiAc stock solution, 500 μl of sterile water, and 4.0 ml of PEG 50% (w/v) stock solution. Since the PEG solution is denser than water, it is important to vortex vigorously to ensure proper mixing.
3. Centrifuge the suspension of cells in 1.0 ml of sterile water (prepared as in *Protocol 3*) at top speed in a microcentrifuge for 10 sec and then pour off the supernatant.

4. Resuspend the cells in 300 μl of 100 mM LiAc. Take up the suspension into a pipette tip and adjust the volume control on the micropipette until all of the suspension is within the tip. Read the volume and add sufficient 100 mM LiAc to bring the titre to 2×10^9 cells/ml.
5. Incubate the cell suspension in a waterbath at 30°C for 15 min.
6. Boil a working aliquot (100 μl) of carrier DNA for 10 min and chill in ice/water.
7. Dispense 5 μl of boiled carrier DNA (50 μg) and 5 μl of plasmid DNA (100 ng) into 1.5 ml microcentrifuge tubes for individual transformations. Include a control with no plasmid.
8. Remove the cell suspension from the waterbath and vortex.
9. Pipette 50 μl of cell suspension (1×10^8 cells) into each transformation tube and vortex briefly.
10. Pipette 300 μl PEG/LiAc into each tube and vortex briefly.
11. Incubate all of the transformation tubes in a waterbath at 30°C for 30 min.
12. Heat shock the transformation tubes in a waterbath at 42°C for 20 min.
13. Microcentrifuge at top speed for 10 sec and remove the PEG/LiAc supernatant with a micropipette.
14. Pipette 1.0 ml of sterile water into each tube. Resuspend the cells by stirring with a 200 μl pipette tip (attached to a micropipette) and suck the suspension up and down to disperse any clumps.
15. Vortex to ensure even suspension of the cells.
16. Make 100-fold dilutions of the suspensions in each transformation tube containing plasmid by pipetting 10 μl into 990 μl sterile water in a microcentrifuge tube. Vortex each dilution.
17. Pipette 100 μl samples of the dilutions on to selective medium, and spread the samples over the agar surface with a sterile glass spreader. Plate 100 μl samples of the cell suspension from the control tube (i.e. no plasmid) on to selective medium.
18. Incubate the control and transformation plates at 30°C for 3–4 days.
19. Calculate the number of transformants per μg plasmid DNA. For example if the plates have an average of 500 transformants, there are 5×10^6 transformants/μg plasmid DNA, *viz.*: 500 × 10 (100 ng plasmid DNA) × 100 (100-fold dilution) × 10 (100 μl sample) = 500 × 10 000 = 5×10^6.

5. Procedures for optimizing the protocol for specific strain/plasmid combinations

The experimental parameters which have the greatest effect on transformation efficiency are:

(a) concentration of single-stranded carrier DNA: the 'effective' concentration of the single-stranded carrier DNA varies from preparation to preparation and it may be necessary to determine the optimal volume for use in the transformation reaction

(b) duration of heat shock: the efficiency of the heat shock will depend on the thermal characteristics of the incubation tube and the volume of the transformation reaction. A large volume transformation may require a longer heat shock. Strains which are hypersensitive to 42°C may require either shorter exposure or a lower temperature (37–40°C).

These parameters can be optimized by following *Protocols* 5 and 6.

Protocol 5. Titration of carrier DNA

1. Prepare a 100 ml culture of cells for high efficiency transformation as described in *Protocol 3* and complete steps **1–6** of *Protocol 4*.
2. Set up 10 transformation tubes containing 100 ng of plasmid DNA and add carrier DNA to give a range from 5 to 15 μl (50–150 μg). Set up a control with no plasmid and 5 μg of carrier DNA.
3. Continue from step **8** to the end of *Protocol 4*.
4. Select the volume of carrier which gives the highest number of transformants.

6. Maximizing the number of transformants

The preceding protocols will have optimized the transformation of the desired strain with the desired plasmid. The absolute number of transformants is determined by the quantity of plasmid and the number of cells. We have obtained 4.4% transformation using 5 μg plasmid DNA in the 'standard' procedure (*Protocol 4*): 4.4×10^6 transformants per tube! If you require very large numbers of transformants either for specific library work, or to detect rare phenotypes using the Fields and Song 'two-hybrid system' (18), try *Protocol 7*.

Note: high plasmid concentrations in the transformation reaction increase the probability that more than one plasmid molecule will be taken up by any one transformant. This complicates subsequent analysis if the transformation involves a complex library.

Protocol 6. Duration of heat shock (HS) at 42°C

1. Prepare a 100 ml culture of cells for high efficiency transformation as described in *Protocol 3* and complete steps **1–6** of *Protocol 4*.
2. Set up 10 transformation tubes with 5 μl of carrier DNA, 5 μl of plasmid DNA (20 ng/μl), and 50 μl of cells. Label the tubes as follows. (1) Control 1: no plasmid, HS 20 min; (2) control 2: no HS; (3) HS 5 min; (4) HS 10 min; (5) HS 15 min; (6) HS 20 min; (7) HS 25 min; (8) HS 30 min; (9) HS 35 min; (10) HS 40 min.
3. Carry out steps **8–11** of *Protocol 4*.
4. Heat shock the transformation tubes for the times indicated. Terminate the heat shock by placing the tubes in ice water for 15 sec.
5. Continue from step **13** to the end of *Protocol 4*.
6. Select the duration of heat shock which gives the highest number of transformants.

Protocol 7. Maximizing the number of transformants

Reagents

- plasmid DNA at 2 μg/μl and 100 ng/μl

Method

1. Prepare a 50 ml culture of cells for high efficiency transformation as described in *Protocol 3* and complete steps **1–6** of *Protocol 4*.
2. Set up seven transformation tubes containing the amount of carrier DNA which you have determined to be optimal.
3. Add plasmid DNA to give 100 ng, 1 μg, 2.5 μg, 5 μg, 7.5 μg, and 10 μg per transformation tube, keeping the volume of plasmid DNA constant at 5 μl. Include a control with no plasmid DNA.
4. Carry out steps **8–15** of *Protocol 4* using the optimized conditions for your strain/plasmid combination.
5. Dilute the transformation tubes containing plasmid 1000-fold in sterile water and plate 100 μl samples on to selective medium.
6. Incubate the control and transformation plates at 30°C for 3–4 days.
7. Calculate the number of transformants per μg plasmid DNA and determine the plasmid concentration which gives the best return in terms of number of transformants in relation to plasmid input. For example, if you get 2×10^6 transformants with 1 μg plasmid and 5×10^6 with 10 μg, then it is more efficient to carry out a number of transformations with the smaller amount of plasmid.

7. Quick and easy transformation procedure

If your only concern is getting a plasmid into a particular strain then *Protocol 8* is sufficient. This protocol works for strains grown overnight on plates of YPAD, or SC-minus media, and it can also be used for 'spur of the moment' transformations of cultures which have been stored on plates in the refrigerator for several days or weeks.

Note: transformation of yeast cells from very old plates is not reliable; it is better to sub-culture the strain on to fresh medium and grow overnight if you want to use this protocol rather than *Protocol 4*.

Protocol 8. Quick and easy transformation procedure

1. Scrape about 50 μl of cells from the plate and suspend them in 1.0 ml of sterile water in a 1.5 ml microcentrifuge tube.
2. Microcentrifuge the cells and wash once in 1.0 ml of sterile water and once in 0.5 ml 100 mM LiAc.
3. Estimate the volume of the cell pellet and resuspend the cells in an equal volume of 100 mM LiAc.
4. Add 5 μl of carrier DNA (50 μg) and up to 5 μg of plasmid DNA to 50 μl of the cell suspension and vortex to mix.
5. Add 300 μl of PEG/LiAc and vortex.
6. Heat shock the transformation tube at 42 °C for 20 min.
7. Microcentrifuge at top speed for 10 sec and remove the PEG with a micropipette.
8. Resuspend the cells in 1.0 ml of sterile water and plate 250 μl samples on to four plates of selective medium.
9. Incubate the plates at 30 °C for 3–4 days.

8. Remedies for problems which can affect the efficiency of LiAc/ssDNA/PEG transformation

The above protocols have been developed using the yeast strains AB1380 (22) and LP2752-4B (11) and the plasmids YCplac33 and YEplac195 (23). In the course of our endeavours to optimize the efficiency of the procedure we have encountered a number of factors which limit or reduce the number of transformants.

8.1 Strains and plasmids

We have found that some strains do not transform well with *Protocol 4*, but even so we still obtain between 1×10^4 and 1×10^5 transformants per μg plasmid DNA. We have also found that some plasmids give 10-fold lower yields than the YCplac series (19), but the number of transformants, 1×10^5 per μg plasmid DNA, is adequate for most purposes. Optimization of the concentration of carrier DNA (see *Protocol 5*), the duration and the temperature of the heat shock (see *Protocol 6*), and of the concentration of plasmid DNA (see *Protocol 7*), should give some improvement.

8.2 Evaporation of PEG

The concentration of PEG in the transformation reaction should be 33.3% (w/v). If the PEG stock solution is not stored in an air-tight container the aqueous component of the solution will evaporate and the resulting increase in the concentration of PEG will give a noticeable decrease in the efficiency of transformation. Make a fresh PEG stock solution and store in a tightly capped container if you suspect that this is happening.

8.3 Reassociation and degradation of the single-stranded carrier DNA

Repeated thawing and freezing of samples of carrier DNA reduces the efficiency of transformation. Repeated boiling has the same effect. Therefore store carrier DNA in working aliquots of 100 μl, sufficient for 20 transformations. These samples can be thawed, boiled, and refrozen at least five times without affecting the efficiency of transformation.

8.4 Preparation of the SC-minus medium

The addition of adenine hemisulphate, amino acids, and other nutrients to Difco yeast nitrogen base (without amino acids) lowers the pH. We have found that the optimal pH for the recovery of transformants is between 5.6 and 6.0. Therefore, check the pH of the medium and adjust as necessary with NaOH.

8.5 Failure to obtain any transformants

The SC mixtures recommended (20, 21) contain all of the supplements required by yeast strains in common use. If you use a partial SC medium and fail to obtain any transformants using these protocols with a specific strain, and you have had success with others, check the genotype of the strain. You may find that the medium lacks a required supplement; amend the SC mixture as required!

Acknowledgements

This work was supported in part by grants from the Medical Research Council of Canada (MT 11373) and the Manitoba Health Research Council to Dr R. D.

Gietz, and from the Medical Research Council of Canada (MT 6112) to Drs P. J. McAlpine and R. A. Woods.

References

1. Ito, H., Fukuda, Y., Murata, K., and Kimura, A. (1983). *J. Bacteriol.*, **153**, 163.
2. Beggs, J. D. (1978). *Nature*, **275**, 104.
3. Hinnen, A., Hicks, J. B., and Fink, J. R. (1978). *Proc. Natl Acad. Sci. USA*, **75**, 1929.
4. Stearns, T., Ma, H., and Botstein, D. (1990). In *Methods in enzymology*, Vol. 185 (ed. R. Wu), pp. 280–97. Academic Press, London.
5. Burgers, P. M. J. and Percival, K. J. (1987). *Anal. Biochem.*, **163**, 391.
6. Schiestl, R. H. and Gietz, R. D. (1989). *Curr. Genet.*, **16**, 339.
7. Gietz, R. D., St Jean, A., and Woods, R. A. (1992). *Nucleic Acids Res.*, **20**, 1425.
8. Gietz, R. D., Weinberg, O., and Woods, R. A. (1992). *Yeast*, **8**, S259.
9. Brzobahaty, B. and Kovac, L. (1986). *J. Gen. Microbiol.*, **132**, 3089.
10. Kang, X., Yadao, F., Gietz, R. D., and Kunz, B. A. (1992). *Genetics*, **130**, 285.
11. Elledge, S. J., Mulligan, J. T., Ramer, S. W., Spottswood, M., and Davis, R. W. (1991). *Proc. Natl Acad. Sci. USA*, **88**, 1731.
12. Gietz, R. D. and Schiestl, R. H. (1991). *Yeast*, **7**, 253.
13. Fleer, R., Nicolet, C. M., Pure, G. A., and Friedberg, E. C. (1987). *Mol. Cell. Biol.*, **7**, 1180.
14. Gietz, R. D. and Prakash, S. (1988). *Gene*, **74**, 535.
15. Guarente, L. (1988). *Cell*, **52**, 303.
16. Murray, A. M. (1992). *Nature*, **359**, 599.
17. Fritz, C. C. and Green, M. R. (1992). *Curr. Biol.*, **2**, 403.
18. Fields, S. and Song, O. (1989). *Nature*, **340**, 245.
19. Meilhoc, E., Masson, J.-M., and Teissie, J. (1990). *Biotechnology*, **8**, 223.
20. Sherman, F. (1991). In *Methods in enzymology*, Vol. 194 (ed. C. Guthrie and G. R. Fink), pp. 3–21. Academic Press, San Diego.
21. Rose, M. D. (1987). In *Methods in enzymology*, Vol. 152 (ed. S. L. Berger and A. R. Kimmel), pp 481–504. Academic Press, San Diego.
22. Burke, D., Carle, G. F., and Olson, M. V. (1986). *Science*, **236**, 806.
23. Gietz, R. D. and Sugino, A. (1988). *Gene*, **74**, 527.

9

Measurement of transcription

PETER W. PIPER

1. Introduction

This chapter discusses the experimental tricks needed to determine if a cloned yeast gene is transcriptionally active under defined physiological conditions, or if DNA constructs newly inserted into yeast cells by transformation are being efficiently transcribed *in vivo*. The techniques it describes rely upon the availability of a DNA probe suitable for quantification of the transcript(s) of interest by DNA–RNA hybridization. It should be read in conjunction with standard methods manuals that describe the experimental details of Northern blotting, probe preparation, blot hybridization and the mapping of transcripts on DNA (1).

Many workers are content merely to measure the *in vivo* levels of mRNA(s) complementary to a probe DNA as an indication of the efficiency with which this DNA sequence is being transcribed. However, steady state levels of mRNA(s) increase not just with an increased rate of their synthesis by RNA polymerase II-catalysed gene transcription, but also with decreases in the rates of their degradation by mRNA turnover. Therefore, for cellular mRNA level measurements to be related to the transcriptional efficiencies of the corresponding genes, the half-lives of these mRNAs in the same cells must also be known. This chapter focuses on how to measure the *in vivo* levels of yeast mRNAs by Northern blot analysis, while Chapter 10 details alternative methods for determining the stabilities of these mRNAs. Recently, *in vitro* extracts from carefully purified *Saccharomyces cerevisiae* nuclei have been developed which allow authentic RNA polymerase II transcription of added DNA templates (2). Since the use of these systems is currently limited to those laboratories studying the detailed molecular mechanisms of transcription, they are not described here.

2. Yeast promoters

The promoter sequences which regulate protein coding genes in yeast are mostly to be found within the 500–700 bp of non-coding DNA immediately 5′

to the coding regions. They consist of at least three elements which regulate the efficiency and accuracy of transcriptional initiation by RNA polymerase II (3–5): upstream activation sequences (UASs), TATA elements, and initiator elements. Many also contain elements involved in the repression of transcription.

UASs have similarities to the enhancers of mammalian promoters. They determine the 'strength' and regulation of yeast promoters by acting as sites for sequence-specific binding of protein *trans*activators of RNA polymerase II. A number of these transcriptional transactivators (e.g. *GAL*4, *GCN*4, and *ADR*1) have been identified in *S. cerevisiae* (3–5). Some UAS elements have been mapped to short regions of DNA, e.g. the binding site of the *GAL*4 *trans*activator protein, needed for gene induction by galactose, is a 17–21 bp dyad-symmetrical DNA sequence (3–5). TATA elements are found 40–120 bp upstream of the initiation site in yeast, unlike in higher eukaryotes where they are a more rigid 25–30 bp upstream (3), and they define the downstream 'window' within which transcription initiation can occur. Within this window, the poorly defined initiator element directs initiation at one site, or a few closely adjacent sites.

Many yeast promoters are complex, with multiple UASs and negative regulatory sites, and sometimes multiple TATA elements associated with different initiation sites (3). Most are regulated to some extent, but the most powerful glycolytic promoters are constitutively active under most conditions of vegetative growth. For expressing proteins toxic to the yeast cell such constitutive promoters are therefore unsuitable, it being preferable to use a tightly regulated promoter so that the yeast can be grown to high cell density and then induced for product at the desired biomass. *GAL1* is the most commonly used regulated promoter, although genetic tricks have to be used to circumvent its limited efficiency when it is present in multiple copies in normal yeast strains (5). A number of the best characterized regulated promoters of *S. cerevisiae* are listed in *Table 1*.

Several groups measure the 'strength' of a promoter sequence by linking this sequence to the protein-coding region of a suitable 'reporter' gene—an open reading frame that encodes a readily assayable enzyme. This promoter–gene fusion is then inserted into yeast on a suitable vector. Both β-galactosidase, as encoded by *Escherichia coli lacZ* (7, 8), and β-glucuronidase (9) can be used as 'reporters' of promoter activity in *S. cerevisiae*, since this yeast does not normally synthesize either of these enzymes. Unfortunately there are disadvantages to this approach:

(a) It will not monitor control regions outside of the 5′ untranslated region, for example the 'downstream activator sites' present in the coding regions of certain yeast glycolytic genes (5).

(b) These reporter enzymes, once made inside the *S. cerevisiae* cell, are extremely stable. They are therefore unsuitable for monitoring *transient*

Table 1. Examples of wild-type *S. cerevisiae* promoters and their level of regulation (3–6)

Promoter	**Host genotype**	**Strength** [a]	**Regulation (3–6)**
PGK, GAP, TPI	wild-type	++++	10- to 20-fold induction by glucose (*PGK*).
GAL1	wild-type	+++	1000-fold induction by galactose in absence of glucose; severely limited in multiple copies by shortage of GAL4 transactivator, hence improved by *GAL4* overexpression.
ADH2	wild-type	++	100-fold repression by glucose; limited in multiple copies by shortage of ADR1 transactivator, hence improved by *ADR1* overexpression.
PHO5	wild-type	+/++	500- to 200-fold repression by phosphate
CUP1	wild-type	+	20-fold induction by Cu^{2+}
*MF*α*1*	wild-type	+/++	constitutive in α haploid cells; strongly repressed in a haploids or **a**/α diploids
HSP26, HSP82	wild-type	++	induced 30- to 50-fold by shift of temperature from 25°C to 39°C.

[a]Relative amounts of these mRNAs as a fraction of total mRNA when the promoter is 'switched on' (see Section 3.3).

changes in promoter activity, such as those frequently found with metabolic change, stress (10–12), pheromone treatment, or the different stages of the cell cycle (11).

Northern blot analysis, in contrast, does not suffer from these drawbacks and is ideally suited to monitor both transient and sustained alterations to mRNA levels. This chapter therefore concentrates on how to prepare undegraded RNA from yeast suitable for quantitative measurements of mRNA levels by Northern analysis.

3. Preparation of undegraded RNA

3.1 Precautions to minimize RNase contamination

To isolate undegraded RNA from yeast meticulous attention should be paid to the following points:

(a) All materials should be prepared free of RNase contamination by adopting the precautions in *Protocol 1* of Chapter 10. Glassware and plasticware should be autoclaved; also all the solutions for *Protocols 1–3* below should be prepared in water that has been previously treated with diethylpyrocarbonate (DEPC) and autoclaved, as in *Protocol 1* of Chapter 10.

(b) Gloves should be worn at all times, and immediately changed if they have come into contact with any surfaces (e.g. door handles, equipment control buttons, Gilson pipettes, etc.) where there may be RNase.

(c) From cell breakage through to the first ethanol precipitation (steps **4-11** in *Protocol 1* below) the steps should be conducted reasonably quickly. Working fast and at high pH minimizes RNA degradation due to release of the highly active, non-specific RNase present in the yeast vacuole. Like other vacuolar activities, this RNase has an acid pH optimum and increases in activity when cells are starved of nitrogen. Any problems it causes could potentially be avoided by using proteinase A-deficient *pep4* (*pra1*) strains since the aspartyl proteinase A is needed for activation of the precursor form of vacuolar RNase *in vivo* (13). However, this RNase should not cause problems when RNA is extracted by *Protocol 1* of this chapter or *Protocol 3* of Chapter 10.

3.2 Rapid RNA isolation

Protocol 1 has been used satisfactorily for a number of years in the author's laboratory to provide RNA from *S. cerevisiae, Schizosaccharomyces pombe*, and *Hansenula polymorpha* (*Pichia angusta*). A similar, yet slightly different, procedure is given in *Protocol 3* of Chapter 10. Other workers report satisfactory results when extracting cells in the presence of high concentrations of guanidium thiocyanate (14). By adopting the precautions stated in this and in the previous section, we find that undegraded RNA can be readily and routinely extracted without the necessity to use such chaotropic agents for RNase inactivation.

Both *Protocol 1* (below) and *Protocol 3* of Chapter 10 (Brown) rely on efficient vortexing causing glass beads to 'knock together' inside the extraction tube, mechanically breaking the yeast cells as they collide. If these beads are omitted from *Protocol 1*, only the small RNAs of the cell (mostly tRNA and 5S rRNA) are isolated, larger RNAs and DNA remaining trapped within the rigid yeast cell wall. The purpose of the ammonium acetate precipitation (step **11**) is to prevent cellular mononucleotides being carried through to the final RNA sample, these mononucleotides being partially ethanol-precipitable from 1 M sodium or potassium acetates, but not 1 M ammonium acetate.

Protocol 1 effectively isolates total yeast cell nucleic acid, but since most (>95%) of this is RNA, the higher molecular weight DNA component does not interfere with Northern blot analysis, and there is generally no need to treat samples with RNase-free DNase before proceeding to Northern blotting. Indeed, the ability of *Protocol 1* to yield samples containing *both* DNA and RNA can be a useful property, as it is possible to use *separate* aliquots of these samples for:

(a) measuring mRNA levels by Northern blot analysis

(b) determining the copy number of recombinant plasmids (after further manipulations involving RNase digestion, restriction endonuclease digestion, and Southern blotting; see ref. 15).

Protocol 1. Extraction of yeast RNA

Reagents and materials

- Siliconized glass tubes: Corex 15 ml and 30 ml tubes are suitable, larger tubes not providing such a good vortex action to their contents when applied to a whirlimix. Before use, soak glass tubes overnight in 1 M HCl, dry, and rinse briefly with dimethyldichlorosilane solution (BDH cat. no. 33164) (**caution**: handle this solution with care in a fume hood; wear gloves and do not breathe vapour). After this silanization step, extensively rinse the tubes with water and dry them at 200°C overnight
- glass beads (0.4 mm diameter; BDH 'Glass beads for GLC', cat. no. 15029 are ideal). Soak these for 1 h in 1 M HCl, rinse thoroughly in distilled water, and dry at 200°C overnight.
- RNA extraction buffer: 20 mM Tris–HCl pH 8.5, 10 mM Na_2EDTA, 1% (w/v) sodium dodecyl sulphate
- aqueous phenol: melt solid phenol by placing the bottle at 60°C overnight. Add 0.02% (w/v) 8-hydroxyquinoline, then saturate with DEPC-treated water and bring the pH to 8.0 with 1 M Tris. Store in the dark at 4°C.
- 6 M ammonium acetate
- 3 M sodium acetate
- TE buffer, pH 7.5 (see Appendix 1)
- absolute ethanol, precooled to −20°C.
- 70% (v/v) ethanol
- chloroform

Method

1. Place the desired volume of culture in a 15 or 30 ml Corex tube. Alternatively, cells stored −20°C under ethanol (as described in *Protocol 2* of Chapter 10) can be used. Harvest the cells by centrifugation at 5000 *g* for 5 min and discard the supernatant.
2. Resuspend cells in 10 ml of ice-cold RNase-free water, and harvest by centrifugation as in step **1**.
3. Leave the tubes containing cell pellets on ice until ready to proceed to steps **4–11**. While they are still on ice add 1–2 g glass beads. It is convenient to process the samples two at a time through steps **4–11**. Equip yourself with two whirlimixers, so that you can vortex the samples in pairs, holding one tube firmly in each hand.
4. *Rapidly* remove the tubes from the ice bucket, add 4 ml (if using 15 ml tubes) or 8 ml (if using 30 ml tubes) each of RNA extraction buffer and phenol, then firmly cover the tops of the tubes with Parafilm. Both the RNA extraction buffer and phenol should be at room temperature.
5. Vortex vigorously and continuously for 5 min at room temperature.
6. Centrifuge at 5000 *g* for 5 min.
7. Carefully remove the upper (aqueous) phase and transfer it to a fresh tube containing an equal volume of phenol–chloroform (1:1 (v/v)). Take care not to take any of the interface.
8. Vortex for 1 min, then centrifuge at 5000 *g* for 5 min.

Protocol 1. *Continued*

9. Transfer the upper (aqueous) phase to a fresh tube containing an equal volume of chloroform.
10. Vortex for 1 min, then centrifuge at 5000 *g* for 2 min.
11. Transfer the upper (aqueous) phase to a fresh tube, add ammonium acetate to a final concentration of 1 M, and two volumes of −20°C absolute ethanol. Mix thoroughly and place at −20°C for at least 20 min (samples can be stored indefinitely at this stage).
12. Centrifuge the tubes at 10 000 *g* for 15 min at 4°C. Pour off the supernatant and drain the tube upturned on paper tissue.
13. Resuspend the white precipitate in 1 ml of TE buffer.
14. Add sodium acetate to a final concentration of 0.2 M, plus 2.5 volumes of −20°C absolute ethanol. Place at −20°C for 20 min.
15. Repeat step **12**.
16. Gently wash the ethanol precipitate with 70% ethanol at 0°C (thereby removing residual salt from the sample). Drain the tube upturned on paper tissue and then vacuum–desiccate until the pellet starts to turn opaque at the edges (slightly damp RNA pellets are easier to redissolve than completely dry ones).
17. Resuspend the precipitate in 100–500 μl of TE buffer. Determine the integrity and concentration of each RNA sample as described in *Protocol 2*. RNA solutions can be stored at −20°C or −80°C and are not damaged by repeated freezing and thawing.

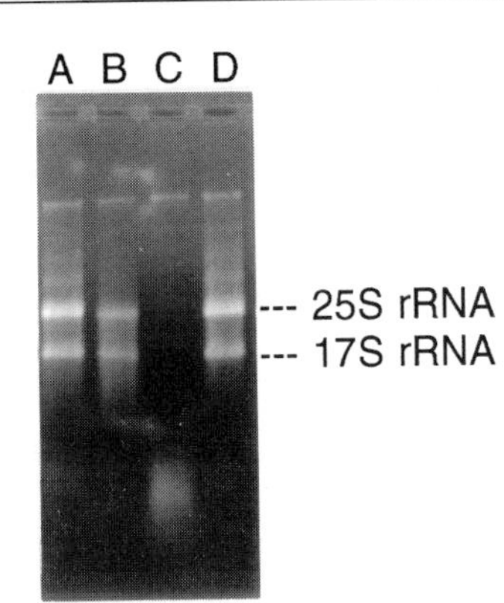

Figure 1. Agarose gel electrophoresis can be used as a rapid (< 1 h) check of the integrity of yeast RNA samples. 1 μg samples of RNAs were electrophoresed on a 1% agarose gel in 0.5 × TBE buffer (see *Protocol 2*). When viewed on a UV light box, high-quality samples of RNA show a 25S rRNA band which is distinctly more intense than the 17S rRNA band (as in samples A and D). With very slight degradation these rRNA bands appear of more equal intensity (B), while with severe degradation all the fluorescence is with the bromophenol blue dye (C).

Protocol 2. Determining the concentration and integrity of RNA samples

Equipment and reagents

- UV spectrophotometer
- small horizontal gel electrophoresis apparatus of the kind used for rapid analysis of restriction digests of DNA
- low wattage power pack
- low EEO agarose (e.g. Sigma Type 1A, cat. no. A-0169)
- 10 × TBE buffer (to minimize the number of buffer solutions in the laboratory it is convenient to use the same 10 × TBE buffer as is used for DNA sequencing gels: 162 g Tris base, 27.5 g boric acid, 9.5 g Na_2EDTA made to 1 litre, pH 8.8)
- gel loading buffer (10% (v/v) glycerol, 0.5 × TBE, 0.01% (w/v) bromophenol blue)

Method

1. Read the absorbance of 1:100 and 1:1000 dilutions of each RNA sample at 260 nm and 280 nm. The yield of RNA can be calculated using the relationship: 1 absorbance unit at 260 nm (1 cm light path) = 42 μg/ml RNA. For a good RNA preparation, the 260 nm/280 nm absorbance ratio should be about 2.0.
2. Melt 1% (w/v) agarose in 0.5 × TBE. Allow the solution to cool until it is slightly too hot to hold comfortably, then add 0.2 μg/ml ethidium bromide (**caution**: this is a mutagen and must be handled and disposed of accordingly) and pour the gel in the horizontal gel electrophoresis apparatus. Allow gel to set (10–20 min).
3. Remove the well-former from the gel and submerge gel in 0.5 × TBE buffer.
4. Remove 1–2 μg aliquots from each RNA sample, add 5–10 μl of gel loading buffer, and load in the wells of the agarose gel.
5. Run the samples at not more than 10 V/cm until the bromophenol dye has migrated 4–5 cm.
6. View the gel on a UV light box (**caution**: wear eye protection), and assess the integrity of each RNA as described in the legend to *Figure 1*. The fluorescence of the 25S rRNA should be approximately twice as strong as that of the 17S rRNA (the latter, at 1650 nucleotides in *S. cerevisiae*, is only half the length of 25S rRNA at 3360 nucleotides). If this is not the case the samples will not give Northern blot data suitable for quantitative mRNA measurements.

3.3 Enrichment for polyadenylated RNA

Individual yeast mRNAs vary greatly in cellular abundance. In fermentative *S. cerevisiae*, mRNAs for certain glycolytic enzymes are the most abundant, constituting up to 5% of the total mRNA (5). Other mRNAs are of moderate abundance, e.g. the *ADH1* transcript (0.7% of total mRNA (16)), while still others are of relatively low abundance, e.g. the transcripts of *CYC1*, *TRP5*, and *CYC7*, which, when derepressed, make up about 0.06%, 0.06%, and 0.003% respectively of total mRNA (16). The use of radioactive DNA probes of very high specific activity (with ^{32}P-labelling to 10^9 d.p.m./μg) allows detection of these low abundance transcripts on Northern blots of total cellular RNA. However, such low abundance transcripts are more readily detected if more mRNA can be loaded on to the gels used for Northern blot preparation. So as not to overload these gels it is first necessary to remove the bulk of the rRNA from the mRNA samples as described in *Protocol 3*.

All yeast mRNAs have a post-transcriptionally added poly (A) tail at the 3′ end. Highly heterogeneous in length (25–80 adenosines) (17), this poly(A) can be used to enrich RNA samples for mRNA prior to Northern blot analysis by selective binding to oligo(dT)-cellulose or poly(U)-Sepharose. *Protocol 3* results in a 3- to 5-fold enrichment for mRNA sequences. It exploits the ability of oligo(dT)-cellulose to bind quantitatively those RNAs with a 3′ poly(A) tail longer than 25 nucleotides (17). rRNA is not retained. The mRNA is subsequently eluted from the oligo(dT)-cellulose (step **6**) and concentrated by ethanol precipitation (step **7**).

Protocol 3. Enrichment for polyadenylated RNA

Equipment and reagents

- Oligo(dT)-cellulose (type III; Collaborative Research, Inc.)
- 1.5 ml microcentrifuge tubes
- 1× binding buffer: 0.5 M NaCl; 10 mM Tris–HCl, pH 7.5, 1 mM Na_2EDTA, 0.05% sodium dodecyl sulphate.
- wash buffer: 0.2 M NaCl; 10 mM Tris–HCl, pH 7.5, 1 mM Na_2EDTA, 0.05% sodium dodecyl sulphate
- 2 × binding buffer
- elution buffer: 10 mM Tris–HCl, pH 7.5, 1 mM Na_2EDTA, 0.05% sodium dodecyl sulphate
- 6 M ammonium acetate
- 3 M sodium acetate
- TE buffer pH 7.5 (see Appendix 1)
- absolute ethanol, precooled to −20°C

Method

1. Resuspend the oligo(dT)-cellulose in elution buffer and allow it to settle, discarding the fine particles. Repeat four times.
2. Transfer samples of the resin (20 mg) to 1.5 ml microfuge tubes. Equilibrate with binding buffer by two cycles of resuspension in 1 ml of binding buffer, then 2 min centrifugation at maximal speed in a microcentrifuge, discarding the supernatant.
3. Heat samples of total cell RNA (100–500 μg in 0.5 ml of TE buffer) at 65°C for 2 min. Add an equal volume of 2 × binding buffer.
4. Add these RNA samples to the microfuge tubes containing the binding buffer-equilibrated oligo(dT)-cellulose. Vortex briefly at intervals over 15 min while maintaining at room temperature.

5. Centrifuge as in step **2**. Discard supernatant. Wash the oligo(dT)-cellulose twice with binding buffer (1 ml), then twice with wash buffer (1 ml) by resuspension and centrifugation.
6. Elute poly(A) + RNA from the oligo(dT)-cellulose by adding 450 μl of water. Incubate at 37 °C for 5 min, then recentrifuge as before.
7. Decant eluate into a fresh 1.5 ml tube, add 45 μl of 3 M sodium acetate and 1 ml of − 20 °C ethanol. Precipitate RNA at − 20 °C for longer than 20 min. The precipitate, 5–15% of the original 100–500 μg RNA, should be visible as a small pellet. This pellet should be vacuum-desiccated, then redissolved in denaturation buffer (1) for Northern blot analysis.

3.4 Designing Northern blotting experiments

RNA may be denatured with either formaldehyde or glyoxal for Northern blotting—refer to a standard manual on molecular biology procedures, such as ref. 1. This section discusses important aspects of the design of Northern blotting experiments, while Section 4 of Chapter 10 describes how to quantify the mRNA signals on blots.

3.4.1 Control to ensure that hybridization has been under conditions of DNA excess

For the hybridization signal on Northern blots to reflect mRNA levels accurately (and therefore to be of use in mRNA quantification) it is essential that the hybridization is carried out under conditions of probe DNA excess. It is easy to see if this condition has been satisfied by applying multiple loadings of a test RNA to the blot—an RNA sample which is known to hybridize strongly to the probe, loaded in parallel tracks at 0.5, 1, 2, and 4 times the amount of all the other RNAs analysed. When the blot is hybridized, the hybridization signal from the tracks of this test RNA should increase in direct proportion to the amount loaded (i.e. the signal should increase 0.5, 1, 2, and 4 times across the blot).

3.4.2 Use of a loading control

Probing the blot for a transcript that does not change in level can be used to show that adjacent wells have been loaded with approximately equal amounts of RNA. Equal intensity of the rRNA bands (as in *Figure 1*) is a good sign that adjacent wells of the gel have been loaded with equivalent amounts of total cell RNA. In principle, blots of total cellular RNA can be probed with a probe containing 25S or 17S rRNA gene sequences. Unfortunately, the high amount of rRNA usually present on these blots means that careful precautions must be taken to perform such probing under DNA excess conditions and therefore to ensure that the hybridization signal accurately reflects rRNA level. Probing for actin mRNA sequences is often used as an indication of uniform RNA loading (10–12), while *Figure 2* shows the use of an alternative transcript (Ty mRNA) as a loading control. We currently probe Northern blots for either the *PMA1* plasma membrane ATPase gene transcript (3.5 kb) or the Ty transposable element transcript (5.7 kb) as an indication of uniform loading. This is because both of these mRNAs are larger than those we are currently studying, thereby obviating the need for a separate hybridization (see Section 3.4.3 and *Figure 2*). Neither actin, nor *PMA1*, nor Ty mRNA levels are completely constant and they have all been observed to change under certain conditions.

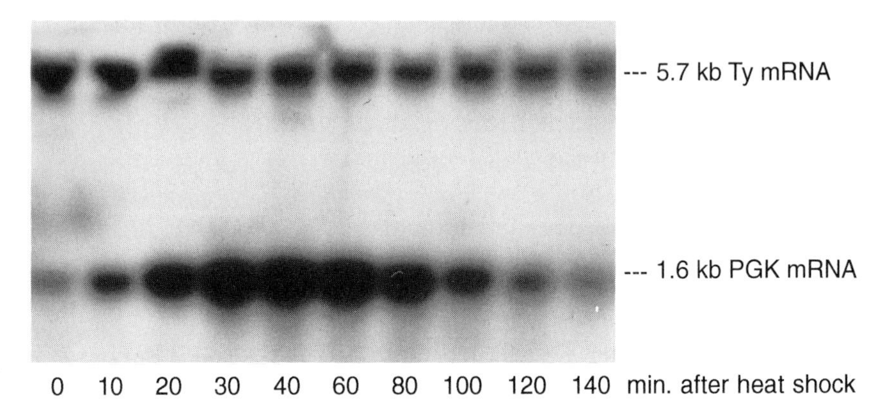

Figure 2. Use of a loading control in in blot hybridization. This blot of *S. cerevisiae* RNA samples is probed for two mRNAs of different sizes. The transcript of the Ty transposable element is a constant-level RNA that serves as a check that parallel lanes of the gel have been loaded with roughly equivalent amounts of RNA. It varies in level between different strains, yet in vegetative cultures of individual strains maintains a reasonably constant level. In this probing of samples of total *S. cerevisiae* RNA isolated at different times after a 25°C to 39°C heat shock, Ty mRNA remains at a constant level whereas PGK (phosphoglycerate kinase) mRNA shows a transient increase. By simultaneously probing for both mRNAs it cannot be argued that the increase in the PGK mRNA signal reflects unequal RNA loading of adjacent gel wells.

3.4.3 Probing for multiple sequences on the same blot

Only a minority of the primary transcripts or pre-mRNAs of *S. cerevisiae* (e.g. those encoding actin and certain ribosomal proteins (10)) undergo splicing in the nucleus. Transcripts of very many natural yeast genes are consequently homogeneous in length. This is an advantage for mRNA quantification, as frequently it allows simultaneous detection of two, or even three, natural *S. cerevisiae* mRNAs of different sizes on the same Northern blot by adding more than one DNA probe to a single hybridization reaction (see *Figure 2*). This is generally not possible in higher eukaryotes, where the transcripts of a single gene are frequently highly heterogeneous in length due to an appreciable fraction existing as incompletely spliced or incompletely polyadenylated forms. Even in *S. cerevisiae*, transcripts of highly heterogeneous length will often result if a gene has no efficient transcriptional terminator.

3.4.4 Choice of blotting membrane

RNA can be blotted on to nitrocellulose filters (e.g. Amersham Hybond-C or Schleicher and Schüll BA85). While these are rather brittle and have an unfortunate tendency to disintegrate with several manipulation steps, they are suitable if blots are only to be probed once. Nylon-based membranes (e.g. Amersham Hybond-N, DuPont GeneScreen, or Pall Biodyne) are generally preferred by most workers. In principle, their strength should allow multiple probings of the same blot. However, despite the manufacturer's claims, we find that Hybond-N is not suitable for multiple probings of the same Northern blot as stripping of the first probe from these membranes also removes substantial amounts of the bound RNA. Northern blots prepared on GeneScreen membranes can be sequentially stripped of probe and rehybridized up to five times (12).

Acknowledgements

I thank the members of my research group for helping to develop these protocols. Work in my laboratory has been supported by SERC, the British Council, the Nuffield Foundation, and the Ciba Geigy Fellowship Trust.

References

1. Sambrook, J., Fritsch, E. F., and Maniatis, T. (1989). *Molecular cloning. A laboratory manual*, 2nd edn. Cold Spring Harbor Laboratory Press, Cold Spring Harbor, NY.
2. Moncollin, V., Stalder, R., Verdier, J.-M., Sentenac, A., and Egly, J.-M. (1990). *Nucleic Acids Res.*, **18**, 4817.
3. Verdier, J.-M. (1990). *Yeast*, **6**, 271.
4. Struhl, K. (1989). *Annu. Rev. Biochem.*, **58**, 1051.
5. Romanos, M. A., Scorer, C. A., and Clare, J. J. (1992). *Yeast*, **8**, 423.
6. Cheng, L., Hirst, K., and Piper, P. W. (1992). *Biochim. Biophys. Acta*, **1132**, 26.
7. Guarente, L. and Ptashne, M. (1981). *Proc. Natl Acad. Sci. USA*, **78**, 2199.
8. Rose, M., Casadaban, M. J., and Botstein, D. (1981). *Proc. Natl Acad. Sci. USA*, **78**, 2460.
9. Schmitz, U. K., Lonsdale, D. M., and Jefferson, R. A. (1990). *Curr. Genet.*, **17**, 261.
10. Herruer, M. H., Mager, W. H., Raue, H. A., Vreken, P., Wilms, E., and Planta, R. J. (1988). *Nucleic Acids Res.*, **16**, 7917.
11. Rowley, A., Johnston, G. C., Butler, B., Werner-Washburne, M., and Singer, R. A. (1993). *Mol. Cell. Biol.*, **13**, 1034.
12. Adams, C. C. and Gross, D. S. (1991). *J. Bacteriol.*, **173**, 7429.

13. Jones, E. W., Zubenko, E. S., and Parker, R. R. (1982). *Genetics*, **102**, 655.
14. Chirgwin, J. M., Przybyla, A. E., McDonald, R. J., and Rutter, W. J. (1978). *Biochemistry*, **18**, 5294.
15. Piper, P. W. and Curran, B. P. G. (1990). *Curr. Genet.*, **17**, 119.
16. Zalkin, H. and Yanofsky, C. (1982). *J. Biol. Chem.*, **257**, 1491.
17. Piper, P. W. and Aamand, J. L. (1989). *J. Mol. Biol.*, **208**, 697.

10

Measurement of mRNA stability

ALISTAIR J. P. BROWN

1. Introduction

The level of an mRNA in the cell is determined both by its rate of synthesis and by its rate of decay. Since mRNA half-lives range from about 1 min to over 1 h in the yeast, *Saccharomyces cerevisiae* (1–4), differential rates of mRNA decay have a fundamental influence upon the overall pattern of protein synthesis in this model eukaryote. However, the structural features which determine the stability of an mRNA in yeast and the mechanisms by which they mediate their effects remain largely obscure. Multiple parameters influence the stability of an mRNA. They include the 3′ poly(A) tail, the 5′-cap, mRNA translation, and various structural determinants that lie within an mRNA (5–10). Also, many of these parameters are inter-related. For example, the inactivation of the poly(A)-binding protein leads to a decrease in mRNA translation (11), and some mRNA destabilizing elements are known to depend upon translation for their function (6). An increasing number of groups are demonstrating how significant contributions can be made to the field by combining molecular, genetic, and biochemical approaches with careful RNA analysis (1–11).

Detailed protocols for two of the most frequently used methods for measuring yeast mRNA half-lives are summarized. The first method exploits a yeast strain which carries a temperature-sensitive mutation in a subunit of RNA polymerase II, *rpb1–1* (12), while the second relies upon the use of a transcriptional inhibitor, thiolutin (4) or 1,10-phenanthroline (3). The relative merits of these methods compared with some less frequently used procedures are discussed. However, since all of these mRNA half-life measurements depend upon the isolation of high quality yeast RNA over short time-courses, protocols for the preparation of RNase-free materials, cell storage, and yeast RNA purification are also described in this chapter. The reader should refer to standard manuals for Northern blotting and hybridization procedures (13).

2. RNA isolation procedures

2.1 Preparation of RNase-free materials

The key to the isolation of high quality RNA is the preparation of materials from which exogenous contamination by RNase has been removed (*Protocol 1*). This preparation takes time, and yet RNA isolation should be done quickly. Therefore, it is advisable to prepare extra materials in case of a mishap.

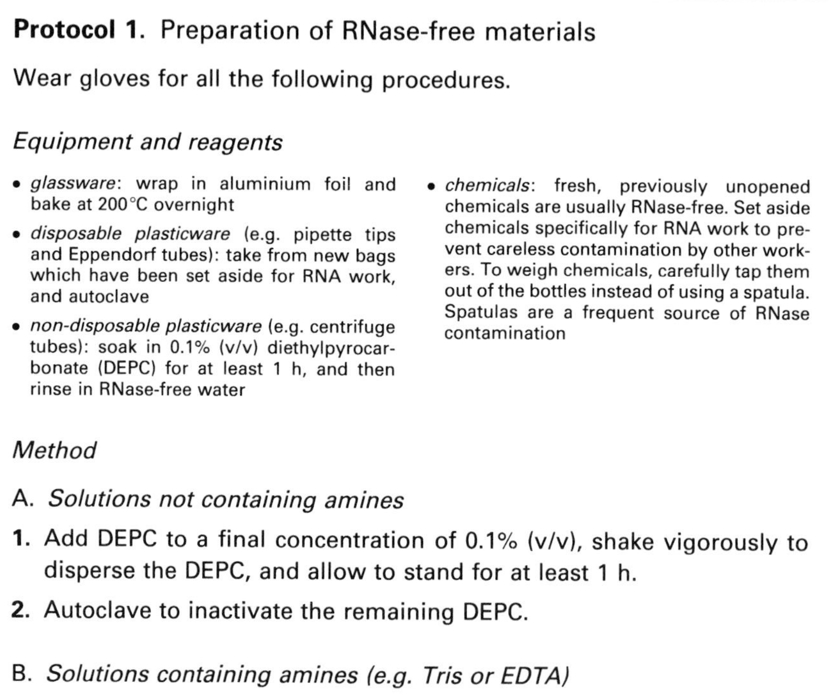

Protocol 1. Preparation of RNase-free materials

Wear gloves for all the following procedures.

Equipment and reagents

- *glassware*: wrap in aluminium foil and bake at 200°C overnight
- *disposable plasticware* (e.g. pipette tips and Eppendorf tubes): take from new bags which have been set aside for RNA work, and autoclave
- *non-disposable plasticware* (e.g. centrifuge tubes): soak in 0.1% (v/v) diethylpyrocarbonate (DEPC) for at least 1 h, and then rinse in RNase-free water
- *chemicals*: fresh, previously unopened chemicals are usually RNase-free. Set aside chemicals specifically for RNA work to prevent careless contamination by other workers. To weigh chemicals, carefully tap them out of the bottles instead of using a spatula. Spatulas are a frequent source of RNase contamination

Method

A. *Solutions not containing amines*

1. Add DEPC to a final concentration of 0.1% (v/v), shake vigorously to disperse the DEPC, and allow to stand for at least 1 h.
2. Autoclave to inactivate the remaining DEPC.

B. *Solutions containing amines (e.g. Tris or EDTA)*

1. These solutions should not be treated with DEPC. Instead, add RNase-free chemicals to DEPC-treated distilled water.
2. Adjust pH of solutions using concentrated acid or alkali. Drop small samples of the solutions on to pH papers to prevent RNase contamination of the main solutions from an electrode.
3. Adjust to the desired final volumes using volumetric markings on RNase-free bottles.

2.2 Storage of cells before RNA preparation

It is possible to store cells before RNA preparation (see *Protocol 2*). This facilitates the execution of the numerous RNA preparations that are required over short time-courses for mRNA half-life determinations. It should be noted that the storage of cells under ethanol at −20 °C for periods longer than 24 h reduces the yield of RNA.

Protocol 2. Storage of cells for RNA preparation

Reagents

- absolute ethanol, precooled to −20 °C
- extraction buffer: 0.1 M LiCl, 0.01 M dithiothreitol (add fresh), 0.1 M Tris–HCl, pH 7.5

Method

1. Dispense the desired volume of culture into a large centrifuge tube.
2. Immediately add two volumes of precooled absolute ethanol and mix. Store at −20 °C for a maximum of 24 h.
3. Harvest the cells by centrifugation at 5000 *g* for 5 min at 0 °C; discard the supernatant.
4. Resuspend the cell pellet in 10 ml extraction buffer at 0 °C, centrifuge at 5000 *g* at 0 °C for 5 min, and discard the supernatant.
5. This cell pellet can be stored for long periods at −70 °C.
6. Prepare RNA according to *Protocol 3*.

2.3 Preparation of RNA

The RNA preparation method described here (see *Protocol 3*) was adapted from the procedure described by Lindquist (14). The diameter of the glass beads (0.4 mm), the ratio of liquid to glass beads, and efficient vortex mixing are all critical to achieve optimum cell breakage and hence good RNA yields. The RNA preparation will not be successful if SDS is omitted from the phenol extraction, or if dithiothreitol is omitted from the extraction buffer.

Protocol 3. Preparation of RNA

Equipment and reagents

- glass beads (0.4 mm diameter): soak in concentrated HNO_3 for 1 h, rinse thoroughly in distilled water, dry thoroughly, and heat-bake at 200 °C overnight
- extraction buffer: 0.1 M LiCl, 0.01 M dithiothreitol (add fresh), 0.1 M Tris–HCl, pH 7.5
- 10% (w/v) SDS
- 3 M sodium acetate
- RNase-free, distilled water
- phenol: melt phenol at 60 °C, bring to pH 8 with 1 M Tris. Add hydroxyquinoline to a final concentration of about 0.02% (w/v). Saturate the phenol phase with RNase-free H_2O and mix. Store at 4 °C. Use only high quality phenol that is clear upon melting. Pink or brown phenol should not be used. (**Caution**: phenol is highly corrosive and toxic, so wear gloves and facemask)
- TE buffer pH 7.5 (see Appendix 1)
- chloroform
- ethanol

Protocol 3. *Continued*

Method

1. For each sample (equivalent to 25 ml of mid-exponential culture), prepare a tube containing 7 g of glass beads, 0.5 ml of 10% SDS, 2.5 ml of phenol, and 2.5 ml of chloroform.
2. Rapidly thaw the cell pellet and resuspend in 2.5 ml of extraction buffer by vortexing.
3. Immediately add the cell suspension to the tube prepared in step **1**, then vortex vigorously and continuously for 5 min.
4. Centrifuge the sample at 5000 *g* for 5 min in a swing-out rotor.
5. Carefully remove the upper aqueous phase and transfer it to a fresh tube containing 2.5 ml of phenol and 2.5 ml of chloroform. Avoid taking any of the interface.
6. Vortex for 1 min, then centrifuge the sample at 5000 *g* for 5 min.
7. Repeat steps **5** and **6**.
8. Transfer the aqueous phase to a fresh tube containing 3 ml of chloroform, avoiding the interface.
9. Vortex for 30 sec, then separate the phases by centrifugation at 5000 *g* for 2 min.
10. Repeat steps **8** and **9**.
11. Transfer the aqueous phase to a fresh tube, add sodium acetate to a final concentration of 0.2 M and two volumes of absolute ethanol, and mix thoroughly. Store overnight at −20°C. At this stage the RNA is stable and may be stored indefinitely.
12. Centrifuge the sample at 10 000 *g* for 10 min at 4°C. Pour off the supernatant and drain the tube. Vacuum-desiccate for 15 min to remove residual ethanol.
13. Resuspend the ethanol precipitate in 1 ml TE buffer.
14. Add 70 µl of 3 M sodium acetate and 2.2 ml of ethanol. Mix thoroughly and store overnight at −20°C.
15. Centrifuge the sample at 10 000 *g* for 10 min at 4°C and pour off the supernatant.
16. Gently wash the ethanol precipitate with 70% ethanol at 0°C, drain the tube, and vacuum-desiccate for 15 min.
17. Resuspend the precipitate in 100 µl of TE buffer; store the RNA preparation at −70°C.
18. Take 10 µl of the sample, dilute to a final volume of 1 ml, and read the absorbance at 260 nm and 280 nm. The yield of RNA can be calculated using the relationship 1 A_{260}=42 µg RNA. The A_{260}/A_{280} ratio should be about 2.0 for a good RNA preparation.

The integrity of each RNA preparation should be assessed by agarose gel electrophoresis. The electrophoresis can be performed using any standard Tris/borate or Tris/acetate agarose gel system. However, the use of formaldehyde/agarose gels is recommended here (see *Protocol 4*) because the activity of exogenous RNases is significantly reduced under these denaturing conditions. In intact RNA preparations, the 25S, 18S, and tRNA bands (and possibly the 5.8S and 5S RNAs) are clearly visible and distinct, with little evidence of smearing (unless the gel is overloaded). Visual comparison of the intensities of ethidium bromide-stained ribosomal RNA bands in each lane will indicate roughly whether attempts to load equal amounts of RNA in each lane have been successful. If not, adjustments should be made to achieve equal loading for Northern analysis of the RNA samples.

Protocol 4. Electrophoretic analysis of RNA

Reagents

- TE buffer pH 7.5 (see Appendix 1)
- MOPS buffer (10 ×): 0.2 M morpholinopropanesulphonic acid, 0.05 M sodium acetate, 0.01 M Na_2EDTA, pH 7.0. Store in a dark bottle, do not autoclave
- MMF: 500 μl of formamide, 162 μl of formaldehyde (37%), 100 μl of 10× MOPS, and 238 μl of H_2O
- ethidium bromide: 0.1 mg/ml. **Caution**: ethidium bromide is a carcinogen. Wear gloves and dispose of ethidium bromide appropriately
- gel loading dye: 0.02% (w/v) bromophenol blue in 50% (v/v) glycerol
- agarose

Method

1. Take the volume of each sample that corresponds to 10 μg of RNA and make up to 10 μl with H_2O.
2. Add 35 μl of MMF and 2 μl of ethidium bromide (0.1 mg/ml), mix, and incubate at 60°C for 15 min.
3. Add 10 μl of gel loading dye and mix. Leave on ice until ready to load agarose gel.
4. Prepare the formaldehyde/agarose gel in a fume hood by melting 1.5 g of agarose in 73 ml of H_2O. Cool to 60°C, add 10 ml of 10 × MOPS and 16.2 ml of 37% formaldehyde, mix, and pour gel immediately.
5. Load the sample on to the gel and perform the electrophoresis in 1 × MOPS buffer. Once the samples have run into the gel, circulate the electrophoresis buffer.
6. Visualize the electrophoresed RNA on a UV transilluminator. (**Caution**: wear a face mask.)

3. Measurement of mRNA half-lives

Several methods have been developed for the measurement of mRNA half-lives in yeast. The methods are divisible into two broad groups:

(a) those that exploit *in vivo* radiolabelling kinetics (e.g. approach to steady-state labelling or pulse–chase kinetics)
(b) those where mRNA decay rates are measured following a cessation in transcription.

Transcription can be stopped using several techniques:

(a) transcriptional inhibitors (thiolutin or 1,10-phenanthroline)
(b) temperature-sensitive RNA polymerase II mutant (*rpb1-1*)
(c) regulatable promoters (*GAL1/10*).

No method is perfect. Each method has particular advantages and disadvantages, but in general there is good agreement between the half-life measurements obtained for a specific mRNA using different procedures (3, 4). The different values reported for the half-lives of some mRNAs, e.g. the actin mRNA (3, 4, 15) may be due in some cases to the different temperatures of the yeast cultures used to perform the measurements (4). Ideally, the stability of an mRNA should be measured using at least two of the methods described below.

3.1 *In vivo* radiolabelling procedures

In principle, *in vivo* radiolabelling procedures can be used to measure the half-life of a yeast mRNA via a 'pulse–chase' or by 'approach to steady-state' (1, 4, 5). Pulse–chase methods involve the labelling of transcripts *in vivo* using a pulse of radiolabelled precursor, usually [^{3}H]adenine or $^{32}PO_4$. The half-lives of specific mRNAs are then measured by determining the rate at which radioactivity decays from the transcript during the chase period when the radiolabelled precursor has been removed from the medium. Alternatively, mRNA half-lives can be measured by determining the time taken for the specific activity of an mRNA to reach steady-state following the addition of radiolabelled precursor to the medium.

Both methods involve the preparation of RNA from radiolabelled cells, and the hybridization of this RNA with an excess of unlabelled, sequence-specific DNA probe which has been immobilized on a membrane filter. The methods suffer the disadvantage that large amounts of radioactivity must be used to ensure that mRNAs are labelled to a sufficiently high specific activity to allow detection following hybridization. Also, pulse–chase methods are dependent upon the rapid and effective clearance of radiolabelled precursors from intracellular pools at the beginning of the chase period, but this can prove problematic (4). Nevertheless, *in vivo* labelling procedures are arguably the least invasive methods in terms of disturbing the growth of the yeast cell and therefore, despite the fact that the procedure is labour intensive, 'approach to steady-state' methodology continues to be used by some groups (5).

3.2 The use of transcriptional inhibitors

Various transcriptional inhibitors have been exploited for mRNA half-life measurements, but currently the most widely used are thiolutin (4, 16) and 1,10-phenanthroline (3, 17). Under the appropriate conditions, thiolutin inhibits mRNA synthesis to less than 5% of wild-type levels (4, 16), and 1,10-phenanthroline inhibits transcription to less than 10% of wild-type levels (3). The protocol described here (see *Protocol* 5) has been adapted from previously published procedures (3, 4, 16).

Briefly, RNA is isolated over a time-course following the addition of either thiolutin or 1, 10-phenanthroline to exponentially growing cultures of yeast, and the half-life of a specific mRNA is then determined by following the decay of the mRNA by quantitative Northern analysis (see *Figure 1*). Therefore, mRNA half-life measurements using transcriptional inhibitors are simple and convenient to perform. Furthermore, multiple mRNAs can be analysed simultaneously, and therefore experiments involving the measurement of mutant mRNAs or artificial mRNA fusions can be *internally* controlled using a 'standard' mRNA. Also, relatively small amounts of radioactivity are required to prepare the hybridization probes, and the half-life of the *intact* mRNA is measured by Northern analysis. These represent significant advantages over the *in vivo* radiolabelling procedures described above (see Section 3.1).

Clearly the use of a transcriptional inhibitor is invasive in that it has a dramatic effect upon cell growth. The use of any inhibitor suffers the disadvantage that it is virtually impossible to exclude any artefacts that might arise through secondary effects of the drug upon other cellular processes. For example, both 1,10-phenanthroline and thiolutin appear to induce a stress response in yeast (18). Furthermore, 1,10-phenanthroline transiently inhibits mRNA translation (3), and this drug (a zinc ion chelator) may also block the *UPF*-specific pathway for mRNA degradation since the *Upf1* protein appears to contain a zinc finger (19). This pathway seems to be specific for mRNAs that carry a premature translational termination codon, since the degradation of most wild-type mRNAs tested is not affected by mutations in the *UPF1* gene (20). Therefore, it would appear more appropriate to use thiolutin to study the degradation of aberrant mRNAs. However, the use of both 1,10-phenanthroline and thiolutin remains a convenient back-up for the analysis of wild-type mRNAs, since mRNA half-life data generated using these drugs are generally consistent with measurements performed using other techniques (3, 4).

It should be noted that while thiolutin is not available commercially, Pfizer have generously provided material to some groups (4, 16). 1,10-phenanthroline can be purchased from Sigma.

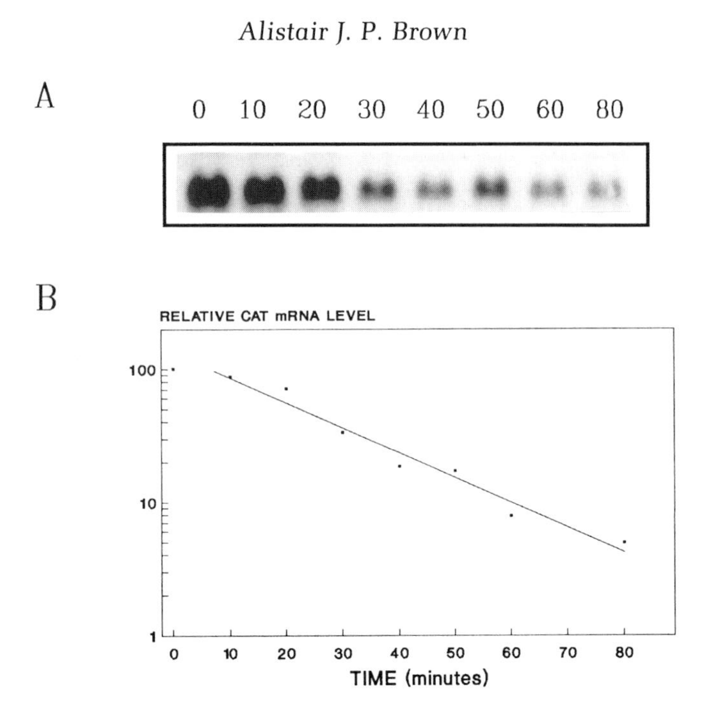

Figure 1. mRNA half-life measurements using 1,10-phenanthroline. (A) RNA was isolated from a culture of the yeast strain, RY137, at 0, 10, 20, 30, 40, 50, 60, and 80 min following the addition of 1,10-phenanthroline to a final concentration of 100 μg/ml. Approximately equal amounts of RNA were subjected to Northern analysis and probed for a heterologous *Escherichia coli cat* mRNA expressed in yeast using the *TEF1* promoter (23). (B) The signals on the Northern blot were quantified by 2D-radioimaging to determine the abundance of the *cat* mRNA at each time-point. The half-life of the *cat* mRNA (16 min) was calculated statistically.

Protocol 5. mRNA stability measurements using transcriptional inhibitors

Reagents

- YPD (see Appendix 1)
- 1,10-phenanthroline (1000 ×): 100 mg 1,10-phenanthroline/ml in ethanol (Sigma P 9375)
- thiolutin (500 ×): 1.5 mg thiolutin/ml in dimethyl sulphoxide (obtained from Pfizer)

Method

1. Grow yeast cultures to mid-exponential phase (A_{600} = 0.4–0.6) in 250 ml of YPD at 30°C with shaking at 250 r.p.m. using a 2 litre flask.
2. Add 1,10-phenanthroline to a final concentration of 100 μg/ml *or* thiolutin to a final concentration of 3 μg/ml.

3. Remove 25 ml portions of the culture at various times following the addition of the transcriptional inhibitor and immediately store cells according to *Protocol 2*. Where possible, the length of the time-course should be designed to reflect the half-life of the mRNA of interest. For example, for mRNAs with short half-lives (1–15 min), time-points of 0, 2.5, 5, 7.5, 10, 15, 20, and 30 min are appropriate. For mRNAs with longer half-lives (15–60 min), an extended time-course of 0, 10, 20, 30, 40, 50, 60, and 80 min is more appropriate (see *Figure 1*).
4. Prepare RNA from each time-point according to *Protocol 3*, and check the integrity of the preparations by gel electrophoresis (see *Protocol 4*).
5. Using equal amounts of total RNA, measure the relative abundance of the mRNA of interest in each preparation by quantitative Northern analysis. Alternatively, measure mRNA levels by dot blotting or an RNase protection assay.

3.3 The use of the RNA polymerase II mutation, *rpb1-1*

The use of the temperature-sensitive RNA polymerase II mutation, *rpb1-1*, which was originally described by Young's group (12), has rapidly become one of the most popular methods of measuring mRNA half-lives in yeast (4, 6, 8, 10, 15, 16, 19–21). In this procedure transcription is inhibited by transferring the *rpb1-1* yeast strain from the permissive temperature (25°C) to the non-permissive temperature (37°C). Within 5 min of this mild heat-shock, transcription is reduced to less than 10% of wild-type levels (12). The half-life of a specific mRNA is then measured by determining the rate at which it decays by Northern analysis of RNA samples prepared over a suitable time-course following the heat-shock (see *Figure 2*). *Protocol 6* is adapted from the previously published procedures of Jacobson's group (4, 16).

This method is rapid and convenient. Also, it has many of the advantages described above for the use of transcriptional inhibitors (i.e. multiple mRNAs can be analysed simultaneously, experiments can be *internally* controlled using 'standard' mRNAs, *intact* mRNAs are analysed by Northern analysis, and relatively small amounts of radioactivity are required). However, it should be noted that the stability of some mRNAs might be affected by the mild heat-shock required to activate the *rpb1-1* mutation (17). The heat-shock can be achieved in several ways. For example, in *Protocol 6*, the heat-shock is performed by adding an equal volume of pre-heated medium to the exponentially growing culture of the *rpb1-1* yeast strain. Although this is an efficient means of achieving a rapid increase in culture temperature, it has been observed that the addition of fresh medium can transiently affect the levels of some mRNAs in exponentially growing cultures of *RPB1* yeast strains (J.-J. Mercado, R. Smith, F. Sagliocco, A. J. P. Brown and J.-M.

Gancedo, submitted). Nevertheless, despite these possible side-effects and the fact that the method is limited to the use of *rpb1-1* strains, this remains one of the most attractive methods for measuring mRNA half-lives in yeast.

Protocol 6. mRNA stability measurements using *rpb1*-1 mutants

Reagents

- YPD (see Appendix1)
- GYNB (see Appendix 1)
- The following temperature-sensitive RNA polymerase II mutants and control strains, which were originally described by Young's group (12), are available from the Yeast Genetic Stock Center: RY136 (*MAT*α, *ura3*), RY137 (*MAT***a**, *ura3, his4, lys2*), RY260 (*MAT***a**, *ura3, rpb1-1*), and RY262 (*MAT*α, *ura3, his4, rpb1-1*)

Method

1. Grow the appropriate strain to mid-exponential phase (A_{600} = 0.4–0.6) in 130 ml of YPD or GYNB (containing the appropriate supplements) at 25°C with shaking at 250 r.p.m., using a 2 litre flask.
2. Prewarm 130 ml of the same medium to 49°C.
3. To start the time-course, add the prewarmed medium (49°C) to the culture (25°C), and immediately place the flask in a shaking waterbath at 37°C (time = 0 min).
4. Immediately remove 40 ml of the culture and store the sample according to *Protocol 2*.
5. Remove 40 ml portions of the culture at various times following the addition of the transcriptional inhibitor and immediately store cells according to *Protocol 2*. For mRNAs with short half-lives (1–15 min), timepoints of 0, 2.5, 5, 7.5, 10, 15, 20, and 30 min are appropriate. For mRNAs with longer half-lives (15–60 min), an extended time-course of 0, 10, 20, 30, 40, 50, and 60 min is more appropriate (see *Figure 2*).
6. Prepare RNA from each sample according to *Protocol 3*, and check the integrity of the preparations by gel electrophoresis (see *Protocol 4*).
7. Using equal amounts of total RNA, measure the relative abundance of the mRNA of interest in each preparation by quantitative Northern analysis. Alternatively, use dot blotting or an RNase protection assay.

3.4 The use of specific promoters to stop transcription

An infrequently used, but practicable method of halting the transcription of an mRNA with a view to measuring its half-life is to construct a hybrid gene in which the synthesis of the mRNA of interest is driven by a tightly regulated promoter, for example the *GAL1* promoter (16, 21). The transcriptional

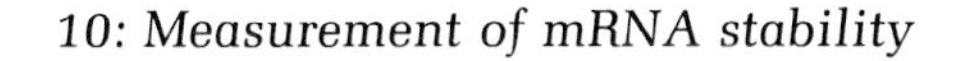

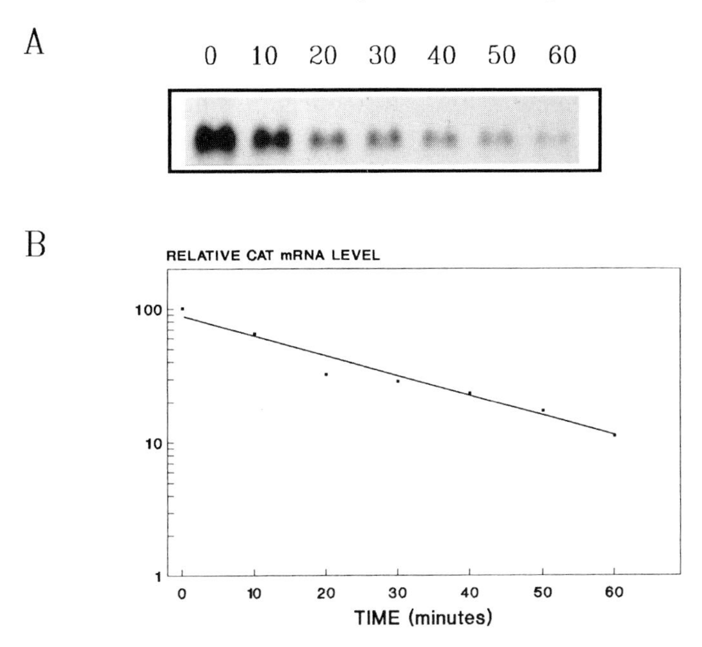

Figure 2. mRNA half-life measurements using an *rpb1-1* mutant. (A) RNA was isolated from a culture of the *rpb1-1* strain, RY262, at 0, 10, 20, 30, 40, 50, and 60 min following the increase in temperature from 25°C to 37°C. Approximately equal amounts of RNA were subjected to Northern analysis and probed for the same heterologous *cat* mRNA as shown in *Figure 1*. (B) The signals on the Northern blot were quantified by 2D-radioimaging to determine the abundance of the *cat* mRNA at each time-point. The half-life of the *cat* mRNA (20 min) was calculated statistically.

repression of this promoter can be conveniently achieved by the addition of glucose to the growth medium, with only a minimal perturbation of cell growth (relative to the use of the *rpb1*-1 mutation or transcriptional inhibitors). However, Parker and co-workers point out that the change in carbon source might affect the stability of some mRNAs (16), and this has indeed been shown to be the case (22). Several alternative promoters might be used if regulation via the carbon source proves inappropriate (such as the *PHO5* promoter which is repressed by the addition of phosphate to the medium).

4. mRNA quantification

All methods for measuring the half-life of an mRNA require the accurate quantification of relative mRNA abundances across a time-course of some sort. Subjective estimates based upon autoradiographic signals are insufficient because these can be misleading due to the non-linear response of X-ray film under some conditions, and this problem can be compounded by unequal RNA loadings in individual lanes on Northern gels. Therefore, the estimates of RNA concentration in each preparation must be as accurate as possible, and accurate quantification of the hybridization signals must be achieved using one of the following methods:

(a) *Scintillation counting*. Individual bands on the filter can be cut out and subjected to scintillation counting, but this suffers at least two disadvantages. Firstly, the filter cannot be reused, and hence separate filters must be prepared for experimental and control mRNAs. This reduces the accuracy of the measurements. Secondly, it can be difficult to align the filter with the X-ray film, and hence to cut out the bands accurately.

(b) *Densitometer scanning*. This method, which measures the intensity of the signal on the X-ray film, allows the filter to be reprobed for additional mRNAs, but the accuracy is entirely dependent upon a linear relationship between the amount of radioactivity bound to the filter and the autoradiographic signal. Unfortunately, this is only the case for a relatively narrow signal range which depends upon the type of film and upon the autoradiographic conditions.

(c) *2D-Radioimaging*. The ideal method involves direct 2D-radioimaging from the filter, for example using an AMBIS 2D-Radioanalytical System (LabLogic). This type of apparatus measures the radioactivity being emitted from each part of the filter. These data can be imaged on a computer and then used to quantify accurately the amount of radioactivity in specific bands on the Northern blot. Having filed this information on the computer, the filter can be repeatedly stripped, reprobed, and requantified. Hence the relative abundance of many mRNAs can be measured accurately from the same filter thus reducing the errors involved in the preparation of multiple filters.

In addition, controls must be performed to ensure that the experiment has been performed under quantitative conditions. For example, control lanes on a Northern blot (which contain a dilution series of an appropriate RNA preparation) should reveal a linear relationship between the hybridization signal and the amount of RNA loaded. All experimental points should lie within this calibration curve. A linear response will only be achieved if the hybridization has been performed under conditions of probe excess. For abundant mRNAs (which can comprise up to about 0.5% of total RNA in the cell), the provision of sufficient probe to generate a molar excess in a Northern hybridization requires careful consideration. Frequently, the most convenient means of achieving this is to mix the radiolabelled probe with additional, unlabelled probe. The resultant reduction in the specific activity of the probe leads to a decrease in the hybridization signal, but more importantly, improved quantification of an abundant mRNA.

Acknowledgements

The data in *Figures 1* and *2* were kindly provided by Francis Sagliocco who was supported by a grant from the EEC (SCI 0143–C [SMA]). Work on mRNA stability in my laboratory has also been supported by the Science and Engineering Research Council, Whitbread PLC, and the British Council.

References

1. Kim, C. H. and Warner, J. R. (1983). *J. Mol. Biol.*, **165**, 79.
2. Losson, R., Fuchs, R. P. P., and Lacroute, F. (1983). *EMBO J.*, **2**, 2179.
3. Santiago, T. C., Bettany, A. J. E., Purvis, I. J., and Brown, A. J. P. (1986). *Nucleic Acids Res.*, **14**, 8347.
4. Herrick, D., Parker, R., and Jacobson, A. (1990). *Mol. Cell. Biol.*, **10**, 2269.
5. Losson, R. and Lacroute, F. (1979). *Proc. Natl Acad. Sci. USA*, **76**, 5134.
6. Parker, R. and Jacobson, A. (1990). *Proc. Natl Acad. Sci. USA*, **87**, 2780.
7. Gerstel, B., Tuite, M. F., and McCarthy, J. E. G. (1992). *Mol. Microbiol.*, **6**, 2339.
8. Vreken, P. and Raue, H. A. (1992). *Mol. Cell. Biol.*, **12**, 2986.
9. Lowell, J. E., Rudner, D. Z., and Sachs, A. B. (1992). *Genes Dev.*, **6**, 2088.
10. Muhlrad, D. and Parker, R. (1992). *Genes Dev.*, **6**, 2100.
11. Sachs, A. and Davis, R. W. (1989). *Cell*, **58**, 857.
12. Nonet, M., Scafe, C., Sexton, J., and Young, R. (1987). *Mol. Cell. Biol.*, **7**, 1602.
13. Sambrook, J., Fritsch, E. F., and Maniatis, T. (1989). *Molecular cloning. A laboratory manual*, 2nd edn. Cold Spring Harbor Laboratory Press, Cold Spring Harbor, NY.
14. Lindquist, S. (1981). *Nature*, **293**, 311.
15. Moore, P. A., Sagliocco, F. A., Wood, R. M. C., and Brown, A. J. P. (1990). *Mol. Cell. Biol.*, **11**, 5330.
16. Parker, R., Herrick, D., Peltz, S. W., and Jacobson, A. (1991). In *Methods in enzymology*, Vol. 194 (ed. C. Guthrie and G. R. Fink), pp. 415–23. Academic Press, London.
17. Herruer, M. H., Mager, W. H., Raue, H. A., Vreken, P., Wilms, E., and Planta, R. J. (1988). *Nucleic Acids Res.*, **16**, 7917.
18. Adams, C. C. and Gross, D. S. (1991). *J. Bacteriol.*, **173**, 7429.
19. Leeds, P., Wood, J. M., Lee, B. S., and Culbertson, M. R. (1992). *Mol. Cell. Biol.*, **12**, 2165.
20. Leeds, P., Peltz, S. W., Jacobson, A., and Culbertson, M. R. (1991). *Genes Dev.*, **5**, 2303.
21. Brown, A. J. P. (1989). *Yeast*, **5**, 239.
22. Lombardo, A., Cereghino, G. P., and Scheffler, I. E. (1992). *Mol. Cell. Biol.*, **12**, 2941.
23. Vega Laso, M. R., Zhu, D., Sagliocco, F. A., Brown, A. J. P., Tuite, M. F., and McCarthy, J. E. G. (1993). *J. Biol. Chem.*, **268**, 6453.

11

Production of foreign proteins at high level

GÉRARD LOISON

1. Introduction

The yeast *Saccharomyces cerevisiae* is one of the most commonly used host cell systems for the production of foreign proteins for research and for industrial or medical use. Although the state of knowledge on yeast molecular biology is increasing day by day, a large part of the yeast biotechnologist's work remains based on empirical approaches rather than on assertive predictions. Insertion of a foreign coding sequence into a standard multiple-copy expression vector for yeast is not always sufficient to ensure a high level of accumulation of the foreign protein. Various strategies have been tested by different groups to overcome limitations in protein production; mostly these depend upon empirical mutagenesis-screening. Reviewing all these cases is beyond the scope of this chapter. Interested readers should consult the recent and comprehensive review of Romanos *et al.* (1), which is entirely devoted to foreign gene expression in *S. cerevisiae* as well as in other yeasts. I will rather limit myself to emphasizing some basic pieces of advice and outlining some successful strategies we have experienced ourselves.

In *S. cerevisiae*, high-level production of heterologous proteins is achieved using replicative expression vectors that are maintained at high copy numbers in the nucleus. Some considerations on yeast transformation and selectable markers, relevant to optimizing protein production, are developed in Sections 2 and 3, respectively. Sections 4 and 5 deal with vectors and expression blocks, whereas protein stability problems and basic analytical methods are examined in Sections 6 and 7, respectively. It should be stressed that the success in producing heterologous proteins at high level generally results from combined studies in two different fields: molecular biology and cell physiology. Physiology applied to gene expression and protein production is a vitally important consideration which is not examined in this chapter.

2. Transformation methods

Choosing a transformation procedure is relevant to optimizing gene expression since plasmid copy numbers in recombinant cells depend on the technique used and on the type of selection applied to obtain transformants. Various methods for transforming yeast cells have been described (1, also see Chapter 8); these include:

- PEG-$CaCl_2$ treated sphaeroplasts (2)
- lithium acetate (LiAc) treated intact cells (3)
- electroporation (4)
- conjugation with *E. coli* (5)

The sphaeroplasting method (2), although somewhat more cumbersome than the other more recently described techniques using whole cells, offers the advantage of easily yielding transformed colonies that can display ultra-high plasmid copy numbers. With the help of a defective selectable marker, it is possible to select for plasmid-enriched colonies (see Section 3). When transforming yeast cells, avoid the addition of carrier DNA in order to minimize the risk of mutation side-effects. Keep in mind that some PEG batches can be much more efficient than others with regard to the number of transformants per μg of DNA, and also that PEG treatment can induce cell fusion, resulting in a variable, and generally low, percentage of transformed cells that have doubled their ploidy (2).

3. Selectable markers

3.1 Introduction

The first and most commonly used markers for the selection of transformants were *LEU2, TRP1, URA3,* and *HIS3*, used with corresponding auxotrophic mutants as recipient strains. It is advisable to use only mutants that do not measurably revert to wild-type; non-reverting mutants can be constructed by introducing multiple mutations in a particular gene. For example, the *leu2*-3-112 and *his3*-11-15 alleles are frequently found in laboratory strains. Such strains can be obtained from the Yeast Genetic Stock Center or from other collections. An example of how to obtain multiple mutations in one gene is given in *Protocol 2* (see Section 3.4.2). Alternatively, a more recent and more rapid technique for constructing a non-reverting mutated gene consists in replacing the chromosomal gene with an *in vitro*-disrupted allele (6).

3.2 Defective *LEU2* marker

This cloned version of *LEU2* lacks essential promoter sequences and is therefore very poorly expressed. Direct selection with this marker can easily be obtained with the sphaeroplasting method but is hazardous with other techniques. The use of the sphaeroplasting method with standard replicative vectors containing the *LEU2*-d marker facilitates the isolation of these transformants which display very high plasmid copy numbers, especially when sphaeroplasts are regenerated in leucine-deficient medium. Note that when the transformation plasmid carries two selectable markers, one of them defective, the selection pressure for the expression of this defective marker must be applied during the sphaeroplast regeneration step. At this stage, yeast cells seem to be capable of amplifying the hybrid vector to some extent, probably as the result of unequal partitioning of plasmid copies during early mitoses. This property appears transient, as yeast cells do not significantly amplify mitotically stable plasmids afterwards, unless plasmid and recipient strains have been especially designed for this purpose (7). Strong selection pressure applied against transformants that harbour standard plasmid DNA at insufficiently high copy number will favour plasmid rearrangement events (8). In some cases, especially when the plasmid is a constitutive expression vector for yeast, high plasmid copy number can be deleterious to the host cell. Transformants will then grow with difficulty, a situation which favours genetic drift and/or selection of plasmid rearrangement events. This problem can be circumvented by using a tightly repressed promoter for foreign gene expression (see Section 5.1).

3.3 Defective *URA3*-d marker

The widely used *URA3* marker offers several advantages over *LEU2*:

- *Ura3* cells are easily selectable (9)
- *Ura3* cells do not grow in synthetic minimal medium (SD) supplemented with casaminoacids (CAA) unless a pyrimidine source is added to the medium (see Appendix 1). Yeast cells grow much more rapidly in SD medium when amino acids are provided. SD+CAA, therefore, constitutes a very convenient medium for selecting ura$^+$ transformants. An equivalent medium for the selection of leu$^+$ transformants requires separate addition of each amino acid (minus leucine) to SD (10).

However, the wild-type *URA3* gene (11) is not a convenient selectable marker for isolating transformants that display ultra-high plasmid copy numbers. Standard replicative plasmids that contain wild-type *URA3* as selectable marker, such as pFL1 (equivalent to YEp24) (12), are commonly maintained at around 30 and 10–15 copies per cell in transformants constructed by sphaeroplast and LiAc procedures, respectively (12, 13). Furthermore, cultures of such ura$^+$ transformants grown in selective medium generally contain a significant percentage of plasmid-cured cells (sometimes as high as 15% in cultures grown in SD+CAA). Apparently, the overproduction of the *URA3* mRNA, a consequence of the gene dosage, allows cured cells to undergo several doublings before being arrested by pyrimidine starvation. A truncated *ura3*-d marker combines the advantages of *URA3* and *LEU2*-d:

- it is easy to select transformants on SD+CAA, even those obtained with transformation techniques performed on whole cells
- *URA3* mRNA is not overproduced in transformants
- selection of high plasmid copy number transformants (50–150 copies per cell)

Here again, as with *LEU2*-d, this selection works optimally in the absence of any plasmid-linked overexpression that handicaps the host cell. It is therefore advisable to avoid strong constitutive promoters and prefer those that are tightly regulated for expressing foreign genes. In this case, choosing rapidly growing candidates among the transformed colonies will minimize the risks for subsequent genetic drift.

A truncated allele of *ura3* can easily be constructed and used in place of the wild-type *URA3* marker. In the cloned *Hin*dIII segment of yeast genomic DNA that encompasses *URA3*, there is a unique *Pst*I site localized 19 bp upstream from the ATG initiation codon. The upstream 0.4 kb *Hin*dIII–*Pst*I subfragment that contains the *URA3* promoter signals can be substituted with a synthetic double-stranded oligonucleotide to yield a promoter-defective version of *URA3*. An example of a sequence for this double-stranded oligo is as follows (8):

5′-AGCTTGGTACCCAACTGCACAGAACAAAAACTGCA-3′
ACCATGGGTTGACGTGTCTTGTTTTTG-5′

In the resulting *URA3* construct, all the sequences localized upstream to −47 with respect to the initiation ATG have been deleted; only some of the proximal mRNA start site sequences have been retained, they are flanked upstream by a *Kpn*I cloning site which can be used for inserting additional upstream sequences, if necessary.

3.4 The *ura3 fur1* autoselection system

3.4.1 Introduction

Ura$^+$ cells obtained from plasmid-transformed *ura3* mutant cells are selected in media devoid of any source of pyrimidine, so the choice of the selective growth medium has to be restricted to synthetic media. The use of a *ura3 fur1* strain as a host makes it possible to grow the plasmid-transformed ura$^+$ cells selectively in practically any growth medium for yeast, including complex rich media such as YPD (see Appendix 1). Growing yeast cells in a rich medium is one of the parameters that can help to improve the productivity of polypeptide biosynthesis in a bioreactor. However, the use of a rich growth medium often complicates the procedure for the purification of polypeptides that are secreted into the medium, so it is generally restricted to the production of cell-associated proteins.

3.4.2 Construction of *ura3 fur1* host strains for the production of foreign proteins

There are various methods for constructing plasmid-transformed *ura3 fur1* strains. One consists in selecting spontaneous *fur1* derivatives from *ura3* cells transformed with the *URA3* plasmid (1, see *Protocol 1*). Another is based on the possibility of rescue of untransformed *ura3 fur1* cells by adding large quantities of uridine to the growth medium. Since doubly impaired *ura3 fur1* strains grow in YPD supplemented with 0.4 g/l uridine, with a generation time of about 6 h, they can be used directly as recipients for ura^+ transformation. Standard YPD medium can be used for the selection of transformants.

Protocol 1. Selection of spontaneous *fur1* mutants

Reagents

- YPD (see Appendix 1)
- YPEG (see Appendix 1)
- SD (see Appendix 1)
- Bacto-casaminoacids (Difco) are stored as a 10% sterile aqueous solution
- 5-fluoro-orotic acid (5-FOA), 5-fluoro-uracil (5-FU), and uracil are obtained from Sigma
- other supplements, as specifically needed by the yeast strain
- a *ura3* mutant strain
- DNA of a yeast plasmid that carries a *URA3* gene as a selectable marker (about 10 μg in TE buffer)

Method

1. Transform the *ura3* $FUR1^+$ cells to ura^+ with plasmid DNA.
2. Purify the transformants by streaking for single colonies on solid SD medium supplemented with 0.5% Bacto-casaminoacids (and any other supplements required by the strain).
3. Choose one of these transformed strains according to specific criteria such as optimal growth characteristics, optimal plasmid copy number, or integrity of the plasmid structure.
4. Plate about 10^8 cells of this transformed strain on to solid YPD medium containing 10 mM (1.3 g/l) 5-FU and incubate at 30°C for 4–6 days.
5. Purify several 5-FU-resistant (5-FU^r) colonies by subcloning on solid YPD + 10 mM 5-FU.
6. Check the ability of subcolonies to grow on YPEG. Discard colonies that do not grow (these are respiration-deficient petite mutants).
7. Test the ability of these subcolonies to yield ura^- cells after serial culture on YPD; ura^- colonies can either be screened by replica plating on uracil-deficient SD medium or selected on solid SD supplemented with 50 mg/l uracil, 1.5 g/l 5-FOA, and other supplements according to the strain's requirements.
8. Choose a strain that does not yield ura^- cells.

Protocol 2. Construction of *fur1* alleles containing multiple mutations

Reagents

- two *ura3* mutants of opposite mating type
- a mitotically unstable plasmid that is maintained under selective pressure in yeast and contains a selectable *URA3* marker e.g. an *STB*-lacking *ARS* vector carrying a complete *URA3* gene (see Section 4.1)
- solid YPD medium (see Appendix 1)
- solid SD medium (see Appendix 1)
- 5-FU, 5-FOA, uracil (see *Protocol 1*)
- uridine (Sigma)
- sporulation medium (see Appendix 1)

Method

1. Start from two haploid strains, one *MAT***a** and the other *MAT*α, both carrying the same 'non-reverting' *ura3* allele, for example *ura3*–52 (a Ty insertion present in various strains of the Yeast Genetic Stock Center).
2. Transform cells of each strain to ura^+ with a mitotically unstable plasmid.
3. Purify two transformants of each mating type by streaking for single colonies on SD plates, then select *fur1* derivatives from each transformant, as described in *Protocol 1*.
4. Mate different plasmid-transformed *ura3 fur1* strains, two by two. Sporulate the diploids.

For each cross, proceed as follows:

5. Plate about 200 000 asci directly on to solid YPD. Incubate at 30°C for 3 days.
6. Replica plate the cell-layer on to SD containing 40 mg/l uracil and 1.5 g/l 5-FOA (recombinant plasmid-cured *ura3* $FUR1^+$ colonies are selected).
7. Purify about 25 *ura3* $FUR1^+$ strains, then induce them to sporulate.
8. Plate the bulk of the asci directly on to SD supplemented with 0.4 g/l uridine and 0.4 g/l 5-FU. Diploid colonies that are ura^- $5\text{-}FU^r$ are selected. Purify diploids and test their frequency of reversion to the ability to utilize uracil as the sole source of pyrimidine.
9. Choose one diploid that does not revert measurably. This strain can be grown in YPD medium supplemented with uridine (0.4 g/l). The strain can be sporulated and haploid segregants germinated and grown on the same medium. *Ura3 fur1* cells can be used as hosts for transformation to ura^+; transformants are selectable on YPD, while non-transformed cells only form abortive colonies on this medium.

3.4.3 Construction of an *in vitro* directed deletion inside the *FUR1* locus

The sequence of the *FUR1* locus has been recently published, and yeast strains that carry deletions inside this locus have already been constructed by cloning and gene replacement techniques (14). Such a strain that displays a deletion inside the *FUR1* locus can be mated with a *ura3* mutant and the *ura3 fur1* progeny isolated on solid YPD enriched with uridine (0.4 g/l).

4. Vectors

4.1 Introduction

The most commonly used expression plasmids are *Escherichia coli*–yeast shuttle vectors which contain at least the *ARS* and *STB* (*REP3*) loci from the yeast 2 μm plasmid (1, 7; also see Chapter 2). Two different types of standard vectors should be distinguished. One class includes vectors, named ARSTAB in this chapter, that lack functional *REP* genes. ARSTAB vectors are defective for equipartitioning functions, and therefore need to be propagated in Cir^+ cells. The second class consists in vectors, named REP^+ hereafter, which have retained functional *REP* genes. It is better to propagate REP^+ vectors in cells that lack the 2 μm plasmid. Both types of vector can give very satisfactory structural and segregational stabilities.

4.2 Expression vectors for Cir^+ cells

The mitotic partition of ARSTAB vectors relies on *trans*-acting functions afforded by 2 μm genes. Therefore, ARSTAB plasmids are much more stably maintained in Cir^+ cells (i.e. cells that harbour the 2 μm plasmid, in contrast to Cir° cells); they behave like typically unstable *ARS* vectors in normal Cir° cells. Since ARSTAB vectors contain only relatively small segments of the 2 μm genome, they are generally much easier to handle and are widely used. Examples of such widely used vectors are pJDB207 (15) and pFL1 (also named YEp24) (12). The pJDB207-based vectors that contain the *LEU2*-d gene can be maintained at very high copy numbers, especially if leu^+ selection is applied during the sphaeroplast regeneration step. Substituting the wild-type *URA3* gene with the defective allele *ura3*-d in pFL1-like plasmids permits the selection of very high copy number transformants (see Section 3.3).

It is worth noting that plasmids that have retained an inverted repeat (IR) sequence from the 2 μm plasmid, e.g. pJDB207 or pFL1, are themselves substrates for the *FLP* gene product, and reversibly co-integrate with the resident 2 μm in Cir^+ cells. The proportion of co-integrates results from a balance with free forms and usually represents a small percentage (commonly 1–5%) of total hybrid plasmids. Levels of co-integrates can be maintained far below 1% by deleting the IR sequence from the shuttle vector. The target sequence of the *FLP* gene product spans a *Xba*I recognition site, which can be used to modify this sequence. For instance, deletion of the *Xba*I–*Eco*RI terminal subfragment of 2 μm D from the vector abolishes this recombination phenomenon, without any detectable consequence on plasmid copy numbers (see e.g. ref. 13). Finally, it is worth noting that no sequence should be cloned into the 2 μm segment in the plasmids that are propagated in Cir$^+$ cells, as events of conversion or recombination with the resident 2 μm can provoke the formation of rearranged molecules. In particular, the small *Pst*I°–*Eco*RI fragment of the 2 μm that flanks the 3′-end of *LEU2*-d marker should ideally be deleted from pJDB207-derived ARSTAB vectors. This precaution impedes the *in vivo* appearance of shorter forms in which the *LEU2*-d-containing chromosomal fragment has been replaced with the reformed *Pst*I cloning site.

In summary, convenient expression vectors for Cir$^+$ cells should include the following elements:

- a bacterial origin of replication
- a selectable marker for *E. coli* (e.g. the ampicillin-resistance gene of pBR322)
- a short fragment of the 2 μm plasmid with no foreign sequence cloned inside. This portion of the 2 μm plasmid should include the *STB* and *ARS* loci but preferably lack the *FRT* site of the IR. The 1.29 kb *Xho*I–*Pst*I subfragment of the 2 μm (coordinates 2652–3945 of form A, or coordinates 703–1686 of form B; see ref. 16) is a good candidate, since it contains the *ARS* and *STB* loci and also provides a useful transcriptional terminator which protects the *STB* domain from transcriptional inactivation. This *Xho*I–*Pst*I subfragment can easily be isolated from any pFL1-type plasmid that contains the 2 μm *Eco*RI D fragment
- a defective selectable marker for *S. cerevisiae* (e.g. *LEU2*-d or *URA3*-d)
- optionally, an additional non-defective selectable marker for *S. cerevisiae* (to be used instead of the defective marker whenever high plasmid copy number is not desirable for optimal production or is deleterious to the host cell)
- the expression block (see Section 5)

4.3 Expression vectors for Cir° cells

These more complex 2 μm-based shuttle vectors, namely REP$^+$, generally contain the entire sequence of the 2 μm plasmid, including functional *REP1* and *REP2* genes. They are more cumbersome than simpler ARSTAB vectors. As they tend to rearrange by recombining with the resident 2 μm plasmid in Cir$^+$ cells, these vectors should be used only in Cir° host strains. The earliest REP$^+$ vectors are exemplified by pJDB219 (1, 7, 15). This plasmid contains the entire 2 μm B form cloned in the bacterial plasmid pMB9 with disruption of *FLP* and the *LEU2*-d cloned into a *Pst*I site of the 2 μm sequence. The presence of the defective marker favours selection of transformed cells that display ultra-high copy numbers. Many expression vectors used in industrial research are pJDB219 derivatives. More recently described REP$^+$ plasmids retain intact *FLP* and *D* genes, and coexist under both FLP-generated forms (1, 7).

5. Expression block

Expression blocks can arbitrarily be divided into three regions: the sequence that encodes the foreign protein (or a precursor form of it), and the regions flanking the 5′ and 3′-ends of the coding sequence. The upstream region includes the promoter signals, the transcription start-site(s) and the sequence of the mRNA untranslated leader region. The downstream region specifies the mRNA 3′-end including the signals for transcription termination and mRNA polyadenylation. Very often, expression vectors are conceived as standard plasmids, in which the upstream and downstream regions are separated by convenient cloning site(s), designed to receive the sequence to be expressed (1).

5.1 Promoters

Although there have been notable examples of high-level expression of foreign coding sequences using promoters from glycolytic genes such as *PGK, ADH1, GAP491*, or *TFDIII*. (1), it should be stressed that these promoters, being constitutive or, at least, not tightly regulated, are not recommended for use for the expression of foreign genes. The risks of undesirable side-effects, such as copy number reduction, genetic drift, or plasmid rearrangements, are much enhanced when production of the foreign protein occurs continuously, instead of being delayed until the optimal time in the production process, ideally at the end of the culture period. In particular, such side-effects inevitably occur when these promoters are used for production that is deleterious to the host cell (1). For these reasons, tightly regulated promoters are presently of much wider use.

5.1.1 An example of a galactose-regulated promoter: the *GRAP1* sequence

Tightly regulated promoters that have been used for heterologous production have been reviewed recently (1). Among them, those regulated by the *GAL4* transactivator are frequently used. These promoters are repressed in glucose, derepressed in the absence of a fermentable source of carbon and energy, and maximally induced in the presence of galactose (17). *GRAP1* (for galactose-regulated artificial promoter sequence 1) belongs to this class of promoter. *GRAP1* contains the two *UAS* elements from the *GAL7* promoter (*pGAL7*) together with downstream regions of the *ADH2* promoter (*pADH2*). This artificial promoter is even more potent than either of the two strong promoters from which it originates (13). As shown in *Figure 1*, the sequence of this hybrid promoter results from the combination of functionally specialized

5'-ACGCGTCTATACTTCGGAGCACTGTTGAGCGAAGGCTCATTAGAT
Mlu1 *UAS1*

ATATTTTCTGTCATTTTCCTTAACCCAAAAATAAGGGAGAGGGTCC

AAAAAGCGCTCGGACAACTGTTGACCGTGATCCGAAGGACTGGCTA
UAS2

TACAGTGTTCACAAAATAGCCAAGCTGAAAATAATGTGTAGCCTT

← pGAL7 | pADH2 →
TAGCTATGTTCAGTTAGTTTGGCATGCCTATCACATATAAATAGAG
Sph1 *TATA REGION*

TGCCAGTAGCGACTTTTTTCACACTCGAGATACTCTTACTACTGCT
Xho1

CTCTTGTTGTTTTTATCACTTCTTGTTTCTTCTTGGTAAATAGAAT
TRANSCRIPTION

ATCAAGCTACAAAAAGCATACAATCAACTATCAACTATTAACTAT
INITIATION REGION

ATCGATATACACAAATG - 3'
Cla1

Figure 1. Sequence of the *GRAP1* promoter. Only one strand is represented for clarity. The *GRAP1* sequence is given in bold letters; it is flanked with an *Mlu1* site (normal letters, underlined) on the 5′ side. One possible example of sequence for the initiation codon-flanking region is given in normal letters at the 3′-end of the *GRAP1* sequence; it was taken from ADH2 (13). The initiation codon is indicated by an open box. The internal *Sph*I site (underlined) delimits the *GAL7*- and *ADH2*-derived regions, as indicated by the arrows. The *Xho*I and *Cla*I sites (underlined) were created in the *ADH2*-derived sequence by modifying the natural sequences CTCGAA and ATCGTA into CTCGAG and ATCGAT, respectively. UAS elements (UASs 1 and 2) are included in open boxes. The subregions that encompass the TATA box and the transcription start-points are indicated by '*TATA REGION*' and '*TRANSCRIPTION INITIATION REGION*' respectively.

cassettes individualized by unique restriction sites. This modular structure facilitates the creation of novel combinations generated by exchanging one or several cassette(s) with other subsequences.

GRAP1 is regulated in a similar fashion to genuine *GAL* promoters of *S. cerevisae*. It is repressed in wild-type yeast cells grown in glucose-containing media, derepressed when glucose is substituted with non-fermentable carbohydrates, and fully activated when galactose is present either as the sole source of carbon and energy or added to other non-fermentable carbohydrates (13). The regulatory circuits that control *GAL* promoters are complex and have been fully reviewed elsewhere (17). In summary, the *GAL4* gene product, namely GAL4, activates the transcription of *GAL* genes by binding to specific *UAS* elements located in each promoter of these genes. The *GAL80* gene encodes a repressor which inactivates the GAL4 transactivator in the absence of a galactose-derived inducer.

The strength of *GAL* promoters cloned in multiple copies is limited by the amount of the GAL4 transactivator (16). It is possible to improve the strength of a GAL4-activated promoter cloned in multiple copies by overexpressing the transactivator (18). The regulation pattern of this promoter can also be modified simply by changing the control exerted on the biosynthesis of the transactivator (13). An ARSTAB vector for *S. cerevisiae* was designed in which a cDNA encoding the *Aspergillus flavus* urate oxidase was put under the control of the *GRAP1* sequence. This vector, namely pEMR547 (13), contains the *LEU2*-d gene as the selectable marker. Yeast cells transformed to leu$^+$ with pEMR547 and grown in galactose-based medium, produce active urate-oxidase (Uox) at high levels representing 20–25% of cellular soluble proteins. Deleting the *GAL4* gene from the genome of the producing strain provokes a 300-fold decrease in Uox production (still detectable by the Uox activity assay). This low production phenotype can be advantageously corrected by replacing the *GAL4* gene with an *in vitro*-engineered version of that gene, in which the transactivator coding sequence has been put under the control of a tightly regulated strong promoter. For instance, putting the *GAL4* coding sequence downstream of the *ADH2* promoter in our pEMR547-transformed *gal4* mutant results in an Uox superproduction phenotype (see *Figure 2*), where Uox now represents about 60% of soluble proteins in cells that grow with galactose as the sole source of carbon and energy. When these cells are grown in ethanol and glycerol, with or without galactose, Uox accumulates up to 34 or 40% of soluble proteins, respectively. The addition of glucose drastically represses the synthesis of Uox. In fact, the regulation of the *GRAP1* promoter now displays a pattern roughly similar to that of its artificially controlled transactivator, as if the Uox coding sequence itself was placed directly under the control of the *ADH2* promoter. Several different recipient strains can be constructed, in which the genuine *GAL4* gene is replaced by a *GAL4* cassette that is controlled in a specific fashion (for instance, using the *PHO5* promoter in one strain, the *CUP1* promoter in another, and so on). A wide choice of regulatory situations can then be tested by transforming these different strains with the same expression vector.

5.2 The mRNA leader sequence

The region that specifies the mRNA leader sequence generally includes a naturally present, or artificially created, cloning site that is used for inserting

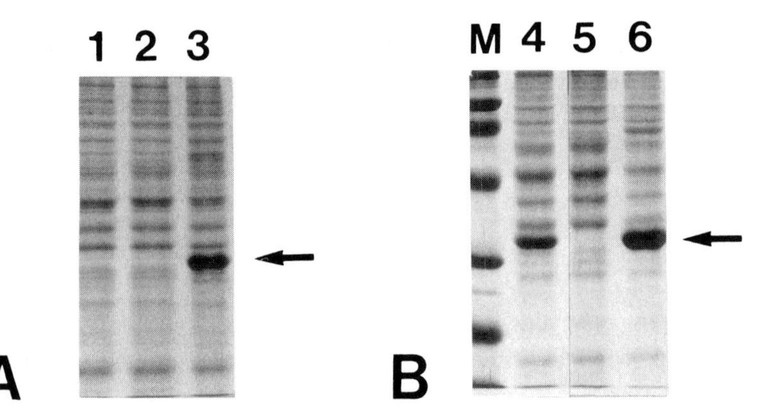

Figure 2. SDS–PAGE analysis of cell-free crude extracts from various transformed strains of yeast. Three congenic derivatives of EMY761 (13) were transformed with pEMR547, a urate-oxidase expression vector. Transformants were grown for 24 h in YPEG (A) or YPGal (B). Cells were lysed in TEA buffer (see *Protocol 4*). Soluble proteins were analysed by 0.1% SDS–12.5 PAGE. Gels A and B were stained with Coomassie brilliant blue. The *GAL4* locus in each strain was as follows: slots 1 and 4, wild-type; slots 2 and 5, deleted; slots 3 and 6, the *GAL4* coding sequence was put under the control of the *ADH2* promoter. Arrows in the right margin point to urate oxidase polypeptide bands. Lane M: molecular weight markers: 200, 97.4, 69, 46, 30, 21.5, and 14.3 kDa.

the coding sequence into the vector. Apparently, yeast is very tolerant regarding the sequence and the length of mRNA leader sequences (19). A common and safe way of designing a leader sequence is to copy that of the highly expressed yeast gene from which the downstream promoter sequences are taken, for instance that of *ADH2* in the case of a *GRAP1*-derived expression cassette (see *Figure 1*). In any case, the following rules must be respected.

(a) Do not generate any 5′-AUG sequence upstream to the initiation codon, as this will drastically inhibit the efficiency of initiation of translation at the correct AUG codon, and, if the additional AUG is placed out-of-frame with respect to the coding sequence, will considerably lower the expression of the foreign gene.
(b) Avoid the presence of any dyad-symmetry region in the mRNA leader sequence, as this also lowers translation efficiency.
(c) The consensus sequence between positions −3 and −1 with respect to the yeast initiator ATG (5′-A A/Y A/U) should preferably be included for optimal expression.

5.3 The foreign protein-encoding sequence

5.3.1 Sequence design

There has been a debate among yeast molecular biotechnologists as to whether the occurrence of rare codons in a coding sequence has any effect on the level of expression. Changing the sequence of an mRNA can no doubt have significant effects on gene expression, but it is also clear that the presence of rare codons in a given gene does not necessarily impede its expression at a high level (18, 1). The biotechnologist who has to design a synthetic sequence should keep in mind that it is the occurrence of secondary structures in mRNAs, rather than just the presence of rare codons, that can be deleterious to the level of expression. No clear data are available on the effect of rare codons on the translation accuracy of highly expressed (foreign) genes in *S. cerevisiae*. For this reason, it seems preferable to choose frequently-used codons. In any case, it is always advisable to anticipate the possibility of having to modify the coding sequence, by introducing convenient cloning sites which will facilitate subsequent gene manipulations.

5.3.2 Secreted proteins

Yeast can be engineered to produce proteins that accumulate intracellularly, as well as proteins that are secreted into the medium (1). Typically, a polypeptide secreted by a eukaryotic cell is synthesized as an amino-terminally extended precursor, which enters and follows the secretory pathway, undergoing various modifications including proteolytic cleavage(s), formation of disulphide bonds, eventually *O*- and *N*-glycosylation before finally giving rise to the mature form(s).

Various signal sequences have been used to direct the secretion of foreign proteins in the culture medium (20, 1). Among these signal sequences, perhaps the most frequently used is the amino-terminal 85 amino acid region of the *MFα1* gene product, namely the pre-prosegment of the precursor for the yeast α-factor pheromone (1, 21). This pre-prosequence ends with a Lys-Arg motif, which is recognized by the *Kex2*-encoded endoprotease. This protease, namely kexin (EC 3.4.21.61), is a membrane-bound enzyme which is localized in the late Golgi compartment and is involved in the processing of some yeast secreted proteins by hydrolysing the peptidyl bond at the carboxy-terminus of specific dibasic doublets (see ref. 22 for a recent review). There have been an increasing number of successful examples of high-level secretion using fusion proteins that start with the *MFα1* pre-prosequence and continue with the mature sequence of the polypeptide to be secreted (20, 1). In most of these constructions, the secreted polypeptide directly flanks the dibasic cleavage site, so that kexin liberates mature forms with the expected amino-terminus. However, the susceptibility of any Lys-Arg motif to cleavage by kexin depends very much on the surrounding context (20–21, 23). The cleavage being generally far from complete *in vivo*, a genetically engineered overproduction of kexin can be of great help for recovering higher amounts of the mature form (20–21). Keeping Glu-Ala and/or Asp-Ala sequences at the carboxy-terminus of the kexin cleavage site also enhances the cleavage efficiency. However, most of these diaminopeptidyl residues remain present at the amino-terminus of the secreted product, as the *STE13*-encoded diaminopeptidylpeptidase that normally clips off these Glu/Asp-Ala doublet is synthesized at limiting levels. It should be noted that the prosequence-extended form itself may be secreted in the medium. This 'pro' form can be detected, together with its matured counterpart, by Western blotting analyses performed on the deglycosylated proteins present in the medium (see Section 7). A deglycosylation treatment is required prior to these visualization experiments as the α-factor prosequence is highly *N*-glycosylated at three sites, the pro form being very polydispersed and hardly detectable otherwise. Methods for studying protein glycosylation in yeast including deglycosylation treatments, have recently been reviewed (24).

The 'pre-pro' region of the α-factor precursor is not the only system that can be used to secrete foreign proteins in yeast. Other presequences can be used whether natural, homologous, heterologous or artificial, alone or in combination with a prosequence; the choice of such a system is very wide. However, the maximal secretion level empirically depends on the choice of the signal system, the sequence of the foreign protein, and the genetics of the host strain. Sometimes, the secretion of a foreign protein comes up against clear limitations. Commonly encountered symptoms of this limitation phenomenon are as follows:

(a) a saturation effect: absence of proportionality between the level of mRNAs and the amount of secreted protein (higher secretion at low gene copy number, inhibition of secretion and intracellular accumulation of precursor forms when produced from many gene copies)
(b) a toxic effect: this effect increases with the mRNA level; the higher the rate of precursor biosynthesis, the more unhealthy the yeast cells become
(c) levels of secreted protein are relatively low

The severity of these problems depends not only on the signal system, but also on the genetics of the host strain and on growth conditions. Classical mutation-screening strategies can bring substantial improvements (1, see also e.g. 25, 26).

Since the secretion of a foreign protein can handicap the host cell, it is, once again, advisable to express the precursor gene under the control of a tightly regulated promoter. A galactose-regulated promoter, such as *GRAP1*, works perfectly. When the plasmid-host strain system works badly, as evidenced by the series of symptoms listed above, note that the yield of secreted protein is generally optimal when the biosynthesis of the precursor is maintained at a low rate. With an expression block involving the *GRAP1* promoter in a wild-type *GAL4* context, this can be achieved by down-regulating the galactose feeding in the bioreactor. It is also possible to limit the copy number of the foreign gene, e.g. by constructing a transformed strain that has integrated several direct repeats of a linearized non-replicative expression vector at a targeted chromosomal locus in the host strain.

5.4 Transcription termination signals and mRNA stability

Providing yeast-specific termination and polyadenylation signals to the foreign sequence is usually necessary for optimal gene expression in yeast. A common strategy consists in downstream flanking of the foreign sequence with the 3′-end region of a highly expressed yeast gene, for example, the 0.4 kb *Bgl*II–*Hin*dIII fragment that contains the end of the *PGK* coding sequence together with mRNA ending signals (27, see also ref. 28 for the sequence). Trimming the 3′-end of the foreign cDNA sequence in order to remove untranslated regions can improve gene expression by stabilizing the corresponding mRNA in yeast (1, 29).

6. Protein stability

Protein stability is a crucial parameter that greatly influences the level of production, especially where intracellularly accumulating proteins are concerned. The foreign proteins that yeast accumulates intracellularly at outstandingly high levels (exceeding 20% of soluble proteins) are stable polypeptides, which in many, if not all, cases have their amino-terminus acetylated. Presumably this protects them against the ubiquitin-dependent degradation system (30, 31). Most of these efficiently produced polypeptides are enzymes that are folded correctly *in vivo*, and consequently are biologically active. Mis-folded polypeptides are generally much more susceptible to degradation. A basic principle should be kept in mind: naturally secreted proteins are generally not folded properly when accumulated intracellularly. The converse is also true. None the less, there are exceptional cases of naturally secreted polypeptides that are more efficiently produced in yeast in a cytoplasmic form, rather than by secretion. Human α-antitrypsin is one good example of a naturally secreted protein that yeast can accumulate at a high level as a soluble, active form in the cytoplasm, whereas secretion of this enzyme is troublesome (32, 25).

Introducing mutations in some protease-encoding genes of the host strain increases the yield of intact molecules at the first steps of extraction and purification. In summary, the list of genes that can advantageously be mutated include the following:

- *pep4*, which encodes the vacuolar aspartyl-protease prA. Mutations in *pep4* induce deficiencies in prA, prB, and cpY activities (33, 34)
- *prb1*, which encodes prB, a vacuolar serine-protease which is mainly responsible for proteolytic degradation in yeast crude extracts. As the prB precursor can be partly activated in a prA-independent fashion, at least *in vitro*, it is therefore preferable to use a strain mutated in both *PEP4* and *PRB1*

- *prc1* which encodes cpY. The action of this vacuolar carboxypeptidase on secreted foreign polypeptides has been documented (1). Note that degradation that affects secreted protein can occur *in vitro*, probably as the result of the action of proteases released by cell lysis
- *kex1*, which encodes the Golgi carboxypeptidase that is involved in the maturations of α-factor and of the killer toxin; some secreted foreign proteins are susceptible to this carboxypeptidase (1)
  KEX2 and *SEX1*: secreted polypeptides can also be susceptible to other proteases of the secretory pathway, e.g. the dibasic and monobasic processing endoproteases, namely the *KEX2* and *SEX1* gene products (21), respectively. In contrast to the previously mentioned proteases, yeast cells need these two for healthy vegetative growth

Strains that are mutated in *pep4*, *prb1*, and *prc1* genes were constructed by E. W. Jones and co-workers (32). Such multiple protease-deficient strains are useful starting points for the construction of strains that produce a foreign-protein, and are available from the Yeast Genetic Stock Center.

7. Basic methods for analysing protein production

Protocol 3. Analysis of total cell proteins extracted under denaturing conditions

Equipment and reagents

- microcentrifuge tubes with the corresponding bench microcentrifuge
- glass beads (0.4–0.5 mm in diameter, Braun)
- 300 μl micropipette
- Pasteur pipettes
- a waterbath (100°C)
- a vortexing machine
- 10% trichloroacetic acid (TCA) solution
- cold acetone
- 10 M NaOH solution
- SDS-gel sample buffer

Method

1. Resuspend about 5×10^8 cells in 300 μl of 10% TCA in a microcentrifuge tube[a].
2. Add glass beads to the level of the meniscus.
3. Vortex vigorously for 5 min.
4. Transfer the crude extract into another microcentrifuge tube by using e.g. a Pasteur pipette, avoid pipetting the beads as much as possible.
5. Centrifuge the crude extract at 10 000 *g* for 20 min. Discard the supernatant.
6. Wash the pellet with 1 ml of cold acetone.

7. Centrifuge at full speed for 20 min and discard the supernatant.
8. Resuspend the pellet in 100 μl of SDS-gel sample buffer (the colour of the suspension turns yellow). Heat at 100°C for 5 min and vortex vigorously in order to redissolve the pellet.
9. Adjust the pH by adding increasing volumes (in μl) of 10 M NaOH until the colour turns blue.
10. Centrifuge the suspension at full speed for 5 min to remove insoluble material.
11. Incubate at 100°C for 10 min.
12. Analyse aliquots of the supernatant by standard SDS–PAGE. Bands can be revealed by silver or Coomassie brilliant blue staining or by Western blotting using antibodies raised against the foreign protein.

[a] Direct cell lysis by boiling in SDS-gel sample buffer, as recommended by some authors, enhances general proteolysis in ordinary strains, presumably by activation of the serine-protease prB. This last method should therefore used only with *prb1* mutants.

Protocol 4. Analysis of cell proteins extracted under non-denaturing conditions

Equipment and reagents

- microcentrifuge tubes
- refrigerated centrifuge
- glass beads (diameter: 0.4 mm)
- 300 μl micropipette
- Pasteur pipette
- a waterbath (100°C)
- ice
- extraction buffer: e.g. TEA (7.5 g/l triethanolamine, 0.38 g/l EDTA, pH 8.9)
- SDS-gel sample buffer

Method

1. Resuspend about 5×10^8 cells in 300 μl of cold (4°C) extraction buffer (e.g. TEA) in a microcentrifuge tube.
2. Add glass beads to the level of the meniscus.
3. Vortex vigorously for five periods of 30 sec each. Cool the suspension on ice for 1 min between each cycle of vortexing.
4. Transfer the crude extract into another microcentrifuge tube by using e.g. a Pasteur pipette; avoid pipetting the beads as far as possible.
5. Harvest the crude extract in a refrigerated centrifuge (about 10 000 *g*, 10 min, 4°C)
6. Analyse both the supernatant ('soluble fraction') and the pellet ('insoluble fraction') by standard SDS–PAGE or by Western blotting. Redissolve the pellet in sample buffer as described in *Protocol 3*, steps **8–12**, except for step **9** which is not necessary in this case.

Protocol 5. Analysing the proteins present in the culture medium

Reagents

- TCA
- deoxycholate
- cold acetone
- SDS-gel sample buffer

Method

1. Add TCA and deoxycholate to the cell-free culture supernatant (e.g. 10 ml) at final concentrations of 10% (v/v) and 0.04% (w/v), respectively.
2. Incubate at 4°C for 30 min.
3. Pellet the precipitate by full-speed centrifugation for 30 min.
4. Wash the pellet with 1 ml of cold acetone and transfer the suspension into a microcentrifuge tube.
5. Incubate at 4°C for 30 min.
6. Centrifuge (about 10 000 g, 4°C, 10 min).
7. Dry under vacuum.
8. Redissolve the pellet in SDS-gel sample buffer, following *Protocol 3*, steps **8–12**.

The presence of *N*- and *O*-glycosylation on the foreign protein can be determined by studying the effect of deglycosylating enzymes, such as endoglucosidase H and α-mannosidase. The O-linked and N-linked oligosaccharides of mannoproteins can be selectively released by alkali-catalysed β-elimination and by hydrazinolysis, respectively (see ref. 24, for further details).

Acknowledgements

I am indebted to many colleagues for useful discussions and experiments, and especially to my co-workers P. Leplatois and B. Le Douarin who also did the experiments shown in *Figure 2*.

References

1. Romanos, M. A., Scorer, C. A., and Clare, J. J. (1992). *Yeast*, **8**, 423.
2. Burgers P. M. J. and Percival, K. J. (1987). *Anal. Biochem.*, **163**, 391.
3. Ito, H., Fukuhada, Y., Murata, K., and Kimura, A. (1983). *J. Bacteriol.*, **153**, 163.
4. Meilhoc, E., Masson, J. M., and Tessier, J. (1990). *Bio/Technology*, **8**, 223.
5. Heinemann, J. A. and Sprague, G. F. (1991) In *Methods in enzymology*, Vol. 194 (ed. C. Guthrie and G. R. Fink), pp. 187–95. Academic Press, London.

6. Rothstein, R. (1991). In *Methods in enzymology*, Vol. 194 (ed. C. Guthrie and G. R. Fink), pp. 281–301. Academic Press, London.
7. Rose, A. B. and Broach, J. R. (1990). In *Methods in enzymology* Vol. 185 (ed. D. V. Goeddel), pp. 234–79. Academic Press, London.
8. Loison, G., Vidal, A., Findeli, A., Roitsch, C., Balloul, M., and Lemoine, Y. (1989). *Yeast*, **5**, 497.
9. Boeke, J. D., Lacroute, F., and Fink, G. R. (1984). *Mol. Gen. Genet.*, **197**, 345.
10. Sherman, F. (1991). In *Methods in enzymology*, Vol. 194 (ed. C. Guthrie and G. R. Fink), pp. 3–20. Academic Press, London.
11. Rose, M., Grisafi, P., and Botstein, D. (1984). *Gene*, **29**, 113.
12. Chevallier, M.-R., Bloch, J.-C., and Lacroute, F. (1980). *Gene*, **11**, 11.
13. Leplatois, P., Le Douarin, B., and Loison, G. (1992). *Gene*, **122**, 139.
14. Kern, L., de Montigny, J., Jund, R., and Lacroute, F. (1990). *Gene*, **88**, 149.
15. Beggs, J. D. (1981). In *Genetic engineering* Vol. 2 (ed. R. Williamson), pp. 175–203. Academic Press, London.
16. Broach, J. R. (1981). In *The molecular biology of the yeast Saccharomyces: life cycle and inheritance* (ed. J. N. Strathern, E. W. Jones, and J. R. Broach), pp. 455–70. Cold Spring Harbor Laboratory Press, Cold Spring Harbor, NY.
17. Johnston, M. (1987). *Microbiol. Review*, **51**, 458.
18. Mylin, L., Hofmann, K. J., Schultz, L. D., and Hopper, J. E. (1990). In *Methods in enzymology*, Vol. 185 (ed. D. V. Goeddel), pp. 297–308. Academic Press, London.
19. Donahue, T. F. and Cigan, A. M. (1990). In *Methods in enzymology*, Vol. 185 (ed. D. V. Goeddel), pp. 366–72. Academic Press, London.
20. Hitzeman, R. A., Chen, C. Y., Dowbenko, D. J., Renz, M. E., Liu, C., Pal, R., Simpson, N. J., Kohr, W. J., Singh, A., Chilsom, V., Halminton, R., and Chang, C. N. (1990). In *Methods in enzymology*, Vol. 185 (ed D. V. Goeddel), pp. 421–40. Academic Press, London.
21. Brake, A. J. (1990). In *Methods in enzymology*, Vol. 185 (ed. D. V. Goeddel), pp. 408–21. Academic Press, London.
22. Bourbonnais, Y., Germain, D., Latchinian-Sadek, L., Boileau, G., and Thomas, D. Y. (1991). *Enzyme*, **45**, 244.
23. Degryze, E., Dietrich, M., Nguyen, M., Achstetter, T., Charlier, N., Charpigny, G., Gaye, P., and Martal, J. (1992). *Gene*, **118**, 47.
24. Balou, C. E. (1990). In *Methods in enzymology*, Vol. 185 (ed. D. V. Goeddel), pp. 440–70. Academic Press, London.
25. Moir, D. T. and Davidow, L. S. (1991) In *Methods in enzymology*, Vol. 194 (ed. C. Guthrie and G. R. Fink), pp. 491–507. Academic Press, London.
26. Chilsholm, V., Chen C. Y., Simpson, N. J., and Hitzeman, R. (1990). In *Methods in enzymology*, Vol. 185 (ed. D. V. Goeddel), pp. 471–82. Academic Press, London.
27. Tuite M. F., Dobson, M. J., Roberts, N., King, R. M., Burke, D. C., Kingsman, S. M., and Kingsman, A. J. (1982). *EMBO J.*, **1**, 603.
28. Hitzeman, R. A., Hagie, F. E., Hayflick, J. S., Chen, C. Y., Seeburg, P. H., and Derynck, R. (1982). *Nucleic Acids Res.*, **10**, 330.
29. Demolder, J., Fiers, W., and Contreras, R. (1992). *Gene*, **111**, 207.
30. Kendall, R. L., Yamada, R., and Bradshaw, R. A. (1990). In *Methods in*

enzymology, Vol. 185 (ed. D. V. Goeddel), pp. 398–407. Academic Press, London.
31. Wilkinson, K. D. (1990). In *Methods in enzymology*, Vol. 185 (ed. D. V. Goeddel), pp. 387–97. Academic Press, London.
32. Cabezón, T., de Wilde, M., Herion, P., Loriau, R., and Bollen, A. (1984). *Proc. Natl Acad. Sci. USA*, **81**, 6594.
33. Jones, E. W. (1990). In *Methods in enzymology*, Vol. 185 (ed. D. V. Goeddel), pp. 372–86. Academic Press, London.
34. Jones, E. W. (1991). In *Methods in enzymology*, Vol. 194 (ed. C. Guthrie and G. R. Fink), pp. 428–53. Academic Press, London.

12

Cell-free translation of natural and synthetic mRNAs in yeast lysates

MICK F. TUITE, DELIN ZHU, and ALAN D. HARTLEY

1. Introduction

All three phases of the translational process—initiation, elongation, and termination—can be faithfully reconstituted in the test tube with defined natural and synthetic mRNA templates. The establishment of such *in vitro* translation systems has underpinned not only studies into the mechanism of protein synthesis in both prokaryotes and eukaryotes, but has also provided a key experimental tool for the elucidation of the genetic code. While the fundamental details of how one prepares a cell-free translation system and uses it to translate mRNA templates have not really changed since Pelham and Jackson first described the mRNA-dependent reticulocyte cell-free lysate (1), progress has recently been made in refining these systems for the study of post-translational processing of secretory proteins (2, 3) and for generating quantitative yields of translation products (4). There have also been impressive developments in the area of *in vitro* transcription of cloned DNA templates (5) to generate both natural and engineered mRNAs for cell-free translation studies. Cell-free systems have been described from a relatively limited number of eukaryotic cells with most researchers still focusing on the rabbit reticulocyte and wheat germ systems primarily due to their commercial availability and reproducibility rather than any innate biological relevance.

The development of a cell-free translation system from the yeast *Saccharomyces cerevisiae*, capable of translating exogenously supplied mRNA templates, was hindered by an apparent inability to 'translate' the technology developed for the reticulocyte and wheat germ lysates to this simple unicellular eukaryote. The arrival in the early 1980s of recombinant DNA technology, coupled with the wealth of the already available classical genetic technologies, highlighted the potential such a cell-free translation system would provide in unravelling both the mechanism and control of translation when used in conjunction with the *in vivo* genetic approach. The first much awaited cell-free translation system from *S. cerevisiae*, capable of initiating translation upon exogenous mRNAs, came from the combined efforts of the McLaughlin and Moldave laboratories in 1979 (6) and has since remained the system of choice for many researchers, although alternative methods of preparing such a system have subsequently been described (7–9). However, extensive use of yeast cell-free translation systems has not been made, probably through a failure to appreciate the attention to detail which must be paid in order to generate an active lysate. In this chapter these fine details will be highlighted, together with protocols for preparing and translating mRNA templates in such a system and ways of analysing the translation products.

2. Preparation of cell-free lysates

The key component of any *in vitro* translation system is the cell-free lysate which must contain all of the necessary translation factors, ribosomes and aminoacyl-tRNA synthetases in a functional state. Such extracts can be effectively prepared from *S. cerevisiae* by any of the following methods:

- by mechanically disrupting whole cells using glass beads
- by rupturing whole cells with a high pressure cell (e.g. French pressure cell or Bio-ex 'X-Press' cell)
- by lysing sphaeroplasts using gentle homogenization

The first two methods, while being rapid, efficient, and largely independent of cell strain or physiological status, generally produce extracts with lower *in vitro* translational activity, particularly when natural mRNAs are used. They also fail to maintain their translational activity during long-term storage. Lysates prepared from sphaeroplasts, on the other hand, while time-consuming to prepare are, in our hands, the most efficient and stable lysates for *in vitro* translational studies with natural mRNAs. Extracts prepared by all three methods do, however, show essentially identical activities with respect to the translation of the synthetic template poly(U) (10).

Active cell-free extracts from a variety of yeast strains have been prepared using sphaeroplast lysis, but it is always important to use strains that grow with doubling times of 2–3 h in rich growth medium (YEPD) at 28°C, and to prepare the extracts from exponentially growing cells (5×10^6 to 2×10^7 cells/ml). Strains that are deficient in the synthesis of one or more endogenous proteases (e.g. strain ABYS1 (11)) have proven particularly good strains to use. It is our experience that, generally, the more slowly growing the strain, the less effective the cell-free extract is at initiating translation on natural mRNAs. The inclusion of a protease inhibitor such as phenylmethylsulphonyl fluoride (PMSF) in the lysis buffers is recommended during the preparation of cell-free lysates by any of the three methods even if a protease-deficient strain is being used.

Active translation extracts can be obtained from *S. cerevisiae* as a 30 000 *g* (S30) supernatant fraction (7, 8). However, our most efficient and stable lysates have invariably been generated by subjecting S30 extracts to a subsequent ultracentrifugation step (100 000 *g* for 30 min) to generate an S100 fraction (see *Protocol 1*). This ultracentrifugation step removes the majority of mRNA-associated ribosomes, i.e. polysomes, but not 80S ribosomes or individual ribosomal subunits, but perhaps more importantly it also appears to remove an as yet undefined particulate inhibitor of translation initiation *in vitro* (6).

Protocol 1. Preparation of an S100 lysate from *S. cerevisiae* using sphaeroplast lysis

Equipment and reagents

- baked (200°C for 3 h) 40 ml Dounce glass–glass homogenizer (Wheaton; supplied by Jencons)
- 30 ml Corex tubes
- YEPD (see Appendix 1)
- YM5-sorbitol (see Appendix 1)
- sterile, double-distilled H_2O
- 2 mM EDTA (pH 8.0), 10 mM β-mercaptoethanol
- 1 M sorbitol
- 1.2 M sorbitol
- Lyticase (partially purified; Sigma L8137) or β-glucuronidase/aryl sulphatase (Boehringer Mannheim, 127060)
- SLB (sphaeroplast lysis buffer) (20 mM Hepes–KOH, pH 7.4, 100 mM KOAc, 2 mM $Mg(OAc)_2$, 2 mM dithiothreitol, 0.5 mM PMSF)

Method

1. Inoculate 50 ml YEPD with a suitable strain (e.g. ABYS1 (11)) and incubate at 30°C until stationary phase (1–3 $\times$ 10^8 cells/ml) has been reached (usually after 48 h). Store this starter culture at 4°C until required.
2. Use the starter culture to inoculate six 1 litre batches of YEPD (usually a 1:10 000 dilution) and incubate with vigorous aeration at 30°C until a cell density of 1–2 $\times$ 10^7/ml has been reached (which usually takes 16–20 h). Harvest the cells by centrifugation (4400 *g*, 5 min, 20°C), resuspend the cell pellets in sterile distilled water, and pool to give a total volume of approximately 200 ml. Pellet the cells by centrifugation as before and determine the wet weight of the cell pellet. Yield should be 12–15 g wet weight of cells per 6 litre culture.
3. Resuspend the cell pellet in 100 ml of 2 mM EDTA (pH 8.0), 10 mM β-mercaptoethanol and incubate with gentle mixing at room temperature for 30–45 min. Harvest cells by centrifugation as in step **1** above, resuspend the cell pellet in 200 ml sterile 1 M sorbitol, and pellet the cells again by centrifugation.

Protocol 1. *Continued*

4. Resuspend the cell pellet in 200 ml of 1.2 M sorbitol then either add 1 mg per g of cells of partially purified Lyticase or 0.5 ml per 1 litre of starting culture of β-glucuronidase/aryl. Incubate at 20–25°C with gentle mixing until more than 95% of the cells have been converted to sphaeroplasts.
5. Pellet the sphaeroplasts by low speed centrifugation (3000 *g*, 10 min, 20–25°C), then gently resuspend in 200 ml of YM5-sorbitol (see Appendix 1) using a sterile 10 ml wide bore pipette. Incubate at 30°C for 60 min with gentle mixing, then pellet the sphaeroplasts as before using low speed centrifugation. At this stage the sphaeroplast pellet may be stored for several days at −70°C.
6. Gently resuspend the sphaeroplast pellet in 4 ml ice-cold SLB, then transfer to a baked 40 ml Dounce glass–glass homogenizer (Wheaton) precooled to 4°C on ice.
7. Homogenize the sphaeroplasts by repeated strokes of the homogenizer pestle until more than 90% of the sphaeroplasts have been lysed. This typically takes 100–150 strokes depending on the tightness of fit of the pestle.
8. Transfer the lysate to a sterile 30 ml Corex tube, then centrifuge at 30 000 *g* for 10 min at 4°C. Remove the supernatant with a glass Pasteur pipette being careful to avoid the flocculent material (cell debris) at the bottom of the tube and the lipid layer at the top of the tube. Transfer to a suitable 10 ml ultracentrifugation tube. This is the S30 fraction.
9. Centrifuge the S30 fraction at 100 000 *g* for 30 min at 4°C in a fixed angle rotor (e.g. Beckman 50Ti rotor). Carefully transfer the supernatant to a fresh sterile 15 ml Corex tube using a glass Pasteur pipette, being careful not to disturb the 'fluffy' layer on top of the pelleted polysomes and the lipid layer at the top of the tube. This is the S100 fraction. Retain the polysome pellet for preparation of RNA; see *Protocol 4*).

To remove the low molecular weight components from the S100 this high speed fraction is passed over a Sephadex G-25 (medium) column (see *Protocol 2*) and the eluting fractions are assayed for their A_{260} and for their translational activity with both natural and synthetic mRNA templates (see Section 4.1). As is illustrated in *Figure 1* the choice of fraction is critical, since late-eluting fractions, i.e. those immediately following the A_{260} peak fraction, appear to be almost totally incapable of initiating translation on natural mRNA while showing optimal translational efficiency with the synthetic template, poly(U). This loss of initiation activity is actually due to the co-elution of a small molecular weight, heat-stable inhibitor of translation initiation (10).

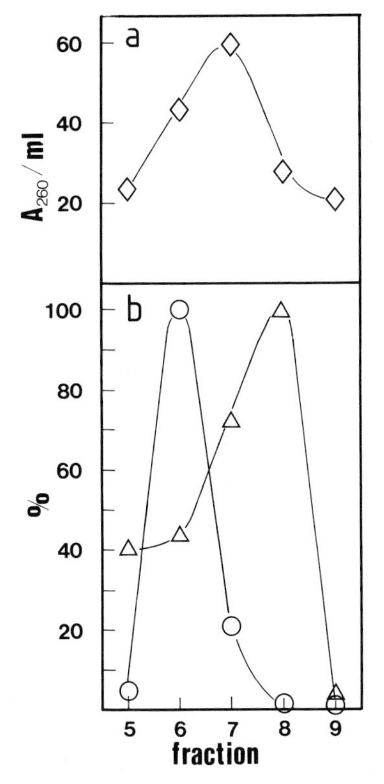

Figure 1. Translational activity of yeast S100 fractions eluting from a Sephadex G-25 column. (a) Absorbance of eluting fractions (A_{260} units/ml). (b) Translation elongation activity (as measured by the translation of poly(U); △) and translation initiation activity (as measured by the translation of total yeast mRNA; ○). The data presented are the percentage of the greatest value for each assay after subtraction of the background incorporated in the absence of added template.

Protocol 2. Chromatographic fractionation of a yeast S100 lysate on Sephadex G-25

Equipment and reagents

- S100 fraction (see *Protocol 1*)
- SLBG buffer (20 mM Hepes–KOH, pH 7.4, 100 mM KOAc, 2 mM Mg $(OAc)_2$, 2 mM dithiothreitol, 0.5 mM PMSF, 20% (v/v) glycerol)
- Sephadex G-25 (medium grade)
- 40 cm × 1.75 cm dimension column
- fraction collector
- liquid nitrogen or a dry ice/acetone bath

Method

1. Prepare a Sephadex G-25 (medium grade) column with the approximate dimensions of 30 cm × 1.75 cm. Pre-equilibrate at 4°C with SLBG buffer.

Protocol 2. *Continued*

2. Add 2–3 ml of the S100 fraction to the top of the column and develop the column at a flow rate of no more than 0.5 ml/min, collecting 1 ml fractions in a suitable fraction collector. This operation should be carried out in a cold (4°C) room.
3. Determine the A_{260} of each of the fractions, retaining only those fractions with absorbance values greater than 15; this in practice means discarding the first 20–25 ml eluting from a 30 cm × 1.75 cm column bed.
4. Prepare 200 μl aliquots of both the peak fraction and the two fractions on either side of the peak fraction. These fractions should have absorbance values in the range 30–60 A_{260} units/ml. The collected fractions should be kept at 4°C for as short a time as possible before being 'snap frozen' using either liquid nitrogen or a dry ice/acetone bath.
5. Assay the *in vitro* translational activity of both the peak fraction and the two fractions either side of the peak (*Figure 1*; see *Protocol 6*).

Active translation lysates can also be isolated from yeast cells disrupted by using glass bead lysis (see *Protocol 3*). The requirement for a protease inhibitor, e.g. PMSF, to be added to the lysis buffer appears to be more critical with this procedure than that detailed in *Protocol 1*, probably reflecting the increased chance of disruption of vacuolar membranes by this method compared with sphaeroplast lysis.

Protocol 3. Preparation of an S100 lysate from *S. cerevisiae* using glass bead disruption of whole cells

Equipment and reagents

- 30 ml Corex tubes
- YEPD (see Appendix 1)
- sterile double-distilled H_2O
- 1 M sorbitol
- YM5-sorbitol (see Appendix 1)
- SLB (see *Protocol 1*)
- 0.40–0.45 mm diameter glass beads (BDH), acid-washed and baked at 200°C for 3 h (BDH)

Method

1. Grow a 1 litre culture of the appropriate strain (preferably a multiply protease-deficient strain e.g. ABYS1 (11) in YEPD and harvest as described in *Protocol 1*, steps **1** and **2**.
2. Resuspend the cell pellet in 50 ml sterile water, pellet by centrifugation (4400 *g*, 5 min, 20°C), and resuspend the washed cell pellet in 50 ml of 1 M sorbitol. Incubate at 20–25°C for 60 min.

3. Harvest the cells by centrifugation (see step **2** above) and resuspend the cell pellet in 150 ml YM5-sorbitol. Incubate at 20–25°C for a further 60 min with gentle mixing.
4. Harvest the cells by centrifugation (see step **2** above) being careful to drain all liquid from the cell pellet. Resuspend cell pellet in ice-cold 0.5 ml SLB and transfer to a sterile 30 ml Corex tube kept on ice.
5. Add sterile, acid-washed glass beads to just below the meniscus making sure that the cell/glass-bead slurry is free to swirl—if there are too many glass beads this will lead to inefficient cell breakage. Agitate vigorously on a high speed vortex mixer for three periods each of 30 sec, with the cell slurry being kept on ice for 30 sec between each vortexing.
6. Remove as much liquid as possible from the glass beads using a Pasteur pipette and transfer to a 30 ml Corex tube. Then add a further 1 ml of fresh ice-cold SLB buffer to the cells and glass beads, vortex for 15 sec, then remove as much liquid as possible adding it to the previously removed sample.
7. Prepare an S30 fraction by centrifugation of the pooled lysate followed by an S100 fraction by ultracentrifugation as described in steps **8** and **9** (*Protocol 1*).

3. Preparation of yeast mRNA templates

3.1 Yeast mRNA structure

Yeast mRNAs are typically eukaryotic in their organization, consisting of a 5′ methylated cap structure covalently attached to a 5′ untranslated leader (5′UTR) preceding the translation initiation codon which in *S. cerevisiae* mRNAs is invariably AUG. Yeast mRNAs, with few exceptions, are polyadenylated (poly(A)$^+$) at their 3′ ends although the number of adenylate residues (30–50) is generally lower than seen in mRNAs from higher eukaryotic cells (12). Both the 5′ cap and the 3′ poly(A)$^+$ tail are important for achieving optimal translation of an mRNA in a yeast cell-free system (13).

Cell-free lysates depleted of endogenous mRNA can be programmed with a variety of mRNA templates, the origin and complexity of the template being defined by the objective of the translation study. Synthetic templates, e.g. poly(U) used for studies on translation elongation and fidelity (14, 15), are commercially available (e.g. from Boehringer Mannheim), although non-standard synthetic templates can be readily synthesized *in vitro* using the enzyme polynucleotide phosphorylase (16). Usually *in vitro* translation studies utilize either complex mixtures of natural mRNAs (see Section 3.2) or defined mRNAs synthesized from cloned genes using *in vitro* transcription reactions (see Section 3.3).

3.2 Preparation of yeast mRNA

Total RNA can be prepared directly from yeast cells using glass bead disruption of either whole cells or sphaeroplasts followed by organic extraction and ethanol precipitation (17). In either case it is important to use exponentially growing cells rather than stationary phase cells since the latter contain very few mRNAs that can be efficiently translated. RNA preparations from exponentially growing cells contain a wide variety of RNA species with mRNA only representing 1–2% of the total. To enrich for mRNA, the polysome pellet generated during the preparation of the S100 fraction (see step **9**, *Protocol 1*) is an ideal starting material since it contains primarily mRNA and ribosomal RNA with only a trace of tRNA. The procedure for the extraction of RNA from polysomes is given in *Protocol 4* and is based on one originally described by Gallis *et al.* (18).

Protocol 4. Preparation of yeast RNA from polysomes

Equipment and reagents

- SLB (see *Protocol 1*)
- polysome sample (see *Protocol 1*, step **9**)
- LETS buffer (100 mM Tris–HCl, pH 7.5, 100 mM $LiCl_2$, 1 mM EDTA, pH 8.0) prepared using diethylpyrocarbonate (DEPC)-treated water
- DEPC-treated double-distilled H_2O
- RNAsin (Promega) or RNAguard (Pharmacia) RNase inhibitors
- 1.0% (w/v) SDS
- 30 ml Corex tubes (baked at 200°C for 3 h)
- phenol:chloroform:isoamyl alcohol (prepared as a 50:50:1 (by volume) mix)
- 3 M NaOAc (pH 5.5)
- absolute ethanol stored at −20°C
- ethanol (75% (v/v)) stored at +4°C

Method

1. Following step **9** (*Protocol 1*) briefly wash the polysome pellet with 2 × 1 ml of SLB buffer, then store the resulting polysome pellet at −20°C until required.
2. Thaw the polysome pellet on ice, then resuspend in 1 ml of ice-cold LETS buffer. A commercially available RNase inhibitor (e.g. RNAsin or RNAguard) may be added at this stage if desired although generally this is not necessary, particularly for those with experience in handling RNA. Resuspension of the pellet can be facilitated by use of a baked (200°C for 3 h) glass rod.
3. Adjust the A_{260} of the suspension to 50 A_{260} units/ml, then add 0.1 vol. of 1.0% SDS solution. Transfer to a baked (200°C for 3 h) 30 ml glass Corex centrifuge tube.
4. Add an equal volume of phenol:chloroform:isoamyl alcohol (50:50:1) and carefully mix for 10 min at room temperature.

5. Separate the organic and aqueous phases by centrifugation (10 000 *g*, 10 min, 4°C), then re-extract the aqueous phase repeatedly until no white proteinaceous interface remains after centrifugation (usually after two or three extractions).
6. Transfer the aqueous phase to a fresh sterile 30 ml Corex centrifuge tube, then add 0.1 vol. of 3 M NaOAc, pH 5.5, and 2 vol. of cold (−20°C) absolute ethanol. Mix thoroughly, then place at −70°C for 20 min or −20°C for 2–24 h to allow the nucleic acid to precipitate.
7. Recover the nucleic acid precipitate by centrifugation (30 000 *g*, 10 min, 4°C). Wash the resulting pellet carefully with ice-cold 75% (v/v) ethanol and then air dry, preferably in a vacuum desiccator.
8. Resuspend the nucleic acid pellet in 1 ml of ice-cold DEPC-treated water, allowing 10 min on ice for resuspension to occur. Measure A_{260} to estimate the concentration of RNA (use 1 A_{260} unit = 50 μg RNA). Adjust the sample to a concentration of 2 mg/ml, aliquot, and store at −70°C until required.

For the majority of *in vitro* translation studies with total mRNA populations we find that it is not necessary to prepare the poly(A)$^+$ fraction. If this fraction is required, however, it can be prepared from the total RNA sample (*Protocol 4*) using chromatography on either oligo(dT)-cellulose or poly(U)-Sephadex according to standard procedures (17).

3.3 *In vitro* transcripts

It is often desirable to translate one specific mRNA rather than a heterogenous collection of mRNAs, particularly where one wishes to relate defined structural features of an mRNA with its *in vitro* translatability. There are relatively few examples of naturally occurring mRNAs that can be readily purified to almost total homogeneity; virus-associated RNAs (especially plant viral RNAs) and α- and β-globin mRNAs, which represent in excess of 60% of the total translatable population of mRNAs in a reticulocyte, are two well exploited examples. However, with the presently routine ability to transcribe efficiently cloned DNA sequences *in vitro* using viral polymerases and their corresponding promoters, it is possible to generate sufficient quantities of a single species of mRNA from any one of a vast number of available cloned gene sequences. If such sequences are from higher eukaryotic cells then they must be cDNAs to avoid any intron sequences being included in the transcript. Yeast genes, with few exceptions, lack introns (19).

A wide variety of transcription vectors are available which are suitable for generating transcripts for *in vitro* translation studies. The vector shown in *Figure 2*, pHST7, is a derivative of plasmid pHSTO (20) and is one such vector we have used routinely to synthesize a variety of mRNAs for transla-

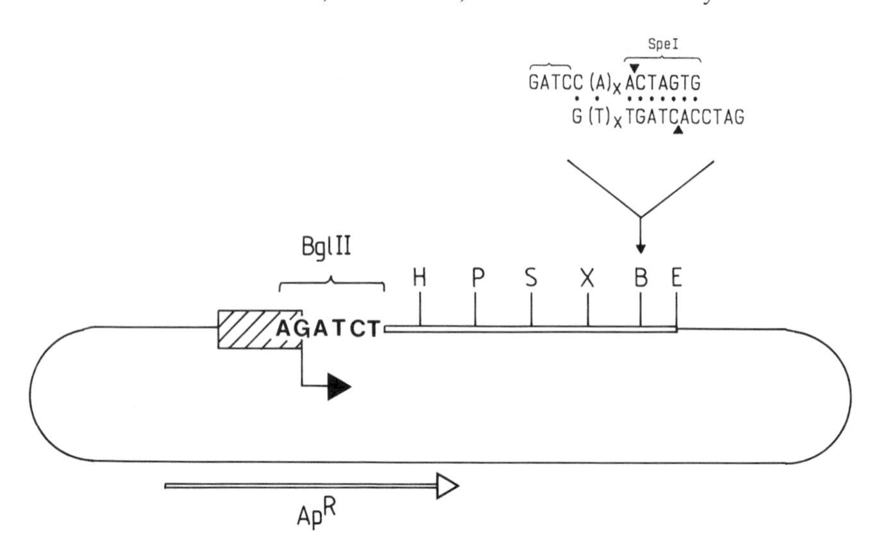

Figure 2. pHST7: an *in vitro* transcription vector. The hatched box is the T7 promoter immediately adjacent to the *Bgl*II site in which transcription initiates. Only some of the restriction sites in the polylinker are shown; abbreviations are B, *Bam*HI; *Eco*RI; H, *Hin*dIII; P, *Pst*I; S, *Sal*I; X, *Xba*I. The plasmid is not drawn to scale.

tion studies in yeast cell-free extracts. pHST7 carries the highly efficient bacteriophage T7 promoter (21) which, in conjunction with commercially available T7 RNA polymerase, can be used to transcribe any cloned cDNA sequence. pHST7 has the advantage over most other *in vitro* transcription vectors in that, by use of the *Bgl*II cloning site, one can generate transcripts that do not contain any vector sequences as part of the transcript since transcription is initiated at the G residue within the *Bgl*II recognition sequence 5′-**AGATCT**-3′. This is a very important consideration when translating mRNAs in yeast cell-free extracts since both the length and the primary and secondary structural features of 5′UTRs can have significant negative effects on translational efficiency both *in vitro* and *in vivo* in *S. cerevisiae* (22–24). A further attraction of using the T7 promoter is that it utilizes a single transcription start site (21).

The 3′ end of the *in vitro* transcript is defined by use of restriction enzymes to cleave the vector plus cloned cDNA or gene within the 3′ polylinker sequence e.g. at the *Bam*HI or *Spe*I sites in pHST7 (*Figure 2*), prior to the addition of the T7 RNA polymerase. The generated 'run-off' transcripts are, however, not polyadenylated which results in less efficient translation of the transcript in yeast cell-free extracts (13). This problem can be circumvented by inserting a synthetic poly(A/T) module within the 3′ polylinker as shown in *Figure 2*; in the case of pHST7, cleavage at the *Spe*I site will generate a near authentic polyadenylated transcript. Transcripts can also be 5′ capped *in vitro* by reducing the concentration of GTP to 0.1 mM and including 1 mM m^7GpppG

in the reaction mixture (25), resulting in in excess of 80% of the mRNA synthesized being capped. The efficiency of capping can be readily checked by carrying out parallel *in vitro* transcription reactions, one in the absence of the cap analogue, using [^{32}P]CTP to label the transcripts and one in the presence of the cap analogue. Following termination of the reactions, the *in vitro* synthesized and labelled RNAs can be fractionated on a 20% (w/v) acrylamide–urea DNA sequencing gel; in the presence of the m^{7}GpppG cap the transcript migrates more slowly than the non-capped derivative. Inclusion of the cap analogue in the *in vitro* transcription reaction does, however, result in less efficient transcription per μg of added DNA template. Using the pHST7 vector we routinely obtain 1–20 μg of capped mRNA per μg added DNA template, the variability being largely template-dependent.

A procedure for generating capped transcripts suitable for *in vitro* translation studies in yeast cell-free extracts, and based on the plasmid pHST7, is given in *Protocol 5*. Other transcription vectors can be used but with the provisos discussed above.

Protocol 5. *In vitro* synthesis of mRNA transcripts that are both capped and polyadenylated using the transcription vector pHST7

Equipment and reagents

- *in vitro* transcription vector e.g. pHST7 (see *Figure 2*)
- restriction enzymes e.g. *Bgl*II and *Spe*I (choice of enzyme will depend on vector used)
- components for *in vitro* transcription (see *Table 1*)
- RNase-free DNase I (100 units/ml)
- DEPC-treated double-distilled H_2O
- phenol:chloroform (1:1 (v/v) mix)
- chloroform
- 3 M NaOAc (pH 5.5)
- absolute ethanol (stored at −20°C)
- 75% (v/v) ethanol (stored at +4°C)
- 1% (w/v) agarose gel with suitable electrophoresis tank and running buffer
- RNA electrophoresis standard e.g. Brome Mosaic Virus RNA (Promega)
- scanning densitometer

Method

1. Clone the cDNA or gene sequence to be transcribed into the *Bgl*II site of the pHST7 vector. Complete the construction by inserting a *Bam*HI–poly(A/T)–*Spe*I synthetic module into the 3′ polylinker (*Figure 2*). Purify the plasmid using standard CsCl–ethidium bromide density gradients (26).
2. Linearize 5 μg of the plasmid with *Spe*I (assuming that no *Spe*I sites occur within the cloned cDNA or gene sequence). Purify the linearized DNA template either by elution from an agarose gel or by simply extracting the *Spe*I digestion mixture with phenol–chloroform by standard procedures and precipitating the DNA with ethanol (26).

Protocol 5. *Continued*

3. Set up the *in vitro* transcription reaction mix as shown in *Table 1*. The reaction mix should be assembled at room temperature due to the presence of spermidine in the transcription buffer which can cause DNA to precipitate at low temperatures.
4. Run the reaction for 1 h at 37 °C in a total volume of 50 μl. Higher yields of RNA transcripts can be obtained by adding a further 40 units of T7 RNA polymerase upon completion of the 1 h incubation and letting the reaction proceed for a further 1 h.
5. Remove template DNA from the reaction by adding RNase-free DNase I to a final concentration of 1 unit/μg linearized template DNA and incubating the reaction for a further 15 min at 37 °C.
6. Add 150 μl of DEPC-treated water to the reaction mixture, followed by 200 μl of 1:1 phenol:chloroform. Mix by vortexing for 1 min, separate the phases by centrifugation, then extract the aqueous phase twice with an equal volume of chloroform.
7. Precipitate the RNA from the aqueous phase by adding 0.1 volumes of 3 M NaOAc and 2.5 volumes of absolute ethanol and leaving at −70 °C for 30 min or −20 °C for 2 h. Recover the RNA by centrifugation at 4 °C, wash the pelleted RNA twice with 75% (v/v) ethanol, and air dry. Resuspend in 20 μl of DEPC-treated water.
8. The concentration of RNA transcript is most effectively measured by running 1 μl of the sample on a 1% (w/v) non-denaturing agarose gel with RNA standards of known concentration e.g. Brome Mosaic Virus RNA. Scanning densitometry of a negative obtained by photographing the ethidium bromide-stained gel using standard Polaroid 665 film can be used to estimate the concentration of the transcript relative to the standards. This is only accurate over the range 0–3 μg of sample per lane (10).

4. Cell-free translation of natural and synthetic mRNAs

To allow for efficient translation of mRNAs the yeast cell-free lysate, prepared as described in *Protocols 1* and *3*, must be supplemented with a number of low molecular weight components. These are:

- K^+ and Mg^{2+} ions
- unlabelled amino acids
- ATP and GTP
- nucleotide triphosphate generating system

Table 1. Components for an *in vitro* transcription reaction used to generate 5′-capped mRNA transcripts

Component	**Amount per 50 μl assay**
5 × transcription buffer (200 mM Tris–HCl pH 7.5, 30 mM $MgCl_2$, 10 mM spermidine, 50 mM NaCl)	10 μl
100 mM DTT	5 μl
1 mg/ml acetylated BSA	5 μl
RNAguard (Pharmacia)[a]	40 units
5 mM ATP	1.25 μl
0.5 mM GTP	1.25 μl
5 mM UTP	1.25 μl
5 mM CTP	1.25 μl
5 mM m^7 GpppG	5 μl
Linearized plasmid DNA (1 μg/μl)	5 μl
T7 RNA polymerase	40 units
DEPC-treated H_2O	to 50 μl

[a] This is just one of a number of commercially available RNase inhibitors.

Table 2. Components of a translation cocktail for use in a yeast cell-free translation system programmed with natural mRNAs

Stock component	**Volume (μl/100 μl)**	**Final concentration**[a]
400 mM Hepes-KOH (pH 7.4)–40 mM DTT	15.6	25 mM Hepes–KOH 2.5 mM DTT
100 mM $Mg(OAc)_2$[b]	7.5	3 mM
2 M KOAc	18.7	150 mM
10 × 'energy mix' (5 mM ATP, 1 mM GTP, 250 mM creatine phosphate)	25.0	0.5 mM ATP 0.1 mM GTP 25 mM creatine phosphate
4 mg/ml creatine phosphokinase	5.0	80 μg/ml
2.5 mM cold amino acids[c]	5.0	50 μM each amino acid
5 mM $CaCl_2$	5.0	0.1 mM
DEPC-treated H_2O	18.2	–

[a] The concentration given is that found in the final translation reaction i.e. after all the additions including radiolabelled amino acids and RNA (see *Protocol 6*, step **5**).
[b] If a synthetic polynucleotide is to be translated then a 500 mM $Mg(OAc)_2$ stock should be used in place of the 100 mM stock in order to generate a final concentration of 15 mM.
[c] A stock of all amino acids *except* for the amino acid to be added in a radiolabelled form. The concentration of each amino acid in the stock solution should be 2.5 mM.

The final concentrations of these components in a translation mix is given in *Table 2*. The K^+ and Mg^{2+} salts are generally acetate rather than their chloride equivalent since the former are less inhibitory to translation of natural mRNAs. The optimal concentration of both K^+ and Mg^{2+} should be determined for each target mRNA sample; the figures shown in *Table 2* are for yeast polysomal RNA. All amino acids except the one chosen to be radiolabelled must be added back to the lysate since the total amino acid pool is effectively depleted by chromatography on G-25 Sephadex. Both ATP and GTP are important for the activity of a number of key translation initiation and elongation factors (e.g. eIF-2, EF-1, and EF-3) and ATP is further required for the aminoacylation of endogenous tRNAs.

The efficiency of translation of natural mRNAs and *in vitro* transcripts can be further improved by the addition of one or more of the following components to the reaction mix: 0.25 mM spermidine hydrochloride, 5 mM putrescine hydrochloride, 200 μg/ml total yeast tRNA, and 2 mM glucose 6-phosphate.

The S100 lysate contains a significant level of endogenous mRNA which needs to be removed prior to the addition of the target mRNA. The most effective means of achieving this, without impairing the translational activity of the lysate, is to use micrococcal nuclease (1). Micrococcal nuclease is a Ca^{2+}-dependent nuclease which can be rapidly inactivated by chelating Ca^{2+} present by the addition of ethylene glycol *bis* (β-aminoethyl ether)-*N,N,N′,N′*-tetraacetic acid (EGTA) as described in *Protocol 6*. While it is advisable to use micrococcal nuclease-treated lysates immediately after their preparation, they can be stored at −70°C, although some loss of translational efficiency will be evident.

Protocol 6. *In vitro* translation of natural mRNA in an mRNA-depleted yeast cell-free lysate

Equipment and reagents

- *in vitro* translation cocktail (see *Table 2*)
- cell-free S100 lysate (see *Protocols 1* and *2*)
- SLBG buffer (see *Protocol 2*)
- micrococcal nuclease (500 μg/ml, Worthington Biochemicals)
- 62.5 mM EGTA, pH 7.3
- radiolabelled amino acid; e.g. [^{35}S]methionine (>1000 Ci/mmol)
- sample of RNA to be translated (see *Protocols 4* and *5*)
- DEPC-treated, double-distilled H_2O
- RNase inhibitor e.g. RNAsin (Promega)

Method

This protocol describes the preparation of sufficient mRNA-depleted yeast lysate to carry out 12 × 25 μl *in vitro* translation reactions using a suitable radiolabelled amino acid.

1. Prepare a stock translation cocktail (*Table 2*) keeping all components on ice and mixing gently after all components have been added.
2. Transfer 100 μl of the translation cocktail to a fresh sterile 1.5 ml microcentrifuge tube and add 110 μl of a cell-free lysate (approximately

8–10 mg protein/ml). If less lysate is required make up the volume with the SLBG buffer. Mix thoroughly.

3. Add 10 μl of micrococcal nuclease, mix gently, then incubate at 20°C for 10 min.
4. Stop the micrococcal nuclease activity by the addition of 5 μl of 62.5 mM EGTA, pH 7.3 and mixing gently. Then add the required radiolabelled amino acid in a final volume of 25 μl; for example we use 50 μCi [^{35}S]methionine (>1000 Ci/mmol) for 12 reactions. Keep on ice at all times.
5. (a) Prepare individual 25 μl translation reactions consisting of
   - 20 μl micrococcal nuclease-treated lysate
   - 0–5 μl mRNA sample (in H_2O)
   - 0–5 μl DEPC-treated H_2O

   (b) Incubate at 20°C for 1 h. Although not usually necessary, an RNase inhibitor can be added instead of the water and before the mRNA is added.

The yeast cell-free translation system can be used to translate a wide variety of natural and synthetic mRNAs (27, 28). For each mRNA template, not only must the optimal K^+ and Mg^{2+} ion concentrations for translation be established, but also the optimal concentration of the RNA itself. For synthetic templates such as poly(U), where no AUG codon is encoded by the mRNA, supraoptimal Mg^{2+} ion concentrations must be used (i.e. 12–15 mM versus 2–3 mM for natural mRNAs) to 'force' ribosomes on to the polynucleotide.

One unusual aspect of cell-free translation with a yeast lysate is the optimal temperature for translation. Most eukaryotic cell-free systems are at their most efficient in the temperature range 30–37°C yet, while *S. cerevisiae* grows optimally in the temperature range 28–30°C, the cell-free system fails to translate natural mRNAs at temperatures above 20°C (6, 28). This temperature-dependent inhibition appears to act at the level of initiation rather than elongation since translation of poly(U) occurs equally well over the temperature range 20–37°C (10, 29). Although originally believed to be due to a temperature-activated nuclease (29), recent data suggest that it may be due to heat lability of a component of the translation initiation apparatus (30).

Under the reaction conditions described above incorporation of amino acids usually occurs linearly for a period of 60–90 min, suggesting that multiple rounds of translation on a single mRNA species may occur. Exogenous mRNAs do not appear to be rapidly turned over in the yeast cell-free system unless they lack a poly(A)$^+$ tail and are not 5′ capped (13).

The efficiency of *in vitro* translation can be quantified by determining the level of incorporation of the radiolabelled amino acid into protein in response to the exogenous mRNA. This is achieved by using hot trichloroacetic acid (TCA) to precipitate the proteins in the reaction (see *Protocol 7*). For high specific activity radiolabelled amino acids e.g. [^{35}S]methionine (≥1000 Ci/mmol) usually only 2–5 μl of the 25 μl reaction mix need be counted. For low specific activity amino acids e.g [^{3}H]leucine (≥50 Ci/mmol) usually the whole 25 μl reaction mix should be used. For accurate quantification, translation reactions should be run in duplicate.

Protocol 7. Assaying the efficiency of *in vitro* translation using radiolabelled amino acids

Equipment and reagents

- 5% (w/v) trichloroacetic acid (TCA; stored at +4°C) supplemented with 10 mM of the unlabelled amino acid whose radiolabelled counterpart is being used
- waterbath at 85°C
- glass fibre filters and appropriate filtration apparatus
- scintillation counter

Method

1. After completion of the *in vitro* translation assay (see *Protocol 6*) terminate the reaction by adding 2 ml of cold 5% (w/v) TCA supplemented with 10 mM of the cold (i.e. unlabelled) amino acid whose radiolabelled counterpart is being used in the assay.
2. Put samples on ice for 10 min, then transfer to a waterbath at 85°C for a further 10 min. Replace the samples on ice for a further 10 min.
3. Collect the hot TCA-precipitate by filtration on to glass fibre filters and wash each filter three times with 5 ml of cold 5% (w/v) TCA. Label the filters in pencil *before* filtration.
4. Dry the filters either under an infrared lamp or in a warm oven (50–60°C) and then count in a scintillation counter using an appropriate scintillant.

To prepare *in vitro* translation samples for analysis by SDS–PAGE and autoradiography, the translation reactions should be first incubated with RNase A (final concentration 500 μg/ml) at 37°C for 15 min. The proteins should then be precipitated with 1 ml of ice-cold 80% (v/v) acetone at 4°C for 1–2 h. The resulting precipitate should be collected by centrifugation and resuspended in an appropriate sample buffer prior to SDS–PAGE.

5. *In vitro* translocation systems

The unravelling of the mechanism whereby nascent secretory proteins are translocated from their site of synthesis in the cytoplasm to the lumen of the endoplasmic reticulum (ER) has been greatly facilitated by our ability to reconstitute the process *in vitro*. Such *in vitro* translocation systems require an active cell-free translation system, microsomal membranes, and a defined mRNA template encoding a secretory protein with an amino-terminal signal sequence essential for translocation. Studies with mammalian *in vitro* translocation systems (largely based on the rabbit reticulocyte lysate supplemented with dog pancreatic microsomal membranes (2, 3, 31) have demonstrated that translocation occurs co-translationally i.e. translation must occur in the presence of the microsomal membranes. *In vitro* translocation systems derived entirely from yeast components were first described in 1986 (32–34) and immediately demonstrated that in *S. cerevisiae* not all secretory proteins are translocated co-translationally. For example, pre-pro-α-factor (the product of the *MFα1* gene) can be translocated post-translationally i.e. after *in vitro* translation has been inhibited with cycloheximide (34). Some yeast proteins e.g. pre-invertase (the product of the *SUC2* gene) are, however, translocated co-translationally.

The yeast cell-free translation system described above can be used for *in vitro* translocation assays providing they are supplemented with microsomal membrane fractions. Such microsomal membrane fractions can be generated relatively easily by gently lysing sphaeroplasts and subjecting the lysate to differential centrifugation (see *Protocol 8*). The required mRNA can be generated by *in vitro* transcription (see *Protocol 5* above). This is usually pre-pro-α-factor for studying post-translational translocation processes and pre-invertase for studying co-translational translocation processes.

Protocol 8. Preparation of a microsomal membrane fraction from *S. cerevisiae* that is suitable for *in vitro* translocation studies

Equipment and reagents

- sphaeroplasts prepared from 4 litres of YEPD-grown cells (see *Protocol 1*)
- buffer A (20 mM Hepes–KOH pH 7.5, 500 mM sucrose, 1 mM dithiothreitol, 3 mM $Mg(OAc)_2$, 1 mM EGTA, 1 mM EDTA, 100 units/ml aprotinin, 0.5 mM PMSF, 2 μg/ml each of pepstatin A, chymostatin, antipain, and leupeptin
- 40 ml Dounce glass–glass homogenizer (Wheaton)
- Beckman JS-13 swing-out rotor or equivalent
- 30% (w/v) Percoll in buffer A
- 1 M $CaCl_2$
- micrococcal nuclease (30 000 units/ml; Worthington Biochemicals)
- 1 M EGTA
- buffer B (20 mM Hepes–KOH, pH 7.5, 250 mM sucrose, 50 mM EDTA, 1 mM dithiothreitol)
- Beckman Ti70 fixed angle rotor (or equivalent)
- buffer C (20 mM Hepes–KOH, pH 7.5, 250 mM sucrose, 1 mM dithiothreitol)

Methods

The following protocol is essentially that described by Garcia *et al.* (35).

Protocol 8. *Continued*

1. Prepare sphaeroplasts from 4 litres of YEPD-grown cells (see *Protocol 1*, steps **2–5**). Resuspend the sphaeroplasts in 15 ml of buffer A and lyse in a 40 ml glass Dounce homogenizer (see *Protocol 1*, step **7**).
2. Centrifuge the homogenate in a swing-out rotor (e.g. a Beckman JS-13 rotor) at 8000 r.p.m. for 10 min at 4°C. Transfer the resulting supernatant to a fresh tube and add 15 ml of the buffer A to the remaining pellet. Resuspend the pellet, homogenize as before, and repeat the centrifugation step.
3. Pool the supernatants and repeat the centrifugation step (see step **2** above).
4. Using a 24 ml centrifuge tube add 18 ml of 30% (w/v) Percoll in lysis buffer and overlay with 5 ml of the supernatant. Centrifuge at 76 000 *g* for 1 h at 4°C in a swing-out rotor. This will generate two turbid bands with the upper of the two bands containing the microsomal membranes. Collect the upper band from five or six separate Percoll gradients and pool them.
5. Add $CaCl_2$ and micrococcal nuclease to the pooled fractions giving final concentrations of 1 mM and 300 units/ml respectively. Incubate at 20°C for 15–20 min, then add EGTA to a final concentration of 2 mM.
6. Place the pooled membrane fraction on ice, add 1 vol. (20–25 ml) of ice-cold buffer B, and leave on ice for 15 min.
7. Centrifuge the sample at 43 000 r.p.m. at 4°C in a Beckman Ti70 rotor. The microsomal membrane fraction forms a band just above a transparent Percoll pellet. Carefully remove the band using a Pasteur pipette and resuspend in 0.5 ml of buffer C. Use a homogenizer to assist resuspension.
8. Dilute the microsomal membrane fraction to give 25 A_{280} units/ml using buffer C. The membranes are now ready for addition to an *in vitro* translation reaction or can be stored at −80°C until required.

The microsomal membrane fraction can be added directly to a cell-free translation extract prior to the addition of the mRNA (see *Protocol 6*); it should represent approximately 20% of the final reaction and some modification of the buffers etc. may be necessary to optimize translational efficiency of the mRNA. If one is interested in post-translational translocation of pre-pro-α-factor then the microsomal membrane fraction should be added to a cell-free system that has been previously incubated for 1 h at 20°C and then inhibited with cycloheximide (25 μg/ml final concentration), a potent translation elongation inhibitor. The efficiency of translocation can be assessed by subjecting the [^{35}S]methionine-labelled *in vitro* translation products to SDS–PAGE and autoradiography; translocation across the microsomal membrane results in apparent changes in the electrophoretic mobility of the translation product as a consequence of post-translational modification e.g. signal peptide cleavage. In addition, the translocated proteins should be protease-insensitive since they will be retained within the ER membrane-bound vesicles. Thus the addition of proteinase K or trypsin to final concentrations of 0.1 mg/ml in the translation reaction should not alter the electrophoretic mobility of the polypeptides if they have been successfully translocated.

6. Alternatives to *in vitro* translation: transient expression systems

Cell-free translation extracts are difficult to prepare and yeast extracts are no exception in this respect. Recently, alternative means of getting an mRNA template translated by yeast ribosomes have been developed, namely by 'RNA delivery' into sphaeroplasts (36, 37). For delivering mRNAs into sphaeroplasts either of two strategies can be used: one is essentially the same as that used for DNA-mediated transformation (36, and see Chapter 8) while the other utilizes electroporation (37). There has been no report of mRNA delivery into whole cells. In both delivery systems, expression—i.e. translation of the delivered mRNA—peaks after 2–3 h but studies to date have been restricted to using mRNAs encoding enzymes that can be easily and sensitively assayed in yeast, namely luciferase and β-glucuronidase. Such systems do, however, show a dependence on both a 5′ cap structure and a poly(A)$^+$ tail for optimal translational efficiency (36, 37).

Acknowledgements

Studies in the authors' laboratory developing and utilizing the technology described in this article have been generously supported by the following funding agencies: EEC, SERC, and Glaxo Group Research.

References

1. Pelham, H. R. B. and Jackson, R. J. (1976). *Eur. J. Biochem.*, **67**, 247.
2. Walter, P. and Blobel, G. (1983). In *Methods in enzymology*, Vol. 96 (ed. S. Fleischer and B. Fleischer), pp. 84–93. Academic Press, London.
3. Clemens, M. J. (1984). In *Transcription and translation. A practical approach* (ed. B. D. Hames and S. J. Higgins), pp. 231–70. IRL Press, Oxford.
4. Ryabova, L. A., Ortlepp, S. A., and Baranov, V. I. (1989). *Nucleic Acids Res.*, **17**, 4412.

5. Krieg, P. A. and Melton, D. A. (1987). In *Methods in enzymology*, Vol. 155 (ed. R. Wu), pp. 397–415. Academic Press, London.
6. Gasior, E., Herrera, F., Sadnik, I., McLaughlin, C. S., and Moldave, K. (1979). *J. Biol. Chem.*, **254**, 3965.
7. Szczesna, E. and Filipowicz, W. (1980). *Biochem. Biophys. Res. Commun.*, **92**, 563.
8. Hofbauer, R., Fessl, F., Hamilton, B., and Ruis, H. (1982). *Eur. J. Biochem.*, **122**, 199.
9. Hussain, I. and Leibowitz, M. J. (1986). *Gene*, **46**, 13.
10. Hartley, A. D. (1992). Ph.D. Thesis, University of Kent at Canterbury.
11. Achstetter, T., Emter, O., Ehmann, C., and Wolf, D. H. (1984). *J. Biol. Chem.*, **259**, 13334.
12. McLaughlin, C. S., Warner, J. R., Edmonds, M., Nakazoto, H., and Vaughan, M. H. (1973). *J. Biol. Chem.*, **248**, 1466.
13. Gerstel, B., Tuite, M. F., and McCarthy, J. E. G. (1992). *Mol. Microbiol.*, **6**, 2339.
14. Tuite, M. F. and McLaughlin, C. S. (1984). *Biochim. Biophys. Acta*, **783**, 166.
15. Eustice, D. C., Waken, L. P., Wilhelm, J. M., and Sherman, F. (1986). *J. Mol. Biol.*, **188**, 207.
16. Grunberg-Manago, M. (1963). *Prog. Nucleic Acid Res.*, **1**, 93.
17. Köhrer, K. and Domdey, H. (1991). In *Methods in enzymology*, Vol. 194 (ed. C. Guthrie and G. R. Fink), pp. 398–405. Academic Press, London.
18. Gallis, B. M., McDonnell, J. P., Hopper, J. E., and Young, E. T. (1975). *Biochemistry*, **14**, 1038.
19. Woolford, J. L. (1989). *Yeast*, **5**, 439.
20. Jobling, S. A., Cuthbert, C. M., Rogers, S. G., Fraley, R. T., and Gehrke, L. (1988). *Nucleic Acids Res.*, **16**, 4483.
21. Tabor, S. and Richardson, C. C. (1985). *Proc. Natl Acad. Sci. USA*, **82**, 1074.
22. Baim, S. B. and Sherman, F. (1988). *Mol. Cell. Biol.*, **8**, 1591.
23. van den Heuvel, J. J., Bergkamp, R. J. M., Planta, R. J., and Raué, H. A. (1989). *Gene*, **79**, 83.
24. Vega Laso, M. R., Zhu, D., Sagliocco, F., Brown, A. J. P., Tuite, M. F., and McCarthy, J. E. G. (1993). *J. Biol. Chem.*, **268**, 6453.
25. Nielsen, D. A. and Shapiro, D. J. (1986). *Nucleic Acids Res.*, **14**, 5936.
26. Sambrook, J., Fritsch, E. F., and Maniatis, T. (1989). *Molecular cloning. A laboratory manual* (2nd edn). Cold Spring Harbor Laboratory Press, Cold Spring Harbor, NY.
27. Tuite, M. F., Plesset, J., Moldave, K., and McLaughlin, C. S. (1980). *J. Biol. Chem.*, **255**, 8761.
28. Tuite, M. F. and Plesset, J. (1986). *Yeast*, **2**, 35.
29. Herrera, F., Gasior, E., McLaughlin, C. S., and Moldave, K. (1979). *Biochem. Biophys. Res. Commun.*, **88**, 1263.
30. Mandel, T. and Trachsel, H. (1989). *Biochim. Biophys. Acta*, **1007**, 80.
31. Blobel, G. and Dobberstein, B. (1975). *J. Cell Biol.*, **67**, 852.
32. Hansen, W., Garcia, P. D., and Walter, P. (1986). *Cell*, **45**, 397.
33. Waters, G. and Blobel, G. (1986). *J. Cell Biol.*, **102**, 1543.
34. Rothblatt, J. A. and Meyer, D. I. (1986). *EMBO J.*, **5**, 1031.
35. Garcia, P. D., Hansen, W., and Walter, P. (1991). In *Methods in enzymol-*

ogy, Vol. 194 (ed. C. Guthrie and G. R. Fink), pp. 675–82. Academic Press, London.
36. Russell, P. J., Hambridge, S. J., and Kirkegaard, K. (1991). *Nucleic Acids Res.*, **19**, 4949.
37. Everett, J. G. and Gallie, D. R. (1992). *Yeast*, **8**, 1007.

13

Virus-like particles: Ty retrotransposons

ALAN J. KINGSMAN, SALLY E. ADAMS, NIGEL R. BURNS, ENCARNA MARTIN-RENDON, GEMMA MARFANY, DOUGLAS W. HURD, and SUSAN M. KINGSMAN

1. Introduction

The *Saccharomyces cerevisiae* genome contains a number of movable genetic elements known as retrotransposons. In moving from one genomic location to another they use an RNA intermediate and a reverse transcriptase reaction that is reminiscent of stages in the mammalian and avian retrovirus life cycles (1, 2). Like retroviruses, the yeast retrotransposons package their RNA intermediate into particles. These are known as virus-like particles (VLPs). The particles of one of the yeast elements, Ty1, is the subject of this chapter. They have been used to produce fusion proteins, authentic proteins, and novel vaccines. In addition, they are used to study fundamental aspects of retrotransposon biology.

2. The biology of Ty1

Most laboratory strains of *S. cerevisiae* contain 20–25 dispersed copies of Ty1 in their genomes. The majority of these are 5.9 kb in length and comprise a 5.2 kb internal unique region flanked by long terminal repeats (LTRs) of about 350 nucleotides (*Figure 1*). The major transcript is a 5.7 kb species that starts in the left LTR and terminates in the right. This RNA is both message and intermediate in transposition. There are two open reading frames, *TYA* and *TYB*, within this full-length transcriptional unit. *TYA* and *TYB* overlap by 34 nucleotides and *TYB* is in the +1 translational reading phase with respect to *TYA*. *TYA* is expressed by simple translation to produce a 50 kDa precursor protein known as p1. Protein p1 is subsequently cleaved to produce a 45 kDa species called p2. *TYB* is expressed as a *TYA*:*TYB* fusion protein. In about 5% of translation events the ribosome translates *TYA*, enters the *TYA*–*TYB* overlap region, and frameshifts by +1 into the *TYB* phase. It then proceeds to the end of *TYB* to produce a second precursor protein, p3,

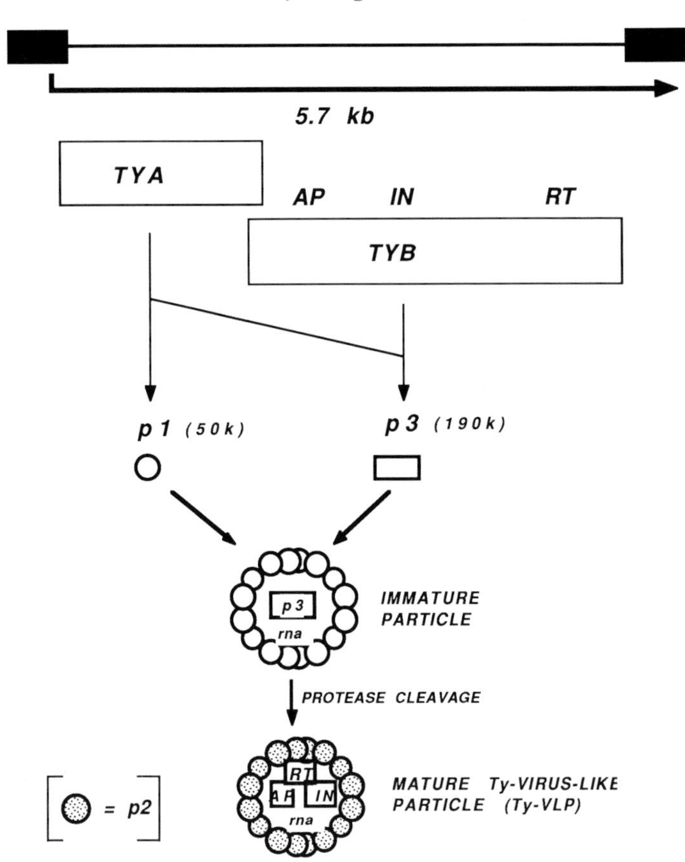

Figure 1. The genetic organization of Ty. For explanation see text. AP, IN, and RT mark positions of homology with retroviral aspartyl proteases, integrases, and reverse transcriptases respectively.

of 190 kDa. Protein p3 comprises protease, integrase, and reverse transcriptase enzymes (1, 2).

The transposition cycle starts with expression of a resident Ty1 element. The p1 and p3 proteins assemble into an immature particle by self-assembly of the p1 proteins and the p1 component of p3. This particle also packages the full-length 5.7 kb RNA. The particle then undergoes a maturation step where a protease encoded by *TYB* and sequestered within the p3 protein cleaves the precursor proteins to produce a mature Ty-VLP. This mature particle is then capable of carrying out reverse transcription and integration of the DNA product, so completing the transposition event (3–5). Isolated Ty-VLPs have been shown to carry out reverse transcription reactions (4, 5) and integration reactions *in vitro* (6).

The assembly of Ty-VLPs is a function of the p1 protein (7). In fact, overexpression of the *TYA* gene alone results in the accumulation of very large numbers of Ty-VLPs within a yeast cell (5, 8). This means that none of the *TYB* functions are required for particle formation. This leads to the notion that it might be possible to express *TYA* fusion proteins that would still assemble into Ty-VLPs (9, 10). This is attractive for a number of reasons. First, Ty-VLPs are extremely easy to purify because of their relatively large physical size (50 nm in diameter). Secondly, particulate structures are frequently more immunogenic than non-particulate structures, making Ty-VLPs a possible carrier for antigens for use as vaccines or to generate laboratory reagents. Indeed the idea that particulate antigens are more potent than simple proteins has led many groups to develop particulate antigen presentation systems from viruses such as polio or hepatitis B or to develop synthetic particulate systems such as ISCOMS (immunostimulatory complexes) and liposomes (11).

3. Ty expression vectors

In order to express Ty fusion proteins in yeast, specialized expression vectors have been developed (see *Figure 2*). The *TYA* gene has been trimmed at the 3′ end so that it encodes only the first 381 amino acids of the p1 protein (9). At codon 381 is a *Bam*HI site for the insertion of additional protein-coding sequences, followed by termination codons in all three reading phases and a transcription terminator. Expression of this truncated *TYA* gene (*TYA*(d)) and any fusions are driven from the efficient phosphoglycerate kinase (*PGK*) gene promoter. This expression cassette, containing the *TYA*(d) gene flanked by the PGK promoter and terminator, has been inserted into a 2 μm-based yeast/*Escherichia coli* shuttle vector to produce the Ty-VLP fusion vector pMA5620. This plasmid uses *LEU2* selection in yeast and is maintained at a copy number of 100–200/cell. Plasmids pMA5621 and pMA5622 are derivatives of pMA5620 in which the *Bam*HI expression sites are in the +1 and −1 reading phases with respect to the *Bam*HI site in pMA5620. This series of vectors therefore allows any coding sequence, from any source, to be expressed in-phase with the truncated *TYA* gene.

The use of the *PGK* promoter results in high-level constitutive expression. In some instances, large amounts of a particular product may be toxic to the cells. An alternative expression strategy is therefore to use a regulated promoter. Plasmids pOGS40, pOGS41, and pOGS42 are derivatives of pMA 5620, pMA5621, and pMA5622 respectively, in which the PGK promoter has been replaced with a hybrid *PGK-GAL* promoter that can be induced by galactose.

Many different hybrid Ty-VLPs have now been produced that carry additional proteins that range in size from 3 to 42 kDa. These include components of HIV encoded by the *env* (9), *pol*, *gag* (12), *tat* (13), *rev*, *nef*, and *vif* genes, influenza virus haemagglutinin, human alpha-interferon (10), feline

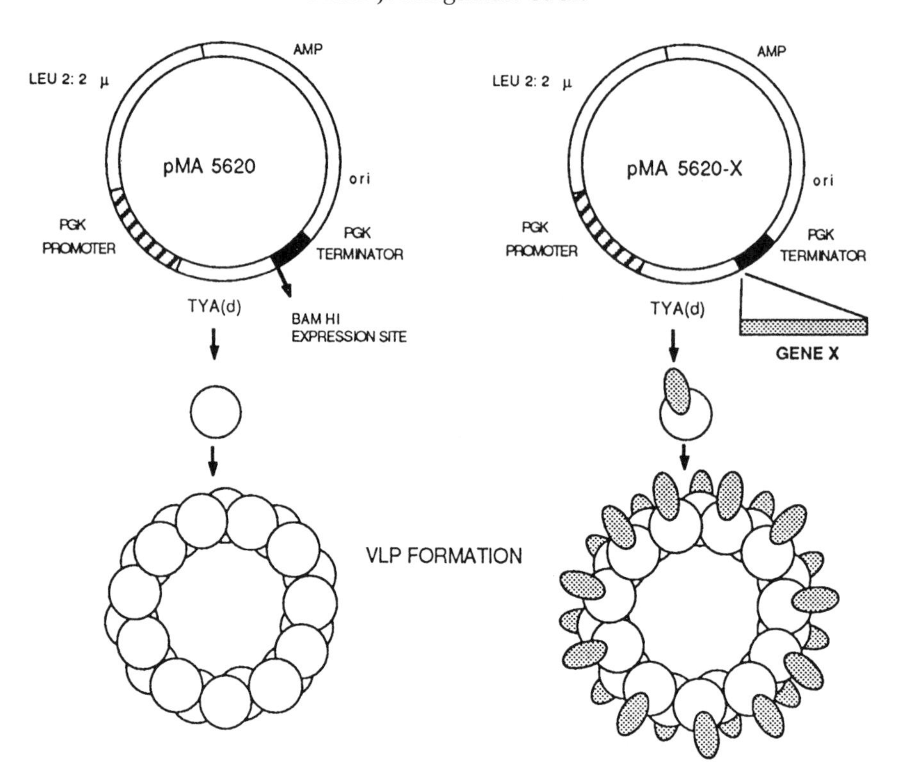

Figure 2. Vectors for the production of Ty-VLPs. The diagram shows one of the Ty-VLP vectors, pMA5620 and a theoretical vector, containing an insertion, pMA5620–X. Below these vectors are diagrams of a non-hybrid and hybrid VLP.

leukaemia virus *env*, and bovine papillomavirus E1 and E2 (A. J. Kingsman *et al.*, unpublished data).

An important feature of the Ty-VLP system is the ease with which pure VLPs can be prepared as a result of their particulate nature. The sedimentation properties of different hybrid Ty-VLPs are similar and this characteristic has been exploited to develop a simple purification process that can be used for any VLP, more or less irrespective of the sequence of the additional protein. The system is therefore extremely versatile, allowing rapid production of a variety of recombinant proteins. Such ease of purification would, for example, make it feasible to survey a genome for regions that encode important antigenic determinants.

Hybrid Ty-VLPs can also be used in many other laboratory applications. For example, they can be used in the production of defined polyclonal and monoclonal antibodies by selecting particular regions of a protein as the added antigen. Conversely, hybrid Ty-VLPs can be used as a rapid primary screen to map monoclonal antibodies raised against non-VLP antigens.

The VLP system can also be manipulated to produce non-particulate pro-

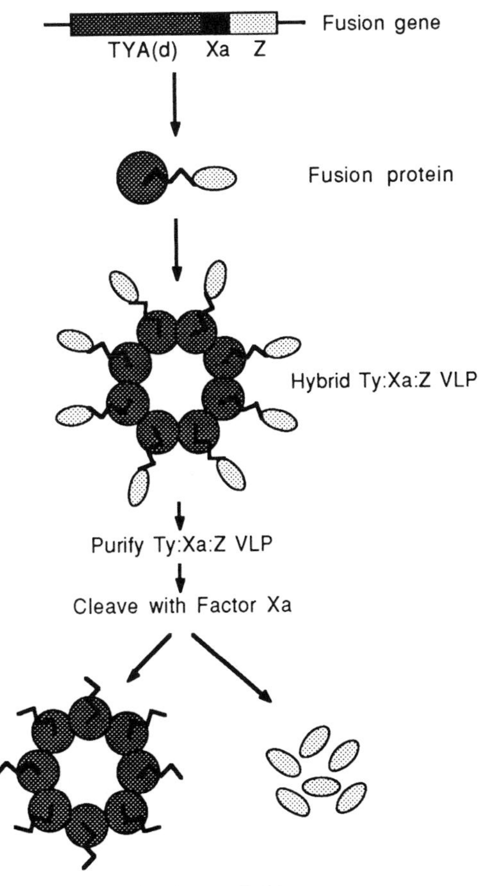

Figure 3. A scheme for cleavable hybrid Ty-VLPs.

teins or protein domains (13, 14). By engineering a protease cleavage site at the carboxy-terminus of the particle-forming protein, p1, hybrid VLPs can be produced with the general structure p1-cleavage site-added protein. Hybrid VLPs have been constructed that contain various proteins downstream of the recognition sequence for factor Xa. Purification of the particles followed by Xa cleavage results in a mixture of three proteins from which the protein of interest can be purified (see *Figure 3*). This technology has been used successfully to purify several HIV proteins, providing sufficient material for structural and functional analyses.

In summary, therefore, Ty-VLPs are isolated for two reasons. The first is to study their biological properties and the second is to use them as a protein preparation system.

4. The purification of Ty-VLPs

The growth of the yeast cultures harbouring the 2 μm-based VLP expression vectors is the same whether one is seeking to use the VLPs for basic studies of Ty or for the production of one's favourite protein. However, growth procedures differ slightly depending on whether one is using a constitutive or inducible promoter.

Protocol 1. Growth of yeast cultures (constitutive promoter)

Reagents

- Synthetic complete glucose (SC-glc) medium (see Appendix 1). The yeast strain used must be a leucine auxotroph for the commonly used VLP vectors and leucine selection is maintained throughout the growth of the cultures
- TEN buffer (see Appendix 1)
- protease inhibitors in TEN buffer (see *Protocol 3*

Method

1. Inoculate 100 ml of SC-glc medium plus appropriate supplements with a 1 ml glycerol stock or overnight culture of the appropriate yeast transformant. Incubate with vigorous shaking in a 500 ml conical flask at 30°C until the cell density reaches 2×10^7 cells/ml.
2. Put 1 litre of SC-glc medium with appropriate supplements into each of two 3 l flasks and inoculate each with 50 ml of preculture. Incubate, as before, overnight until the cell density reaches 5×10^7 cells/ml.
3. Split the total 2 l of culture among 16 individual litres of SC-glc medium, with appropriate supplements, in 3 l flasks. Grow overnight and harvest the cells at a density of 4×10^7/ml by centrifugation at 3500 *g* for 20 min. Wash by resuspending the pellets in 10 ml of water per litre of culture. Transfer to 50 ml tubes (e.g. Falcon) pooling the cells from each 2 l of culture into one 50 ml tube. Centrifuge at 3500 *g* for 5 min. Repeat the washing procedure twice more. Finally wash each pellet with 20 ml of TEN buffer plus protease inhibitors. Remove the supernatant and freeze the cell pellets at −20°C until required.

For small scale preparations the cells can be harvested and washed at stage **2**.

When using a galactose-inducible system the growth protocol is slightly different to accommodate the need to induce with galactose in the absence of glucose. The scheme uses the fact that yeast cells will use glucose before galactose so that a mixed carbon source medium can be used.

Protocol 2. Growth of yeast cultures (inducible promoter)

Reagents

- SC-glc/gal medium (as SC-glc but containing 0.3% (w/v) glucose and 1% (w/v) galactose)

Method

1. After steps **1** and **2** of *Protocol 1*, at stage **3** split the 2 l of culture among 16 individual litres of SC-glc/gal medium and grow for about 24 h. At this time the cells will be at a density of about 4–8 $\times$ 10^7 and will have exhausted the glucose. There will not necessarily be significant growth in the last 10 h.
2. For a smaller scale preparation, inoculate 50 ml of the preculture into 1 litre of SC-glc medium at stage **2** (see *Protocol 1*) and 150 ml from this culture into each of two 1 litre batches of SC-glc/gal medium at stage **3**.

The purification procedure that is used for VLPs depends on the final use of the particles. If the intention is to use the particles either to analyse Ty proteins or Ty enzyme activities using exogenous substrates or to prepare either cleavable or non-cleaved hybrids then the high purity protocol (*Protocol 3*) should be used. If, however, it is important to preserve the endogenous RNA for a reverse transcription reaction using that endogenous template then the fast protocol (*Protocol 4*) should be used.

The high purity Ty-VLP preparation is designed for a 16 l final culture as, in general, relatively large amounts of hybrid VLPs are required. Yields are in the range of 20–50 mg/l and the VLPs are more than 90% pure. For smaller scale preparations, the procedure can be scaled down proportionately.

Protocol 3. High purity Ty-VLP preparation

Equipment and reagents

- acid-washed glass beads: glass beads, 40 mesh, can be obtained from BDH Ltd. Wash the beads in concentrated sulphuric acid, rinse 20 times in distilled water, and then dry and bake them at 150°C for 2 h.
- protease inhibitors: make separate solutions (25 mg/ml) of chymostatin, antipain, leupeptin, and pepstatin A in dimethyl sulphoxide and store in 25 μl aliquots. Make a 25 mg/ml solution of aprotinin in water and store in the same way. Then add 25 μl of each of the above solutions to 1 litre of TEN buffer with 1 ml of fresh 5 mM phenylmethylsulphonyl fluoride dissolved in ethanol. All the protease inhibitors can be obtained from Sigma
- 60% (w/v) sucrose in TEN buffer
- 5–20% linear sucrose gradients: prepare a 12.5% stock sucrose solution in TEN buffer and autoclave. Add 30 ml to Beckman SW28 tubes (or similar) and freeze at −20°C. The day before the gradients are needed, remove them and thaw slowly for about 16 h at 4°C. Add a 2 ml 60% sucrose cushion to the bottom of the tubes with a Pasteur pipette before loading.
- Sephacryl S1000 superfine (LKB): equilibrate in TEN before use
- ultrafiltration units: Millipore immersible CX-30 (30 kDa cut-off)

Protocol 3. *Continued*

Method

All procedures are performed at 4°C using prechilled buffers.

1. Thaw cells in eight 50 ml tubes. Add 4 ml of TEN buffer containing the protease inhibitors to each tube and resuspend the cells. Add 5 ml of acid-washed glass beads to each tube.
2. Vortex the tubes in ten 30 second bursts with 30 sec periods of cooling on ice between each burst. This is a critical step as far as yield is concerned. It is essential to achieve efficient breakage of the yeast cells with the beads. When vortexing ensure that the mixture is forced as far up the tube as possible and monitor cell breakage by phase-contrast microscopy. Damaged cells are phase-dark and intact cells are phase-bright. A bead beater can be used as long as it is calibrated. After breakage centrifuge the suspension at 2000 *g* for 5 min. Remove the supernatant and keep on ice.
3. Add 4 ml of fresh TEN buffer to the pellet and repeat stage **2**. Retain the supernatant on ice.
4. Add 3 ml of fresh TEN buffer and repeat stage **2**. Pool all the supernatants and centrifuge at 13 000 *g* (Sorvall HB4 rotor or equivalent) for 20 min. This clears any remaining debris.
5. Centrifuge the cleared supernatant in a Beckman SW40 at 30 000 r.p.m. for 1 h or SW28 at 28 000 r.p.m. for 1.5 h on to a 2 ml 60% sucrose cushion in TEN buffer. The sucrose cushion prevents the VLPs pelletting which can result in a decreased yield. Collect the cushions with a Pasteur pipette and dialyse overnight against 1 litre of TEN buffer with protease inhibitors. During dialysis a precipitate may form. This does not contain the VLPs and can be removed by centrifugation at 13 000 *g* for 20 min.
6. Centrifuge the dialysate at 13 000 *g* for 20 min and retain the supernatant. Add 2 ml of 60% sucrose in TEN buffer to the bottom of each of six 5–20% sucrose gradients in Beckman SW28 tubes with a Pasteur pipette. Load each gradient with up to 2 ml of the supernatant and centrifuge at 53 000 *g* (25 000 r.p.m. for SW28) for 6 h.
7. Following centrifugation, remove the sucrose cushions containing the VLPs using a Pasteur pipette. Dialyse against 1 litre of TEN buffer for 36 h with five changes of buffer.
8. Centrifuge the dialysate at 13 000 *g* for 20 min and then filter the supernatant through a 0.45 μm filter. Concentrate the filtrate to 16 ml using a Millipore CX-30 filtration unit. Load the sample on to a Sephacryl S1000 superfine 5 cm × 1000 cm column and develop at a rate of 16 ml/h with TEN buffer.

9. Fractions containing VLPs are identified by running aliquots on 10% SDS–polyacrylamide gels. Pool the VLP fractions and concentrate if necessary.

If hybrid Ty-VLPs are used as a means of producing a non-particulate protein via a protease cleavage reaction, such as a factor Xa reaction, then the cleavage reaction is carried out according to standard conditions for the cleaving enzymes (e.g. ref. 15 for factor Xa). The carrier particles can then be removed by centrifugation at 40 000 r.p.m. for 60 min in a Beckman SW41 rotor. It only then remains for the protein of interest to be separated from the cleaving protease.

The fast Ty-VLP preparation protocol is a modification of that developed by Mellor *et al.* (5). It has the advantages of speed and simplicity and can yield very pure samples (>90%). Its disadvantage is that it is somewhat variable in yield. However, this method is essential if the integrity of packaged RNA is required. Growth of cultures is as described in *Protocols 1* and *2* except that the 1 litre cultures are harvested at about 10^7 cells/ml.

Protocol 4. Fast Ty-VLP preparation

Equipment and reagents

- 15–45% sucrose gradients: prepare in Beckman SW28 tubes using autoclaved 15% and 45% sucrose solutions in a gradient maker. The volume of the final gradient should be about 32 ml.
- TEN buffer
- acid-washed and baked glass beads (see *Protocol 3*)
- sonicator

Method

1. Transfer culture to 300 ml centrifuge tubes and chill on ice for 5 min. Pellet cells at 5000 r.p.m. in a Beckman JA-2 rotor for 20 min at 4°C.
2. Wash cells in 10 ml of ice-cold water and resuspend in 4.0 ml TEN buffer. Transfer to 30 ml Corex tubes.
3. Disrupt the cells by vortexing with acid-washed and baked glass beads until more than 70% are broken. Glass beads should come to the shoulder of the Corex tube and vortexing should be carried out in 30 sec bursts followed by 30 sec on ice.
4. Complete the disruption by sonicating the cells in ten 15 sec bursts on ice (6–7 μm peak to peak).
5. Remove debris by centrifugation at 40 000 *g* for 1 h at 4°C.
6. Layer the supernatant on to the 15–45% sucrose gradient in a Beckman SW28 tube and centrifuge at 76 300 *g* for 3 h at 4°C.

Protocol 4. *Continued*

7. Analyse fractions by SDS–PAGE and pool fractions showing the Ty proteins p1 and p2. Fractions can also be analysed by Northern blotting if they are first phenol extracted. If Northern blots are used the 5.7 kb full-length Ty RNA is seen.

5. Reverse transcription and integration *in vitro*

To date Ty-VLPs have been shown to carry out two types of reaction *in vitro*. The first is a reverse transcription reaction (4, 5) using either the endogenous Ty 5.7 kb RNA template and tRNA primer or exogenous template and primers. The second is an integration reaction, again using either endogenous or exogenous substrates (6, 16).

The reverse transcriptase assay is a derivative of that described by Goff *et al.* (17). Some characteristics of this Ty enzyme are shown in *Table 1*. The reaction using the endogenous template/primer produces a range of products of up to about 200 nucleotides presumably representing premature termination events prior to the 'strong stop' position at the 5′ end of the template RNA. A full-length product can be produced only if an exogenous oligonucleotide primer is added (4). Any exogenous template/primer can be used although the most rigorous way of ensuring that one is assaying reverse transcriptase rather than a DNA polymerase is provided by using poly(2′-*O*-methylcytidylate)-oligo(dG) as this is not a substrate for the latter.

Protocol 5. Reverse transcriptase assay

Typically a reaction is carried out in 30 μl containing 10 μl of a sucrose gradient fraction from *Protocol 4*.

Methods

1. Prepare the 30 μl reaction mix to contain, at final concentration:
   - 50 mM Tris–HCl pH 8.3
   - 20 mM dithiothreitol
   - 6 mM $MnCl_2$
   - 2 mM $MgCl_2$
   - 60 mM NaCl
   - 0.1 mM each of dATP, dCTP, and dGTP
   - 0.06 U/ml RNasin (Promega)
   - 0.05% NP40
   - 60 μCi [^{32}P]TTP (400 Ci/mMol) (Amersham)

2. Incubate for 2 h at 25°C and then transfer 15 μl of each reaction to a DE81 filter (Whatman).
3. Wash the filter three times in 5% Na_2HPO_4 for 15 min, once in water, and twice in ethanol. Allow to dry and count in a scintillation counter.
4. Products of the reaction can be recovered by extracting twice with phenol/chloroform (1:1 (v/v)) equilibrated to TEN buffer and containing 0.1% SDS and then precipitating with ethanol.

Table 1. Properties of Ty reverse transcriptase

Variation on standard reaction	**Activity (%)**
None	100
$-Mn^{2+}$ $-Mg^{2+}$	2
− NP40	50
+ Actinomycin D	100
+ RNase	24
10°C	77
30°C	85
37°C	64
42°C	46

Most *in vitro* transposition assays using Ty-VLPs have used a system designed by Brown *et al.* (18) for retroviral integration assays. Briefly, the Ty element is marked with an *E. coli supF* gene and the whole element is packaged into VLPs *in vivo* (6, 16). The VLPs are isolated using the fast procedure (*Protocol 4*) and then added to a transposition reaction mix that contains λgt.*WES* genomes. λgt.*WES* contains amber mutations in the *W*, *E*, and *S* genes that are suppressible by *supF*. This phage can only grow on a non-suppressing *E. coli* host if it acquires a *supF* gene via Ty transposition. Transposition is readily scored, therefore, by counting plaques on a lawn of a non-suppressing host. In order to increase the overall sensitivity of the system the lambda genomes are packaged into phage particles after the reaction, using standard procedures. In addition to this endogenous transposition assay where the genome is incorporated into the VLPs *in vivo*, exogenous substrates can be added to the VLPs. These exogenous substrates are double-stranded DNA molecules having, at least, the terminal 12 nucleotides of the LTR (delta sequences). In these experiments a *supF* gene is included within a fragment of about 1.0 kb if a λgt.*WES* assay is used. Alternatively a drug resistance marker could be used for insertion into bacterial plasmids. The double-stranded DNA fragments used as integration substrates are produced conveniently using the unusual properties of the restriction enzyme *Fok*I

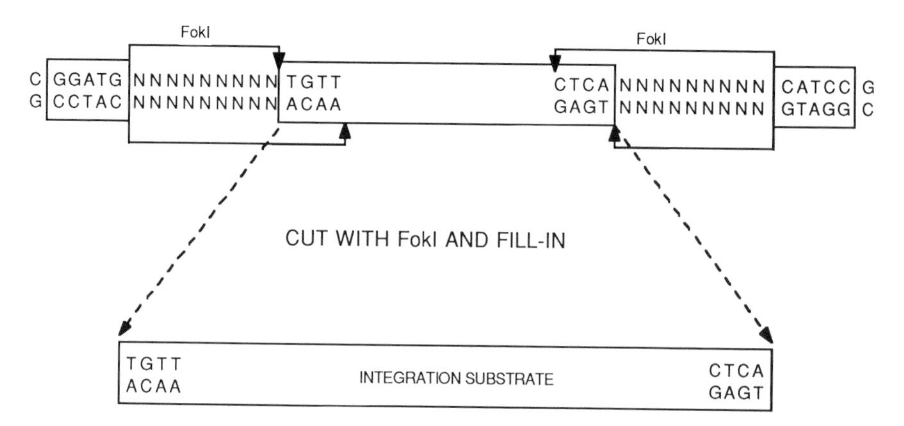

Figure 4. The use of *Fok*I to generate authentic ends of integration substrates.

which recognizes the sequence GGATG but produces a staggered cut 9 and 13 nucleotides away (see *Figure 4*). This means that a cut can be made to produce a fragment with ends unconstrained by the restriction enzyme recognition site sequence.

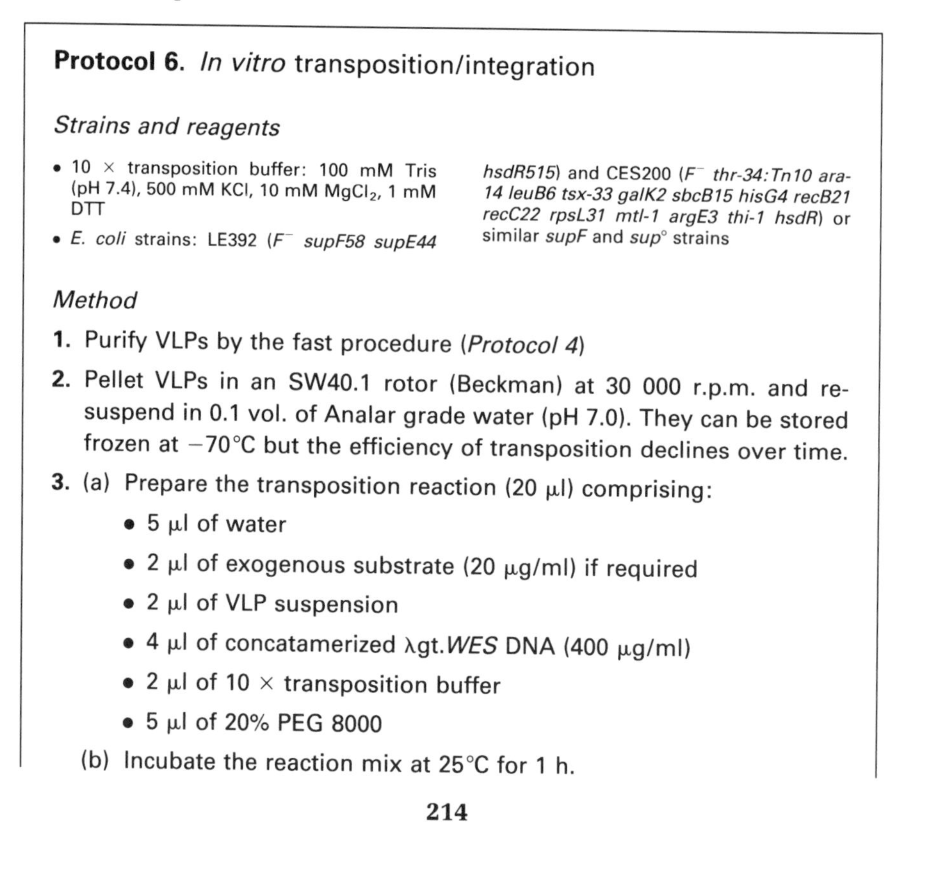

Protocol 6. *In vitro* transposition/integration

Strains and reagents

- 10 × transposition buffer: 100 mM Tris (pH 7.4), 500 mM KCl, 10 mM $MgCl_2$, 1 mM DTT
- *E. coli* strains: LE392 (*F⁻ supF58 supE44 hsdR515*) and CES200 (*F⁻ thr-34:Tn10 ara-14 leuB6 tsx-33 galK2 sbcB15 hisG4 recB21 recC22 rpsL31 mtl-1 argE3 thi-1 hsdR*) or similar *supF* and *sup°* strains

Method

1. Purify VLPs by the fast procedure (*Protocol 4*)
2. Pellet VLPs in an SW40.1 rotor (Beckman) at 30 000 r.p.m. and resuspend in 0.1 vol. of Analar grade water (pH 7.0). They can be stored frozen at −70°C but the efficiency of transposition declines over time.
3. (a) Prepare the transposition reaction (20 μl) comprising:
   - 5 μl of water
   - 2 μl of exogenous substrate (20 μg/ml) if required
   - 2 μl of VLP suspension
   - 4 μl of concatamerized λgt.*WES* DNA (400 μg/ml)
   - 2 μl of 10 × transposition buffer
   - 5 μl of 20% PEG 8000

   (b) Incubate the reaction mix at 25°C for 1 h.

4. Stop the reaction by adding 130 μl of TE buffer (pH 7.4; see Appendix 1), 5 μl of 0.25 M EDTA containing 1 mg/ml of proteinase K, and 4 μl of 10% SDS and incubating at 30°C for 5 h.
5. Extract with phenol, phenol/chloroform (1:1 (v/v)), and chloroform, ethanol precipitate, and then dissolve in 5 μl of water at room temperature. It is critical that the DNA is in solution. **Do not overdry** the ethanol precipitate.
6. Use this DNA in a standard lambda packaging reaction (19). Kits can be obtained from Stratagene or Promega.
7. Plate phage on bacterial strains such as CES200 (*supO*) and LE392 (*supF*) and count plaques. Calculate the transposition events per reaction according to the equation: no. of events = A/B × C/D × E where A is the number of plaques on CES200, B is the number of plaques on LE392, C is the number of lambda genomes in 1.6 μg of DNA (3.52×10^{10}), D is the fraction of the lambda genome able to accept an insertion (0.27), and E is the correction factor for the relative efficiencies of packaging/plating of λgt.*WES* with and without an insertion. Efficiencies vary but typically a few hundred for the endogenous reaction to 10^3–10^4 for reactions with exogenous substrates.

Acknowledgements

Ty work has been supported by the MRC and AFRC at Oxford University and by the DTI at British Biotechnology Ltd.

References

1. Kingsman, A. J. and Kingsman, S. M. (1988). *Cell*, **53**, 333.
2. Boeke, J. D. and Sandmeyer, S. B. (1991). In *The molecular biology of the yeast Saccharomyces* (ed. J. R. Broach, J. R. Pringle, and E. W. Jones), pp. 193–261. Cold Spring Harbor Laboratory Press, Cold Spring Harbor, NY.
3. Boeke, J. D., Garfinkel, D. J., Styles, C. A., and Fink, G. R. (1985). *Cell*, **40**, 491.
4. Garfinkel, D. J., Boeke, J. D., and Fink, G. R. (1985). *Cell*, **42**, 507.
5. Mellor, J., Malim, M. H., Gull, K., Tuite, M. F., McCready, S., Kingsman, S. M., and Kingsman, A. J. (1985). *Nature*, **318**, 583.
6. Eichinger, D. J. and Boeke, J. D. (1988). *Cell*, **54**, 955.
7. Burns, N. R., Saibil, H. R., White, N. S., Pardon, J. F., Timmins, P. A., Richardson, M., Richards, B. M., Adams, S. E., Kingsman, S. M., and Kingsman, A. J. (1992). *EMBO J.*, **11**, 1155.
8. Adams, S. E., Mellor, J., Gull, K., Tuite, M. F., Kingsman, A. J., and Kingsman, S. M. (1987). *Cell*, **49**, 111.

9. Adams, S. E., Dawson, K. M., Gull, K., Kingsman, S. M., and Kingsman, A. J. (1987). *Nature*, **329**, 68.
10. Malim, M. H., Adams, S. E., Gull, K., Kingsman, S. M., and Kingsman, A. J. (1987). *Nucleic Acids Res.*, **15**, 7571.
11. Kingsman, S. M. and Kingsman, A. J. (1988). *Vaccine*, **6**, 304.
12. Gilmour, J. E. M., Senior, J. M., Burns, N. R., Esnouf, M. P., Gull, K., Kingsman, S. M., Kingsman, A. J., and Adams, S. E. (1989). *AIDS*, **3**, 717.
13. Braddock, M., Chambers, A., Wilson, W., Esnouf, M. P., Kingsman, A. J., and Kingsman, S. M. (1989). *Cell*, **58**, 269.
14. Burns, N. R., Craig, S., Lee, S., Richardson, M., Stenner, N., Adams, S. E., Kingsman, S. M., and Kingsman, A. J. (1990). *J. Mol. Biol.*, **216**, 207.
15. Nagai, K. and Thorenson, H. C. (1984). *Nature*, **309**, 810.
16. Eichinger, D. J. and Boeke, J. D. (1990). *Genes Dev.*, **4**, 324.
17. Goff, S., Traktman, P., and Baltimore, D. (1981). *J. Virol.*, **38**, 239.
18. Brown, P. O., Bowerman, B., Varmus, H. E., and Bishop, J. M. (1987). *Cell*, **49**, 347.
19. Murialdo, H., Fife, W. L., Becker, A., Feiss, M., and Yochem, J. (1981). *J. Mol. Biol.*, **145**, 375.

14

The K1 killer toxin: molecular and genetic applications to secretion and cell surface assembly

JEFFREY L. BROWN, TERRY ROEMER, MARC LUSSIER, ANNE-MARIE SDICU, and HOWARD BUSSEY

1. Introduction

The killer phenomenon is widespread among yeasts, having been demonstrated in at least eight genera (1). The best characterized is the K1 killer system of *Saccharomyces cerevisiae* (2–5). K1 killer strains are immune to the toxin they produce, and secrete a small pore-forming toxin that kills sensitive strains. K1 killer strains maintain a double-stranded RNA (dsRNA) viral genome of two distinct species: M_1, a medium-sized dsRNA (1.8 kb) encoding the killer toxin and immunity precursors, and L-A, a larger dsRNA (4.6 kb) required for the replication and maintenance of M_1. These two dsRNAs are packaged separately with a common capsid protein (encoded by L-A) to form virus-like particles that are transmitted cytoplasmically during vegetative growth and conjugation (6).

The selection for various killer phenotypes has led to the identification of a large number of nuclear genes required for the killer phenomenon. *MAK* genes represent a class of 30 or more genes required for the maintenance of killer; *mak* mutants typically lack M_1 dsRNA but retain L-A. A smaller class of *SKI* genes confer a super killer phenotype through an elevated copy number of both L-A and M_1 dsRNA. Several genes from both *MAK* and *SKI* classes have been characterized and encode proteins involved in replication and plasmid maintenance. *KEX* and *KRE* genes define two additional classes required for killer expression and killer resistance respectively. The *KEX1*-encoded carboxypeptidase and the *KEX2* gene product, a subtlisin-like endoprotease, are both required for processing the killer toxin precursor to its biologically active form. When mutated, *KRE* genes cause sensitive cells to become resistant to killer toxin by affecting the cell wall receptor and preventing toxin binding.

In this chapter, we have outlined some of the applications of the *S.*

cerevisiae K1 killer toxin to the identification and characterization of new genes. References 2–5 give more comprehensive reviews on the virology, synthesis, and action of killer toxin.

2. Killer toxin structure and function

2.1 Killer toxin processing

K1 toxin is a heterodimeric protein secreted into the medium after extensive modifications from a larger precursor form (7). The pre-protoxin is a 316 amino acid (42 kDa) protein possessing an amino-terminal leader sequence, a short pro-region of 18 amino acids, followed by the two toxin subunits α and β which are separated by a central glycosylated γ sequence. Proteolytic modifications begin in the endoplasmic reticulum, where the signal peptide is removed by the signal peptidase, and three asparagine residues in the γ subunit are glycosylated. The product of the *KEX2* gene then excises the α and β subunits from the protoxin during transit through the Golgi, by cleaving dibasic residues bordering the central γ subunit, as well as a proline–arginine junction between the pro-region and the α subunit. Processing by the *KEX1* protein removes a pair of basic residues at the carboxy-terminal end of the α subunit, and the α and β subunits, associated through disulphide bonds, form the mature toxin.

Site-directed mutagenesis has identified three functional domains in the mature toxin protein: one required for receptor binding, another for channel formation, and a third to confer immunity to the toxin. The channel-forming domain resides in the α subunit which possesses two hydrophobic stretches, of 20 and 16 amino acid residues, each capable of spanning a membrane. This subunit also appears necessary for both receptor binding and immunity. Mutagenesis of the hydrophilic β subunit reveals only a receptor-binding defect.

2.2 Cell surface interactions

A two-step model for killer toxin action came from the following observations: K1 toxin can kill sphaeroplasts derived from some resistant mutant strains; and mutant forms of the toxin deficient in receptor-binding retain the ability to kill sphaeroplasts. These findings suggest that the toxin initially binds to a receptor at the cell surface, and, in a subsequent and lethal step, inserts into the plasma membrane. The receptor molecule has been identified as β(1→6) glucan through a combination of biochemical and genetic data (8). K1 toxin can bind to a trypsin-resistant, periodate- and β(1→6)-glucanase-sensitive, Zymolyase-solubilized cell wall fraction, which is either altered in structure or reduced in amount in the *kre* mutants. In addition, β(1→6) glucan is a strong competitive inhibitor of toxin action on sensitive cells. Affinity chromatography provides direct evidence of toxin interactions with β(1→6) glucan as judged by a pH-dependent binding of K1 killer toxin to the β(1→6) glucan, pustulan.

2.3 Lethal action

Studies on the physiological effects of K1 toxin on sensitive yeast cells provided early evidence on its mode of action. Toxin-treated cells were found to leak intracellular ATP and K^+ shortly after toxin addition (9). Subsequent studies showed that after exposure to toxin, H^+/amino acid symport is rapidly inhibited. This causes a reduction in the electrochemical gradient across the plasma membrane, ultimately resulting in acidification of the cytoplasm and K^+ efflux. Together, these findings suggested that the toxin acted by perturbing an energized membrane state. More recent patch clamp analysis has directly demonstrated that highly purified toxin preparations are capable of forming ion conductance channels in both sphaeroplasts and artificial liposomes (10). Channel formation in artificial liposomes implies that the K1 toxin has the capacity to form membrane channels autonomously, but whether additional components are required for channel formation *in vivo* is not known.

2.4 Immunity

The lack of sensitivity of K1 killer strains to the toxin they produce is defined as immunity. Mutagenesis studies have mapped the immunity domain to the α subunit of the toxin, but, while several hypotheses explaining this phenomenon exist, our understanding of the process remains incomplete. Expression of an unprocessed toxin precursor is sufficient to confer the immunity phenotype to normally sensitive strains. The finding that strains expressing the killer virus in either a *kex1* or *kex2* mutant background retain immunity argues that the pre-protoxin need not be proteolytically processed to confer immunity. Proposed models depict the pre-protoxin saturating a membrane receptor while in transit through the secretory pathway, or, more simply, interacting with active toxin to interfere with channel formation (5, 11).

3. *KRE* cell wall mutants

The biology of K1 killer toxin and the amenability of yeast genetics has allowed the isolation of a number of mutants resistant to K1 killer toxin. All of the characterized *kre* mutants appear to confer toxin resistance either by disrupting the normal synthesis of the β(1→6) glucan receptor, or by altering the cross-linking of the polymer at the cell surface. The cloning and functional analysis of a number of these genes have begun to make it possible to elucidate a sequential pathway for β(1→6) glucan biosynthesis in *S. cerevisiae* (12).

KRE5 presently defines the earliest step in the pathway, encoding a large soluble endoplasmic reticulum protein, Kre5p, of 156 kDa. Kre5p is essential for β(1→6) glucan biosynthesis and is presumed to act early, since cells harbouring deletions of *KRE5* fail to synthesize the β(1→6) polymer.

KRE6 and *SKN1* encode two highly homologous membrane proteins (80 and 86 kDa, respectively), predicted to be of type II topology. Mutant strains harbouring deletions of both *KRE6* and *SKN1* possess less than 10% of the wild-type levels of β(1→6) glucan. These proteins may provide the core subunits of a membrane-associated glucan synthase complex in conjunction with several additional cytoplasmic or secretory proteins.

The *KRE11* gene encodes a putative cytoplasmic subunit (63 kDa), which when disrupted leads to a 50% reduction in the level of cell wall β(1→6) glucan. The prediction that Kre11p may act as a regulatory component at the synthase level may be strengthened by the lethality of *kre6 kre11* double mutants. Disruption of the *KRE9* locus causes dramatic morphological alterations and a reduction of the β(1→6) glucan levels to 10% of isogenic wild-type strains, also suggesting a function early in the pathway.

The 30 kDa protein encoded by the *KRE1* gene appears to be involved later in β(1→6) glucan maturation. Kre1p most probably acts through a glycosyl-phosphatidylinositol anchor at the plasma membrane to add linear β(1→6) glucan side-chains on to a core glucan backbone, completing the synthesis of the polymer.

KRE2 is a novel killer-resistant gene: null mutations do not directly affect the synthesis of the β(1→6) glucan polymer, but appear instead to disrupt normal cross-linking of the polymer at the cell surface. *KRE2* (*MNT1*) encodes a 51 kDa α(1→2) mannosyltransferase involved in *O*-glycosylation (13, 14), suggesting that the final toxin receptor *in vivo* is β(1→6) glucan cross-linked to glycoprotein in the cell wall.

Together, these genes outline components in a pathway for β(1→6) glucan assembly in yeast, which may serve as a model for cell surface assembly in other systems.

4. Applications

4.1 Heterologous systems

An extension of killer resistance methodology is in the isolation of functional homologues to existing *KRE* genes from other β(1→6) glucan-containing organisms. Cloning and expression studies of several *Candida albicans* genes in *S. cerevisiae*, for example a *C. albicans* homologue to the *S. cerevisiae* chitin synthase 1 gene, exemplify such an approach (15). Screening of a *C. albicans* genomic library, constructed in an *S. cerevisiae* multicopy vector, for functional complementation in a *kre1* background led to the isolation of the *C. albicans* homologue, *canKRE1* (16). Such an approach with pathogenic fungi should allow the isolation of potential anti-fungal target genes, which would be difficult to clone directly. Presumably other functional homologues and/or suppressors of other *kre* genes could be cloned from a variety of different organisms in a similar fashion, providing comparative structural insights into the conserved domains of related proteins involved in glucan biosynthesis.

The heterologous functional complementation of *kre* mutants, while potentially eliminating the work of generating new mutants and sidestepping the genetic limitations of an organism, is not without obstacles. A first approximation of the $\beta(1\rightarrow6)$ glucan structure in the organism of study is necessary before hypothesizing the existence of functional homologues. For example, the characterization of a *C. albicans* $\beta(1\rightarrow6)$ glucan with a very similar structure to the *S. cerevisiae* polymer provided a basis for cloning the *canKRE1* gene. Genomic libraries from *C. albicans* and *Schizosaccharomyces pombe*, constructed in *S. cerevisiae* multicopy vectors, are available, although genomic libraries from other organisms may have to be constructed by the investigator. Isolating heterologous genes may be limited by variables difficult to control, such as expression levels of the putative homologue and the extent to which the foreign protein substitutes functionally. However, even modest complementation can be detectable phenotypically when using a completely resistant, slowly growing *kre* mutant.

4.2 New toxins

Just as the K1 killer toxin has been used to examine the synthesis and assembly of the $\beta(1\rightarrow6)$ glucan receptor, killer toxins using other cell surface polymers as receptors have also found use. For example, the KT_{28} toxin of *S. cerevisiae* binds to cell surface mannoprotein, and several *mnn* mutants defective in *N*-glycosylation are resistant to KT_{28} toxin (17). An exhaustive set of mutants defective in KT_{28} toxin binding may identify new genes involved in glycosylation. Several *S. cerevisiae* mutants resistant to the heterotrimeric ($\alpha\beta\gamma$) *Kluyveromyces lactis* killer toxin have also been isolated (18). The α subunit of the toxin has local homology to chitinases, suggesting that this toxin may initially bind to cell surface chitin. Thus, genes involved in the synthesis of three of the four major *S. cerevisiae* cell wall polymers have been isolated using killer toxin biology. This approach might be profitably extended to other surface components by examining additional protein toxins secreted by other killer yeasts.

5. Killer toxin methodology

5.1 Isolation of yeast K1 killer toxin

The following procedure takes advantage of the secretory nature of the K1 killer toxin of *S. cerevisiae*. By using strains selected for killer toxin overproduction, it is possible to isolate milligram quantities of crude toxin directly from about 20 l of extracellular culture medium. Cells are grown in minimal synthetic medium containing Halvorson salts (19; also see Appendix 1) at 18°C in the presence of 5% glycerol to help stabilize the heat-labile toxin prior to concentration. The final toxin product represents a 1000-fold concentration, and should be buffered to pH 4.7 and kept on ice.

Protocol 1. Large scale toxin preparation

Equipment and reagents

- diploid killer strain T158C/S14a (ATCC 26427)
- minimal medium (MM) and MM with 5% glycerol (to help stabilize the toxin). Prepare 1.6 l by adding 320 ml of 5 × Halvorson salts, 80 ml glycerol, and 1135 ml dH_2O, autoclave, then aseptically add 64 ml of sterile 50% glucose and 1.6 ml of vitamins
- Amicon ultrafiltration unit with PM 10 membrane
- toxin storage buffer (50 mM sodium acetate (pH 4.7), 15% glycerol)

Method

1. Starting with a fresh culture of diploid strain T158C/S14a (ATCC 26427) grown at 30°C on an MM plate (see Appendix 1), inoculate 10 ml of MM, and grow overnight at 30°C until it reaches the stationary phase.
2. Inoculate three 100 ml flasks of MM with 1 ml of overnight culture (from step **1**), and grow at 30°C until the cultures reach stationary phase.
3. Inoculate 12 flasks, each with 20 ml of the overnight cultures from step **2**. The flasks should each contain 1.6 l of MM plus glycerol to 5%. Grow at 18°C, with shaking at 250 r.p.m. for 36 h, until the cultures reach late log phase. (Growth at 18°C is **essential** for high yields of active toxin.)
4. Place the 12 flasks at 4°C and centrifuge the cultures at 1000 *g* for 10 min at 4°C to remove the cells. From this point on, **always** keep the toxin on ice.
5. Concentrate the 20 l of supernatant using an Amicon ultrafiltration unit with a PM 10 membrane at 4°C, under 20 p.s.i. of N_2 gas. (It normally takes about 36 h to complete the filtration.)
6. Disassemble the ultrafiltration unit and repeatedly wash the toxin from the surface of the membrane, using a total of 20 ml of cold toxin storage buffer, into a Petri dish on ice.
7. Spin the 20 ml of toxin in a refrigerated centrifuge at 2500 *g* for 10 min to remove any remaining cells or debris. The toxin may also be sterilized at this point using a disposable 0.45 μm filter.
8. Aliquot the toxin-containing supernatant into 1.5 ml microcentrifuge tubes on ice and store at −20°C. Thaw slowly on ice when ready for use. The toxin can be repeatedly frozen and thawed without significant loss of activity, provided it is *always* kept on ice.

9. The activity of the toxin should be tested by spotting a dilution series of the preparation on a plate seeded with a suitable sensitive strain, as described below (see *Protocol 3*). Normally 20 μl of toxin will result in a 'kill zone' of approximately 15 mm in diameter.

5.2 Toxin competition assays

Competition assays were useful in identifying β(1→6) glucan as the killer toxin receptor (8, 20). It is possible to inhibit the action of killer toxin in a liquid culture by introducing a competitive inhibitor that sufficiently resembles the cell surface receptor. Competitive activity is measured by plating appropriate dilutions of the cells/toxin/inhibitor mix before and after a set incubation time, and calculating the percent survival over a range of inhibitor concentrations. In this way sugars, oligosaccharides, polysaccharides, cell wall fractions, whole cell walls, and heat-killed whole cells can be tested as potential competitive inhibitors. Only minor modifications to *Protocol 2* would be necessary to adapt it to other toxins from other yeasts.

Protocol 2. Characterization of toxin inhibitors

A. *Toxin competition assay*

1. Grow killer-sensitive yeast strain to early log phase in 1× Halvorson buffered liquid YEPD medium at pH 4.7 (see Appendix 1).
2. Mix inhibitor and toxin to a total volume of 25 μl in 1.5 ml microcentrifuge tubes maintained on ice. Prepare duplicate tubes (labelled T_0 and T_1) for each inhibitor at each concentration tested.
3. Add 175 μl of liquid culture to each tube and mix. Immediately plate appropriate dilutions of T_0 samples on YEPD agar, and incubate the T_1 samples at 18°C with shaking for 1.5 h. T_1 samples are then diluted and plated in a similar fashion. (Killer toxin is heat-labile at room temperature but sufficiently stable at 18°C to carry out the killing reactions.)
4. Incubate plates at 30°C for 2 days and score colonies from T_0 and T_1. Determine percent survival as the ratio of c.f.u. per ml from T_1 versus T_0 (×100).

Note: typically we use a range of 0–0.5 mg/ml carbohydrate and graphically present the data as a dose–response curve showing percent survival on a logarithmic scale versus inhibitor concentration. Toxin sufficient to give approximately 1% survival provides a sensitive working range, so that a good inhibitor like pustulan or luteose will enable more than 50% of the cells to survive at concentrations of approximately 50 μg/ml. Cell wall fractions achieve this level of inhibition at concentrations of approximately 300 μg/ml.

Protocol 2. *Continued*

B. *Toxin-activity removal assay*

1. Incubate a fixed amount of toxin with a range of inhibitor concentrations (cells or insoluble carbohydrate) in 1× Halvorson buffered liquid YEPD medium at pH 4.7 (see Appendix 1).
2. Measure the affinity between the inhibitor and toxin by briefly centrifuging the sample and spotting the unbound toxin remaining in the supernatant on to plates seeded with sensitive cells (see *Protocol 3*).
3. Estimate the amount of killer toxin activity removed after incubation with the inhibitor, by comparing the diameter of the 'kill zone' to an equivalent amount of untreated toxin.

5.3 Seeded plate tests

Yeast strains are assayed for sensitivity to the killer toxin by seeding the strain in question into buffered agar medium, and spotting concentrated killer toxin on to the centre of the plate. These tests have been used extensively in the characterization of the *kre* mutants (13, 21–23), and a related test has been used to assay killer toxin production (see *Protocol 4*). The vital stain, methylene blue, is included in the medium to highlight the 'kill zone', appearing as a blue halo of dead cells when a strain is sensitive to the toxin. A reduced zone size, or confluent lawn of cells, will emerge if a strain is toxin-resistant. (It is important **always** to use fresh cells and the appropriate controls each time strains are assayed, since there is some variation between tests. Variation in toxin sensitivity has also been noted for strains grown on different (YEPD or YNB) media.)

Protocol 3. Seeded plate tests

1. With a sterile toothpick, resuspend approximately 1×10^6 cells from a fresh culture in 100 μl of sterile water.
2. Use this slurry to inoculate 10 ml of 1× Halvorson buffered seeded plate medium at pH 4.7 (same as Halvorson-YEPD or Halvorson-minimal medium, but containing 1% agar and 0.001% methylene blue, and aliquoted into sterile 10 ml tubes), molten and held at 45°C. (Avoid repeated heating and cooling cycles, since the low pH tends to soften the agar.) Mix well.
3. Quickly pour the agar into a 60 mm × 15 mm Petri dish and allow to cool to room temperature (this takes 15–20 min).
4. Spot concentrated toxin (20 μl) on to the surface of the solidified medium and place the plate at 18°C overnight. (A fresh culture of a diploid, killer-secreting strain may be substituted if a cell-free toxin concentrate is not available.)

5. Leave plates at 18°C (most sensitive), or the following morning move to a 30°C incubator for 24 h to hasten the growth of the seed, and measure the zone. Twenty microlitres of 1000 × toxin (from *Protocol 1*) normally produces about a 15 mm kill zone. (Incubating the toxin-treated plates at 30°C directly is *not* recommended as this results in the zone size being much smaller than that in plates left at 18°C, due to inactivation of the toxin.)

5.4 Killer processing

Gene products involved in processing the K1 killer toxin have been identified and analysed by assaying toxin production. This is illustrated by the isolation and characterization of the *KEX1* gene, and such an approach could be applied to other gene products involved in killer toxin production. The gene encoding the *KEX1* carboxypeptidase was cloned functionally, by transforming a yeast genomic bank into a *kex1* mutant strain carrying an M_1 dsRNA, but phenotypically killer minus. By plating transformants on Petri dishes seeded with a toxin-sensitive strain, positive clones with a restored capacity to process an active toxin were identified by the kill zone they produced (24).

Kex1p also provides a model where assaying killer toxin production by zone size comparisons can be a useful technique in the functional characterization of a given protein. Both the lumenal active-site and the cytoplasmic carboxy-terminus targeting signal of the Kex1p membrane protein have been studied *in vivo* by evaluating the effect of specific mutations on the ability to process toxin. For example, if important amino acids are removed or mutated from the catalytic domain of the protein, the mutant Kex1p should show a proper Golgi localization, but would be unable to cleave the killer toxin precursor. This will result in either a reduced kill zone or no zone, depending on the severity of the mutations. The same approach may be applied to the study of a specific targeting signal (25). If targeting information is removed or mutated, the catalytic domain of Kex1p will remain functional, but the mutant protein may now be mislocalized and therefore unable to process the toxin to a biologically active form. The zone size assay thus offers a simple functional test of secretory pathway proteins involved in killer toxin processing. The following procedure may also be usefully extended to the characterization of heterologous proteins involved in toxin processing.

Protocol 4. Killer production assay

1. Transform a strain harbouring a K1 killer toxin dsRNA virus with the plasmids expressing the desired mutant forms of the protein to be analysed.
2. Grow transformants to stationary phase in selective medium.

Protocol 4. *Continued*

3. Harvest cells by centrifugation and wash in distilled H_2O. After pelleting the cells again, resuspend in distilled H_2O to a concentration of 5×10^8 cells/ml.
4. Spot 10 μl of cells on to a 1× Halvorson buffered MM plate at pH 4.7 (see Appendix 1) seeded with a lawn of S6 cells (or suitable toxin-sensitive strain).
5. Incubate plates at 18°C for 24 h and examine zones of toxin killing. (10 μl of wild-type toxin-secreting cells normally produce a kill zone of approximately 12 mm in diameter.)

6. Using killer toxin to identify new genes

6.1 Isolating killer-resistant mutants

Selection for mutations leading to killer toxin resistance has been useful in identifying *KRE* genes implicated in cell wall biosynthesis (21, 22). The following simple procedure allows for the selection of mutations with severe growth impairments, and could easily be adapted for use with other toxins, under a variety of conditions. The importance of generating new mutants in a parental strain that has an isogenic partner of opposite mating type and the appropriate genetic markers for subsequent crosses and cloning efforts *cannot be overemphasized.*

Protocol 5. Selection for killer-resistant mutants

1. From freshly grown streaks of a suitable sensitive strain, take individual colonies with a toothpick, mix with 100 μl of Halvorson-buffered YEPD (pH 4.7) and 25 μl of sterile killer toxin.
2. Spread the slurry on to a YEPD/1.2 M sorbitol agar plate, and incubate at 18°C for 1–3 weeks.
3. Prepare 20–40 plates in this way, using a different colony for each plate to obtain independent spontaneous mutations.
4. Pick several colonies of varying size from each plate for further analysis.
5. If possible, cross the resistant mutants with strains bearing mutations of previously identified genes, and test the diploids for complementation.
6. Backcross the remaining mutants, presumably identifying new complementation groups, with an isogenic parental strain of opposite mating type. Analyse the diploids from these crosses for killer resistance to determine whether the mutations are dominant or recessive, prior to sporulation and tetrad analysis to establish if the resistance phenotype segregates as a single gene.

6.2 Cloning genes by functional complementation of killer-resistant mutants

A number of yeast *KRE* genes have been cloned by functional complementation of killer-resistance alleles using single copy or centromere-based genomic libraries; these include *KRE1, KRE2, KRE5, KRE6, KRE9*, and *KRE11*. The same cloning scheme (*Protocol 6*) has also been successfully extended to the identification of *SKN* genes, which are multicopy suppressors of some *kre* mutants. The strategy is based on first isolating 10 000 to 20 000 independent transformants (several genome equivalents) in the mutant allele background on YNB selective medium, then replica plating on to methylene blue-containing YEPD or YNB agar seeded with a diploid killer strain, to test for those plasmids conferring restored toxin sensitivity.

Protocol 6. Cloning killer-resistance genes

1. Start with a non-reverting, toxin-resistance mutant allele, and perform a test transformation (using standard lithium acetate or electroporation techniques) to calibrate the quantity of both gene-bank DNA and competent cells required to obtain approximately 200 colonies per plate. The addition of sheared, single-stranded salmon sperm DNA (1 μg/μl competent cells) as a carrier may be useful to increase lithium acetate transformation efficiencies (26).
2. Perform a large-scale transformation (50–100 plates), taking care to plate the cells evenly and away from the plate edge for ease of subsequent replica plating.
3. After growing the colonies for 4–5 days at 30°C, prepare the methylene-blue killer plates (see *Protocol 3*). Cool the agar-containing medium to 45°C, add 75 μl per litre of an overnight culture of a diploid killer strain, T158C/S14a (ATCC 26427), and very quickly pour plates. Always use plates on the same day that they are prepared. It is wise to test a couple of seeded plates at 30°C overnight for adequate seed growth.
4. Replica plate the transformants (using glass fibre filters or paper towels to absorb liquid from the freshly poured killer-seeded plates under a Whatman no. 1 filter paper) on to the plates seeded with the diploid, toxin-producing cells. (Whatman paper is more convenient than velvet since large numbers can be used and discarded.)
5. Incubate the blue replica plates at 18°C for 3–5 days; during this time the diploid seed will slowly grow, secreting toxin into the medium. Cells with plasmids containing a wild-type copy of the mutant gene should now have restored killer toxin-sensitivity, which shows up as a

Protocol 6. *Continued*

white colony edged with a blue ring of killed cells. Colonies with non-complementing plasmids typically remain toxin-resistant and white.

Note: as in all cloning by complementation, it is important to demonstrate that the complementation is plasmid-based, and reproducible upon re-transformation, since the screen can give a background of false positives with some alleles.

7. Cell wall analyses

7.1 β(1→3) and β(1→6) glucan

The *S. cerevisiae* cell wall represents about 20% of the total dry weight of the cell, and is predominantly composed of glucan (β(1→3)- and β(1→6)-linked), and mannoprotein in approximately equal amounts, and a small amount of chitin (27). The abundance of glucan polymers, and their arrangement in the yeast cell wall, greatly facilitates glucan isolation. *Protocol 7* parts *A* and *B* are based on modifications of procedures of earlier workers (cited in ref. 21), which have utilized the cross-linked attachment of glucan to chitin in the cell wall, rendering it insoluble in alkali. Essentially, repeated extraction of yeast cells with alkali, leaves an insoluble pellet containing β(1→3) glucan, β(1→6) glucan, and chitin. Water-soluble β(1→6) glucan can be released from the alkali-extracted walls after digestion with a β(1→3) glucanase preparation, such as Zymolyase.

Protocol 7. Isolation of alkali-insoluble β(1→6) glucan

A. *Large-scale isolation*

As it is usually necessary to have large quantities of purified material for the accurate structural analysis of carbohydrates (using techniques such as column chromatography, linkage analysis, or NMR), the following simple protocol has been developed for the isolation of yeast β(1→6) glucan. Approximately 250 mg of alkali-insoluble β(1→6) glucan can be purified from 4 litres of a wild-type, stationary phase culture in about 3 days.

Equipment and reagents

- 3% NaOH
- 10 mM and 100 mM sodium phosphate buffer (pH 6.8)
- Zymolyase 74T (ICN)
- 10 mg/ml α-amylase (Boehringer Mannheim)
- dialysis tubing: 2000 and 6000–8000 mol. wt cut-off (Spectra/por, Spectrum Medical Industries Inc.)
- lyophilization equipment

Method

1. Grow cells as 2 l cultures in YEPD-Halvorson (see Appendix 1) (or alternative medium) at 30°C until the cultures reach stationary phase. Treat each 2 l culture as described below.
2. Harvest cells by spinning for 10 min at 1000 *g*; wash once with distilled H_2O.
3. Extract the cells five times with 100 ml of 3% NaOH, each at 70°C for 1 h, making sure that the slurry is mixed well each time.
4. Neutralize the cell walls by washing twice with 100 mM sodium phosphate buffer (pH 6.8) and twice with 10 mM sodium phosphate buffer (pH 6.8). (Note: Tris–HCl or buffers containing carbon compounds must not be used for samples that will be subjected to ^{13}C NMR.)
5. Test the insoluble slurry with litmus paper to make sure that the pH is approximately 7 before digestion in 10 mM sodium phosphate buffer (pH 6.8) with 1 mg/ml Zymolyase 74T in a total volume of 40 ml, containing 0.01% sodium azide. Digest for 16 h (or overnight) at 37°C.
6. Remove the remaining insoluble material by centrifugation (10 min at 4000 *g*), and treat the supernatant with 20 μl of α-amylase (10 mg/ml) for 2 h at room temperature to digest glycogen.
7. Remove any residual protein from the glucan-containing aqueous phase with two 5 ml phenol extractions, followed by four or five extractions with ether. (Initially the organic phases will not completely separate from the aqueous, but the repeated ether treatment overcomes this problem.)
8. Dialyse the 40 ml of glucan-containing aqueous solution against 4 l of distilled H_2O using Spectra/por membranes with a 6000–8000 mol. wt cut-off for 4 h before lyophilization at −40°C until dry.
9. Resuspend the freeze-dried glucan in 5–10 ml of distilled H_2O, and again dialyse against 4 l of distilled H_2O for 30 h using 2000 mol. wt pore-size tubing. A second freeze drying produces a β(1→6)-linked glucan polymer.

B. *Small-scale isolation*

The above large-scale procedure is also routinely used on a modified, smaller scale to estimate quantitatively β(1→3) and β(1→6) glucan levels in yeast. Normal levels of total alkali-insoluble, (β(1→3) plus (β(1→6)), glucan are 150 ± 10% μg/mg dry weight, versus alkali-insoluble β(1→6) levels of 30 ± 5% μg/mg dry weight, for S288C-derived strains on YEPD.

1. Grow cells as 10 or 20 ml cultures until they reach stationary phase, then harvest them and wash once with distilled H_2O.

Protocol 7. *Continued*

2. Resuspend the cell pellet in 1 ml of H_2O, and separate into four microcentrifuge tubes (250 μl each), taking great care to ensure an equal cell distribution. Take one of the four tubes for a dry weight determination, and process the three remaining samples as below.
3. Centrifuge each sample for 10 min at 13 000 *g*, discard the supernatant, and extract the cell walls with 0.5 ml of 3% NaOH at 75°C for 1 h.
4. Repeat this step twice (giving a total of three 1 h extractions).
5. Neutralize the walls by washing once with 1 ml of 100 mM Tris–HCl (pH 7.5) and twice with 1 ml of 10 mM Tris–HCl (pH 7.5). (Check the pH before proceeding, and change tubes if necessary.)
6. Resuspend the walls in 1 ml of 10 mM Tris–HCl (pH 7.5), containing 1 mg/ml Zymolyase 100T, and digest overnight at 37°C.
7. Centrifuge the samples to remove any residual insoluble material, and dialyse 0.5 ml of the supernatant against distilled H_2O for 16 h using Spectra/por tubing with a 6000–8000-D pore size. Save the remaining 0.5 ml to determine the total alkali-insoluble, β(1→3) plus β(1→6) glucan.
8. Determine the carbohydrate remaining in the insoluble pellet (from step 7) to ensure that the Zymolyase has digested the β(1→3) glucan to completion. (Typically less than 10% of the total carbohydrate, approximately 10 μg/mg dry weight, remains in the pellet.)
9. Estimate the carbohydrate present prior to dialysis, as a measure of the total alkali-insoluble, β(1→3) plus β(1→6) glucan; the post-dialysis fraction contains the β(1→6) glucan.

7.2 Cell wall dyes

The use of dyes to visualize specific components of the yeast cell wall provides a rapid test for major cell surface alterations. Two fluorescent dyes commonly used to identify chitin and β(1→3) glucan are Calcofluor White and Aniline Blue, respectively. Calcofluor White (also known as Fluorescent Brightner) is highly specific, providing an intense fluorescence staining of the cell wall chitin while emphasizing bud scars and bud necks (28). Aniline Blue binds specifically to β(1→3) glucan, and the purified fluorochrome is now available (Biosupplies Australia Pty). The specificity of Aniline Blue to yeast β(1→3) glucan is illustrated by an even staining over the entire cell periphery, while being excluded from the chitinous bud scars, which appear as black holes.

Alcian Blue, a non-fluorescent dye, binds to a mannoprotein moiety on the yeast cell surface and is often used as a quick screen for dramatic glycosylation defects (29).

Protocol 8. Yeast cell wall dyes

A. *Calcofluor White or Aniline Blue staining*

1. Make a 1 mg/ml stock solution of the dye in distilled H_2O and store at 4°C in the dark (e.g. wrap in foil).
2. Collect 1×10^5 cells by centrifugation and resuspend in 100 μl of a 10-fold dilution of the dye stock solution.
3. Incubate for a few minutes at room temperature, wash twice with H_2O, and mount resuspended cells.
4. Observe cells under a fluorescence microscope.

B. *Alcian Blue staining*

1. Collect 10^7 cells by centrifugation and wash once with 2×10^{-2} M HCl.
2. Resuspend in 2.5 ml of a 0.1% Alcian Blue solution in 2×10^{-2} M HCl.
3. Mix for 30 sec and wash the sample again with 2×10^{-2} M HCl.
4. Examine for evidence of binding, which is observed in a wild-type strain as a dark blue staining of the cell pellet. Non-binding strains appear light blue.

Note: Alcian Blue is dependent on both low pH and low ionic strength. The dye binds immediately and binding is essentially irreversible.

Acknowledgements

This work was supported by the Natural Sciences and Engineering Research Council of Canada, through Operating and Strategic Grants.

References

1. Young, T. W. (1987). In *The yeasts*, Vol. 2 (2nd edn) (ed. A. H. Rose and J. S. Harrison), pp. 131–64. Academic Press, London.
2. Wickner, R. B. (1991). In *The molecular and cellular biology of the yeast Saccharomyces: genome dynamics, protein synthesis, and energetics*, Vol. 1 (ed. J. R. Broach, J. R. Pringle, and E. W. Jones), pp. 236–96. Cold Spring Harbor Laboratory Press, Cold Spring Harbor, NY.
3. Bussey, H. (1991). *Mol. Microbiol.*, **5**, 2339.
4. Wickner, R. B. (1986). *Annu. Rev. Biochem.*, **55**, 373.
5. Tipper, D. J. and Bostian, K. A. (1984). *Microbiol. Rev.*, **48**, 125.
6. Somers, J. M. and Bevan, E. A. (1969). *Genet. Res.*, **13**, 71.
7. Bussey, H. (1988). *Yeast*, **4**, 17.
8. Hutchins, K. and Bussey, H. (1983). *J. Bacteriol.*, **154**, 161.

9. Bussey, H. and Sherman, D. (1973). *Biochim. Biophys. Acta*, **298**, 868.
10. Martinac, B., Zhu, H., Kubalski, A., Zhou, X., Culbertson, M., Bussey, H., and Kung, C. (1990). *Proc. Natl Acad. Sci. USA*, **87**, 6228.
11. Boone, C., Bussey, H., Greene, D., Thomas, D. Y., and Vernet, T. (1986). *Cell*, **46**, 105.
12. Bussey, H., Boone, C., Brown, J., Hill, K., Roemer, T., and Sdicu, A.-M. (1992). In *New approaches for antifungal drugs* (ed. P. B. Fernandes), pp. 20–31. Birkhäuser, Boston.
13. Hill, K., Boone, C., Goebl, M., Puccia, R., Sdicu, A.-M., and Bussey, H. (1992). *Genetic*, **130**, 273.
14. Häusler, A., Ballou, L., Ballou, C. E., and Robbins, P. W. (1992). *Proc. Natl Acad. Sci. USA*, **89**, 6846.
15. Au-Young, J. and Robbins, P. W. (1990). *Mol. Microbiol.*, **4**, 197.
16. Boone, C., Sdicu, A.-M., Laroche, M., and Bussey, H. (1991). *J. Bacteriol.*, **173**, 6859.
17. Tipper, D. J. and Schmitt, M. J. (1991). *Mol. Microbiol.*, **5**, 2331.
18. Stark, M. J. R., Boyd, A., Mileham, A. J., and Romanos, M. A. (1990). *Yeast*, **6**, 1.
19. Halvorson, H. O. (1958). *Biochim. Biophys. Acta*, **27**, 267.
20. Bussey, H., Saville, D., Hutchins, K., and Palfree, R. (1979). *J. Bacteriol.*, **140**, 888.
21. Boone, C., Sommer, S., Hensel, A., and Bussey, H. (1990). *J. Cell Biol.*, **110**, 1833.
22. Brown, J. L., Kossaczka, Z., Jiang, B., and Bussey, H. (1993). *Genetics*, **133**, 837.
23. Roemer, T. and Bussey, H. (1991). *Proc. Natl Acad. Sci. USA*, **88**, 11295.
24. Dmochowska, A., Dignard, D., Henning, D., Thomas, D. Y., and Bussey, H. (1987). *Cell*, **50**, 573.
25. Cooper, A. and Bussey, H. (1992). *J. Cell Biol.*, **119**, 1459.
26. Schiestl, R. H. and Gietz, R. D. (1989). *Curr. Genet.*, **16**, 339.
27. Ballou, C. E. (1982). In *The molecular biology of the yeast Saccharomyces: metabolism and gene expression* (ed. J. N. Strathern, E. W. Jones, and J. R. Broach), pp. 335–60. Cold Spring Harbor Laboratory Press, Cold Spring Harbor, NY.
28. Roncero, C., Valdivieso, M. H., Ribas, J. C., and Durán, A. (1988). *J. Bacteriol.*, **170**, 1950.
29. Friis, J. and Ottolenghi, P. (1970). *Compt. Rend. Trav. Lab. Carlsberg*, **37**, 327.

15

Immuno-electron microscopy

MICHAEL W. CLARK

1. Introduction

The budding yeast, *Saccharomyces cerevisiae*, well known as a model experimental system for the relative ease of its genetics and molecular biology, has disappointed many modern molecular cell biologists because of the limited cytological techniques that are available for this organism. As a result of the small size of a single *S. cerevisiae* haploid cell, 3–5 μm in diameter, most immunofluorescence observations provide relatively simple mapping information, such as whether the protein is in the nucleus or the cytoplasm, or whether or not it is in the vacuole. Most of the important cytological questions for yeast can be adequately answered only by the higher resolutions obtained with the electron microscope. The search for such efficient and effective cytological methods has thus far provided many usable techniques for mapping specific yeast constituents at this higher resolution. Various cell preparation methods have been reported which demonstrate the antibody mapping of specific yeast proteins with the electron microscope (1–3). These techniques are all variations upon the traditional electron microscopy (EM) methodology of sectioning fixed, resin-embedded samples. The diversity in these EM techniques arises from the different viewpoints taken by the investigators as they overcame problems specific to the interactions of a particular antibody with a particular protein. Thus far, each immuno-EM localization published for yeast has had the EM procedures 'tailored' for the specific protein and the antibody reagents utilized in the investigation. At this time, there is no single procedure for fixation and resin-embedding that works for all antibody-based EM examinations.

This chapter is a guide through the more traditional and somewhat confusing techniques of antibody labelling for fixed, resin-embedded sections of yeast. As presented here, these techniques, when done in the systematic manners proscribed, can yield a high resolution, distinct cellular localization for most proteins without too much agony for the investigator.

1.1 Objectives

The main objective of any cytological fixation and embedding protocol is to minimize the disruption of the natural morphology of the cell during the procedure. When preparing samples for antibody mapping, optimum cytological preservation must be further weighted against providing conditions conducive to an optimum reaction between the antibody and the antigen. Although the specific fixation and embedding procedures that yield good cellular preservation can be made routine, the inherent variability of the reaction of the antibody binding site with the antigenic determinants on the protein endows any immune mapping experiment with a high degree of uncertainty. To overcome much of this uncertainty, I have always approached immuno-EM in yeast through a systematic search procedure. I prepare a series of resin blocks which contain wild-type yeast cells fixed and embedded in the specific manners listed below. Each new antibody is then 'tested' on sections from each block to determine the optimum labelling conditions. Once an optimum fixation and embedding protocol is found, further antibody staining optimizations can be done by varying the antibody and salt concentrations. A good rule to remember when dealing with any antibody-based detection methodology is that 'more is *not* better': too much antibody will increase the non-specific staining, too little antibody only weakens the signal.

1.2 Compromise

With the present state of immuno-EM technology, compromise is the operative word. In this chapter, a fixation and embedding regime will be described in detail which provides good morphological preservation while yielding a reasonable antibody reaction with most proteins of moderate or greater abundance. The sections which follow will then present short discussions on the compromises required and the modifications of protocols that are necessary to intensify the immune signal. If the signal is inherently weak (due to the affinity of the antibody being low or because the protein is a low copy number one), refer to ref. 1 which describes a data acquisition methodology that allows the collection of mapping results from a large number of sections.

2. Materials

Various chemicals, buffers, and equipment that should be acquired and/or prepared, are listed below, together with their potential suppliers.

2.1 Chemicals and equipment

2.1.1 Fixatives

- glutaraldehyde, 70%, EM grade (Polysciences, cat. no. 1201)
- paraformaldehyde, EM grade, powder (Polysciences, cat. no. 0380)
- osmium tetroxide, 4% solution (Polysciences, cat. no. 972A)

2.1.2 Embedding resins
- Poly/Bed 812 embedding kit (Polysciences, cat. no. 8792)
- LR White (Polysciences, cat. no. 17411)

2.1.3 Heavy metal stains
- ammonium heptamolybdate (Polysciences, cat. no. 0085)
- uranyl acetate (Polysciences, cat. no. 0379)
- vanadyl sulphate (Polysciences, cat. no. 1310)

2.1.4 Proteins and enzymes
- β-glucuronidase, type H-2 (Sigma, cat. no. G-2887)
- bovine serum albumin, Pentex (Miles Diagnostics, cat. no. 81–001–3)
- 20 nm colloidal gold–anti-rabbit IgG conjugate (E-Y labs, cat. no. GAF-012)
- 10 nm colloidal gold–protein G conjugate (SPI Supplies, cat. no. 04882-AB)
- Zymolyase 100T (ICN ImmunoBiologicals, cat. no. 320931)

2.1.5 Other chemicals
- Freon 113, trichlorodifluoromethane (Ted Pella, cat. no. 17388)
- propylene oxide (Polysciences, cat. no. 00236)
- sodium cacodylate (Polysciences, cat. no. 01131)
- sodium metaperiodate (Mallinckrodt, cat. no. 1139)

2.1.6 Equipment
- adjustable-temperature drying oven (Polysciences, cat. no. 0294)
- embedding BEEM capsules, conical tips, 00 size (Polysciences, cat. no. 0294)
- multiple grid staining unit (either Polysciences, cat. no. 7332, or SPI Supplies, cat. no. 2480)
- multi-well glass depression plates
- Nalgene filter units, 0.45 μm (Nalgene, cat. no. 245–0045)
- Slimbar hexagonal mesh, nickel grids (400 mesh) (SPI Supplies, cat. no. 2240N)
- Teflon-coated forceps, #4 (SPI Supplies, cat. no. 504T)

2.2 Buffers and staining solutions
- **antibody buffer**: 20 mM Tris–HCl, pH 7.5, 200–500 mM NaCl, 0.1% BSA. Pass this solution through a 0.45 μm filter before use
- **cell wall digestion buffer**: 0.1 M potassium phosphate–citrate, pH 5.8, 1.2 M sorbitol

- **cell wall reducing solution**: 20 mM Tris–HCl, pH 8, 5 mM Na_2EDTA, 25 mM dithiothreitol (DTT), 1.2 M sorbitol
- **fixation buffer**: 0.1 M sodium cacodylate, pH 7.0, 0.5 mM $MgCl_2$, 5 mM $CaCl_2$, 1.2 M sorbitol
- **modified Luft's epon removal solution**: prepare this just prior to use by dissolving seven pellets of KOH in 100 ml of *absolute* ethanol containing *no* water
- **section blocking solution**: 20 mM Tris–HCl, pH 7.5, 200 mM NaCl, 8.0% BSA
- **vanadium stain solution**: mix 20 ml of 1% aqueous vanadyl sulphate with 80 ml of 1% aqueous ammonium heptamolybdate and stir until the dark purple solution is oxidized, becoming a clear, yellow colour. Adjust the pH to 6.0 with 0.1 M NaOH before use (4)

3. Traditional fixation

Before describing the detailed protocols, I will provide some brief explanatory comments on the various steps of the overall immuno-EM procedures.

3.1 Cell growth and harvesting

Yeast cultures to be examined by EM are harvested at an optical density at 600 nm (OD_{600}) of 0.25–0.5, just as the culture has entered the logarithmic phase of growth. Cells harvested at this stage of growth have a cell wall which is more easily digested enzymatically than at later phases of the growth curve. When growing yeast cells in minimal media, be aware that cells treated in this manner contain large vesicles which can interfere with the immune mapping procedure by greatly distorting the cellular morphology. To eliminate these nutrient-dependent cytoplasmic vesicles, grow the cell culture overnight in the required minimal medium to an OD_{600} of 0.25–0.5, then add directly to the culture flask an equal volume of rich medium containing double the required concentration of the necessary nutrients and grow for another 2 h. I have never observed any substantial plasmid loss using this procedure and the cytoplasm is always free of large vesicles (5). To transfer the cells from the growth medium to the primary fixative as rapidly as possible, collect the cells by filtration through a 0.45 μm Nalgene filter unit.

3.2 Primary fixation

The first step in good immuno-EM mapping is a rapid and thorough fixation of the cells on ice to prevent drastic cytological changes occurring. The best primary fixation for yeast cells has been found to result from a combination of two aldehyde fixatives: paraformaldehyde and glutaraldehyde. Further protection of cellular morphology is provided in the primary fixation step by including osmotic stabilizers, such as 1.2 M sorbitol, in the fixation buffer.

3.3 Permeabilization of the cell wall

Sufficient penetration of the viscous embedding resin throughout the yeast cell can only be accomplished by the total permeabilization of the barrier produced by the yeast cell wall. The standard method for cell wall removal is by enzymatic digestion. The lytic enzymes are not as effective on the fixed cell wall components, thus the disulphide bonds that are contained in the cell wall must be reduced to increase the efficiency of digestion. The inclusion of 25 mM DTT in the cell wall reducing solution produces effective reduction even at 6°C for 10–20 min.

3.4 Post-fixation

CAUTION: great care must be taken at this stage of the fixation procedure. Always use OsO_4 in a fume hood: this reagent is quite volatile and can react with any tissue, even your own. In the EM protocols OsO_4 is used to retain membrane integrity and good cellular morphology. Because of the nature of the reagent, which deposits metallic osmium on to the biological components, this fixative can interfere with the antibody–antigen interaction. *Protocol 2* includes a chemical treatment with sodium metaperiodate which removes the metal from the surfaces of the sectioned sample just before exposure to the antibody (6). (See Section 5 for a continued discussion of OsO_4 usage.)

3.5 Dehydration and embedding

Dehydration of the fixed sample, i.e. removing H_2O from it, is a necessary transition step to further allow sufficient penetration of the embedding resin into the fixed sample. For each of the dehydration steps, resuspend the cells in the appropriate solution, incubate the suspension on ice for the appropriate time, pellet the cells by centrifugation, remove the supernatant, and then add the next solution. **Note**: *never* let the cell pellet dry out during the changing of the solutions.

Protocol 1. Fixation and embedding of the cells

A. *Harvesting the cells*

1. Rapidly collect the cells from the growth medium by filtration.
2. While the cells are still in the filter unit wash them once with 10 ml of ice-cold **fixation buffer**[a] without the paraformaldehyde/glutaraldhyde.
3. Resuspend the cells from the unit in 5 ml of ice-cold **fixation buffer** plus the primary fixatives: 3% **paraformaldehyde** and 0.5% **glutaraldehyde,** and transfer the solution to a 15 ml centrifuge tube with a sealable cap.

Protocol 1. *Continued*

B. *Primary fixation*

1. Keep the cells from step **A(3)** on ice and in suspension for 3–5 h.
2. After the primary fixation period add 10 ml of the fixation buffer minus the fixative agents and pellet the cells by centrifugation in the cold.
3. Resuspend the pellet in 10 ml of ice-cold **cell wall reducing solution** and again pellet the cells by centrifugation.
4. Resuspend this cell pellet in 1 ml of the same buffer.

C. *Cell wall permeabilization*

1. Incubate the cells in **cell wall reducing buffer** for 10–20 min at 6°C.
2. Add 12 ml of **cell wall digestion buffer** and pellet the cells by centrifugation.
3. Resuspend the cell pellet in **cell wall digestion buffer** to obtain an OD_{600}/ml of 10–20, then add 0.1 vol. of β-glucuronidase and 0.1 vol. of a 5 mg/ml **Zymolyase** solution. Incubate this mixture for 2 h at 30°C. Keep the cells in suspension throughout this incubation.
4. After this digestion step add 12 ml of **cell wall digestion buffer** and pellet the cells by centrifugation.

D. *Post-fixation*

1. Resuspend the washed cell pellet from step **C(4)** in a 2% solution of OsO_4 prepared in **cell wall digestion buffer**. Keep this solution in suspension on ice for 1 h.
2. With the cap sealed on the centrifuge tube, pellet the cells by centrifugation in the cold; carefully remove the OsO_4 solution and discard it as a hazardous chemical.
3. Resuspend the cell pellet in 14 ml of ice-cold distilled H_2O and pellet the cells by centrifugation.
4. Resuspend this cell pellet in 1 ml of ice-cold distilled H_2O, transfer the solution to a 1.5 ml microcentrifuge tube, pellet the cells by centrifugation, and discard the supernatant in preparation for the dehydration steps to follow.

E. *Dehydration and embedding*

1. Treat the pelleted cells with an increasing series of ethanol concentrations of 50, 70, 80, and 95% (5 min for each concentration).
2. Treat the cells with 100% ethanol, four times for 5 min each.
3. Treat the cells with 100% **propylene oxide**; twice for 15 min each.

4. Resuspend the dehydrated cells in a solution composed of 1:1 **propylene oxide:Poly/Bed 812**, seal the tubes, and agitate this solution overnight at room temperature. The next day, pellet the cells by centrifugation in a microcentrifuge for 5 min.
5. Resuspend the cell pellet in 100% **Poly/Bed 812** and place the tubes unsealed in a fume hood for 6–8 h. This treatment allows the resin to exchange with the propylene oxide remaining in the cells and provides enough time for the propylene oxide to evaporate from the resin. At the end of this incubation the cells must be spun in the microcentrifuge for about 10 min before they form a firm pellet.
6. Resuspend the cell pellet in about 0.5 ml of freshly prepared **Poly/Bed 812**. Place a drop or two of this viscous cell suspension in the tip of a dust-free, oven-baked BEEM capsule. Fill the remaining space in the capsule with fresh **Poly/Bed 812**.
7. Harden the resin with a three-stage temperature regime:
   (a) 35°C overnight
   (b) 45°C for 8 h
   (c) 60°C overnight

[a] Details of the reagents and buffers given in bold are listed in Section 2.1.

4. Section preparation and immuno-gold staining

4.1 Section etching

A major difficulty in using any resin-embedded material for an immune mapping procedure is that the resin, which is needed to support the sectioned tissue, prevents the antibody from gaining access to the majority of the antigens in the sample. It has been calculated that the epon-based resins, when sectioned, produce a surface that is so smooth that an antibody can only react with tissue within 1.0 nm of the surface (7). This would give the antibody access to only about 5% of the cellular constituents that are contained in a 20–30 nm thick ultrathin section. For a reasonable immuno-EM mapping more of the sectioned material can be made available by partial removal of the resin after sectioning. The **modified Luft's solution**, which is a very basic ethanol solution, will solubilize the resin surface of a Poly/Bed 812 ultrathin section. Because of the effectiveness of this solution in resin removal, it is important to use 'silver' (60–80 nm) ultrathin sections when treating sections in this manner. The thicker sections also provide more tissue to be sampled by the antibody. Prepare the modified Luft's solution (a KOH–ethanol solution) just before the section treatment. This KOH–ethanol solution **must** be free of water. (KOH in water is a very caustic reagent, whereas

KOH in *absolute* ethanol is quite mild and will not drastically disrupt the cellular morphology or cause protein denaturation.)

Another method of section etching has been developed which not only removes surface resin from the section but also removes from the exposed tissue the cross-linking moieties and heavy metals used in the fixation procedure. In this etching procedure the sections are treated with a saturated aqueous solution of sodium metaperiodate for 1 h (6). In *Protocol 2*, the metaperiodate etching is done after KOH–ethanol resin removal, so here the sections are only exposed to the metaperiodate for 10–20 min. The incubation periods for these two procedures will vary slightly depending upon the hardness of the resin and the thickness of the section, but the times given below are within a generally useful range. After these etching procedures **DO NOT** let the sections dry. They now have the cellular constituents exposed to the air and should be *immediately* processed through the antibody staining steps.

A few words must be said about the EM grids that are used to transport and stabilize the sections in the electron microscope.

(a) The grids should **not** be made of copper (because the colloidal gold particles would react with the copper and precipitate on to the grids). Instead, use nickel EM grids. They are reasonably cheap and work well.

(b) Because of the extensive manipulation of the section in this immunostaining procedure, care must be taken with the grids on which the sections will be mounted. I routinely use a Slimbar, hexagonal 400 mesh EM grid for section support. The hexagonal mesh configuration provides the section with more support than conventional grids while the open areas of a 400 mesh grid allow more sample observation per section. Also the thinness of the bar of the Slimbar EM grid allows the corners of the section to wrap around the bar, anchoring the section securely on to the grid.

(c) Immediately prior to a sectioning session, rinse the grids in 5 ml of Freon 113 or some other volatile solvent, pour off the excess solution, and invert the breaker with the grids into a Petri dish containing a piece of filter paper. The grids will fall on to the filter paper as they dry. Treating the grids in this way before each session ensures that the grids are free of static, oils, or water which can interfere with the adherence of sections to the grid surface.

4.2 Immuno-gold labelling of the sections

Non-specific staining is a major problem in any immuno–EM procedure, so extra care should be taken with the primary and secondary antibodies that are to be used. The primary antibody, that is the antibody made against the protein of interest, should be the highest affinity antibody that can be obtained; the higher the affinity of the antibody, the less of it is needed in the procedure, thus the less chance there is of non-specific reactions occurring.

Protocol 2. Section etching and blocking

A. *Section etching*

1. Mount 'silver' (60–80 nm) ultra-thin Poly/Bed 812 sections upon a hexagonal, 400 mesh Slimbar nickel EM grid.
2. In the first well of a multi-well glass depression slide, expose the grids to the freshly prepared **modified Luft's solution** for 5–10 min at room temperature.
3. *Immediately* move the treated section into the next well of the depression slide which contains the **antibody buffer** *minus the BSA*. Leave the grid there for 5 min.
4. Move the grid to the next well containing the saturated aqueous sodium metaperiodate. Expose the sections to this solution for 10–20 min.

B. *Blocking the section to prevent non-specific antibody interactions*

1. Rinse the grids individually with a gentle stream of **antibody buffer** *minus the BSA*.
2. Place the grids in the **section blocking solution** for 10 min.

Affinity purified antibodies would be best, but in the affinity purification the highest affinity antibodies can remain on the affinity column and be lost. For a polyclonal serum, first try an IgG fraction of the serum obtained from either a protein A or protein G column. A concentration of 2.5–25 μg/ml of the IgG fraction for the primary staining is a reasonable concentration with which to begin. For affinity purified material, start with a concentration of 0.05–0.5 μg/ml. I routinely incubate the sections in the primary antibody at 6°C overnight in a moist container. This long incubation time allows even the lower affinity antibodies to have a chance to react with the sections.

In working with yeast all secondary antibody conjugates must be approached with caution. It is my experience that commercially available secondary antibodies, even when affinity purified, contain material which reacts with the yeast cell wall. Thus, I first preabsorb all antibody conjugates before use with whole yeast cells to remove much of that contaminating material.

Here are a few specific cautionary comments directed particularly at the colloidal gold conjugates:

(a) In general, colloidal gold particles tend to form aggregates, so just before use centrifuge the colloidal gold dilution to remove such aggregates: for a 20–25 nm particle diameter, centrifuge the solution at 700 *g* for 15 min, while for a 10–15 nm particle diameter, centrifuge the solution at 2000 *g* for 15 min.

(b) This tendency to aggregate will also cause the colloidal gold particle to adhere non-specifically to hydrophobic surfaces (which include many embedding resins as well as cell membranes). To prevent such non-specific interaction of the gold:
i. thoroughly block the sections with 8% BSA in buffer before beginning the staining procedure
ii. always keep at least 0.1% BSA in all antibody solutions
iii. keep the NaCl concentration in the antibody solutions between 200 and 500 mM

5. Systematic modification of the procedure to increase the antibody signal

The procedures described above have the advantages of retaining good cellular morphology while providing a highly contrasted and stable section. High affinity antibodies against moderately to highly abundant cellular proteins will yield good mapping information (1). For lower copy number proteins, however, the above procedure must be modified. Furthermore, as stated in the introduction, no single immuno-EM method has been found to work for all antibody–antigen interactions. Still, by modifying the above procedures in a systematic fashion, appropriate conditions for immuno-EM can be found for a majority of proteins. *Table 1* lists six regimes of fixation and embedding

Table 1. Fixation and embedding screen

	Poly/Bed 812			**LR White**		
Step	**1**	**2**	**3**	**4**	**5**	**6**
A	P1-A3	P1-A3	(P1-A3)[a]	P1-A3	P1-A3	(P1-A3)[a]
B	P1-B,C	P1-B,C	P1-B,C	P1-B,C	P1-B,C	P1-B,C
C	P1-D	–	–	P1-D	–	–
D	P1-E1,2	P1-E1,2	P1-E1,2	P1-E1,2	P1-E1.2	P1-E1,2
E	P1-E3,4	P1-E3,4	P1-E3,4	–	–	–
F	P1-E5-7	P1-E5-7	P1-E5-7	LR white	LR white	LR white
G	P2-A,B	P2-A,B	P2-A,B	P2-B	P2-B	P2-B
H	P3	P3	P3	P3	P3	P3

In the first stage of optimizing the immuno-EM signal for a specific antibody, screen that antibody against ultrathin sections cut from resin blocks prepared according to this battery of fixation and embedding regimes (*Steps A–H*, columns 1–6). P1-E1,2 means *Protocol 1*, section E, steps **1** and **2**, etc. – indicates that no treatment is required at this step.

[a] In these samples the initial fixation is to be done as in *Protocol 1*, section A, steps **1–3**, except that 2% paraformaldehyde and 0.075% glutaraldehyde should be used instead of the concentrations listed in step **3**.

Protocol 3. Antibody staining of the prepared sections

Spin all of the solutions used below in the microcentrifuge for 10 min before use to remove aggregates and debris.

A. *The primary antibody*

1. After the etched sections have been thoroughly blocked (*Protocol 2*, step **B2**) move the grids immediately, without rinsing, to the primary antibody solution. Incubate the grids overnight in this solution at 6°C.
2. Rinse the grids gently by passing 2 ml of **antibody buffer** *minus the antibody* over them and then place in the colloidal gold antibody conjugate solution.

B. *The secondary anti-IgG antibody–colloidal gold*

1. 1.5 h before it is needed, dilute the colloidal gold conjugate to a concentration of 0.5 μg/ml. Add to this solution, 0.05–0.1 OD_{600} units of washed, whole yeast cells.
2. Keep this solution on ice for 1 h, then centrifuge it at the appropriate speed (see Section 4.2) for the diameter of the gold particles to remove gold aggregates and the yeast cells. Use the supernatant as the secondary antibody conjugate.
3. Expose the grids to the secondary antibody conjugate for 1 h at room temperature.
4. Rinse the grids using a Multiple grid staining device in the **antibody buffer** *without BSA* for 5 min.

C. *Cross-linking of the antibody to the section and heavy metal staining*

1. Place the rinsed grids into 1% glutaraldehyde in *antibody buffer without BSA* for 10 min at room temperature.
2. Rinse the grids with 2 ml of distilled water.
3. Place the grids in 1% aqueous uranyl acetate for 5 min in a covered Petri dish.
4. Rinse the grids with 2 ml of distilled water and place in the **vanadium stain** solution for 15 min.
5. Rinse the grids with 2 ml of distilled water and then dry on filter paper. These grids are now ready for viewing in the electron microscope.

conditions one of which should work for any particular antibody–protein interaction. Prepare blocks of a wild-type yeast strain for each of the six regimes. A new antibody can then be screened on sections from this battery of cell preparations to discern the optimum fixation and embedding conditions.

5.1 Reduction of the exposure to the fixatives

By their very nature the fixative reagents interact physically with the cellular constituents, cross-linking them and stabilizing the morphology of the cell. As a result, they can directly affect the interaction of the antibody with the protein to be mapped. In general, to improve an immune signal weakened by the fixation of the cells, simply reduce the amount of fixatives to which the cell constituents are exposed. This reduction in exposure can be done by varying either the reagent concentration or the length of time that the sample is exposed to the fixative. In the described procedures, only the fixative concentrations have been varied. Also notice that all fixations are done in the cold; this not only helps maintain the *in vivo* cellular morphology but also reduces the effectiveness of the fixative agents. Doing the fixations at room temperature without drastically reducing the time interval of exposure will greatly increase the amount of cross-linking in the sample and will decrease the antibody–antigen interactions.

5.1.1 Omission of the OsO_4, post-fixation

The most obvious reagent to eliminate is the OsO_4. This chemical is thought to interact with the polar heads of the fatty acid molecules in the lipid bilayer. The metal thus acts as a membrane preservative in the EM procedure by stabilizing the bilayer. When proteins react with the OsO_4, the resulting osmium metal coating will prevent the antibodies from sampling the protein. The sodium metaperiodate treatment described in the Poly/Bed 812 procedure (see *Protocol 2*, step **A4**) will remove the osmium from the sample, but the conformation of the protein could already have been altered too severely to allow the antibody to react. The remedy for this problem is simple: omit the OsO_4 post-fixation. Although the antibody labelling of many proteins will increase, there will be a very noticeable change in the morphological preservation of the cell. Despite this rather drastic change in the appearance of the sample, many reproducible and interpretable results have been obtained without osmium post-fixation (1–3, 5, 8).

5.1.2 Reduction of the concentrations of the primary fixatives

Sometimes the primary fixatives themselves can prevent adequate antibody labelling. Usually the problem comes from the glutaraldehyde. This fixative is a bifunctional cross-linking reagent which is very proficient in stabilizing cellular constituents, hence its inclusion in the fixation. It is the cross-linking ability of glutaraldehyde which directly interferes with the antibody–protein interactions. The concentration of glutaraldehyde should thus be reduced. Concentrations as low as 0.075% have been used in combination with paraformaldehyde and are still useful in the immuno-EM mapping of nuclear pore proteins (8). With these low concentrations of fixatives, osmotic stabilizers, such as 1.2 M sorbitol, *must* always be present to retain a reasonable morphology of the cell. Any further reduction of the fixatives will cause severe morphological disruptions and will make the mapping data unreliable and difficult to interpret. If the antibody signal is still not sufficient, changing the embedding resin to a water-soluble resin could improve the antibody penetration into the section without chemical etching of the section.

5.2 Changing the embedding resin

The control of hardening, the relative efficiency in sectioning, and the distinct contrast between the tissue and the resin which is common with epon-based resins has kept these resins in use in electron microscopy for over thirty years. Yet, these very characteristics can also be disadvantageous for immuno-EM. A reasonable replacement for epon-based resins is a polymer called London Resin (LR) White. LR White does not require propylene oxide in the final infiltration procedure, so it is possible to go directly from the 100% ethanol into the resin. Also, LR White can be hardened either by heat or by a chemical accelerator. I generally use the accelerator and allow the resin blocks to sit on ice as they harden.

LR White has other advantages over epon-based resins. Since LR White is not as hydrophobic as the epon-based resins, LR White sections have less non-specific adhesion of the colloidal gold particles. Also the surface of the LR White section is not as dense and smooth as an epon-based section, thus antibodies have more access to the fixed sample. Specific labelling is thus, in many cases, higher than with the epon-based resins. However, LR White has a few quirks that translate into major disadvantages during the collection of immuno-EM data. The hardening of the resin is difficult to control, making the actual cutting of sections quite frustrating at times. Because LR White is not very hydrophobic, sections do not adhere well to the EM grid. Generally, plastic-coated supports are suggested for LR White sections (2). Because of the weaker binding to the grid and the retention of some water in the resin, LR White sections will distort and sometimes physically move around if the electron beam is concentrated in a single region of the section for too long. To alleviate partially these random section movements, I routinely irradiate the entire grid with a low intensity electron beam at low magnification. Once you are used to the specific characteristics of LR White sections, the LR White resin can be a useful tool for immuno-EM procedures.

6. Future directions

Ideally, molecular cytologists will need an EM technique which will allow the localization of both the protein and nucleotide sequences on the same sample. Although *in situ* hybridizations have been accomplished at an EM level with isolated yeast nuclei (9), no resin-embedded, sectioned material has the attributes necessary to accommodate both procedures. However, progress is being made in this direction. Drs N. Dvorkin and B. Hamkalo at the University of California, Irvine have now extended the *in situ* mapping procedures from isolated nuclei into the realm of cryo-sectioning for yeast (personal communication). They have developed a preservation regime for the yeast cell that provides a cellular morphology as good as that obtained with resin-embedded sections. This not only allows the antibody mapping protocols to be done at the EM level, but also makes it possible to conduct *in situ* hybridization mappings on the same samples. This cryo-sectioning technique for yeast appears to have the potential to be the single technique which the yeast molecular cell biologist seek.

References

1. Clark, M. W. (1990). In *Methods in enyzmology*, Vol. 194 (ed. C. Guthrie and G. R. Fink), pp. 608–626. Academic Press, San Diego.
2. Wright, R. and Rhine, J. (1989). *Meth. Cell Biol.*, **31**, 473.
3. van Tuinen, E. and Riezman, H. (1987). *J. Histochem. Cytochem.*, **31**, 327.
4. Hayat, M. A. (1975). *Positive staining for electron microscopy*. Van Nostrand-Reinhold, Princeton, NJ.
5. Clark, M. W. and Abelson, J. (1987). *J. Cell Biol.*, **105**, 1515.
6. Bendayan, M. and Zollinger, M. (1983). *J. Histochem. Cytochem.*, **31**, 101.
7. Carlemalm, E., Colliex, C., and Kellenberger, J. (1984). In *Advances in electronics and electrophysics* (ed. P. W. Hawkes), pp. 280–7. Academic Press, New York.
8. Wente, S. R., Rout, M. P., and Blobel, G. (1992). *J. Cell Biol.*, **119**, 705.
9. Dvorkin, N., Clark, M. W., and Hamkalo, B. A. (1991). *Chromosoma*, **100**, 519.

16

Industrial *Saccharomyces* yeasts

MORTEN C. KIELLAND-BRANDT

1. Introduction

Genetic and biochemical work with industrial *Saccharomyces* yeasts usually has close connection with their industrial applications, because basic biological phenomena are generally best studied in organisms that 'behave well' in certain respects. When *Saccharomyces* yeasts are chosen as model organisms, the highly developed genetics are usually one of the key virtues, and one therefore chooses members of a large group of closely related strains of *Saccharomyces cerevisiae* that behave regularly in crosses and carry convenient genetic markers (e.g. laboratory strains and genetic reference strains). These strains are ill suited for the traditional practical applications of *Saccharomyces* yeasts, such as bread-making or brewing, while industrial strains suffer from difficulties when it comes to crosses and various other genetic techniques. The present chapter describes some ways of dealing with these difficulties.

1.1 Breeding and research

When development of improved yeast strains is considered, it is often clear that there is insufficent knowledge concerning the genetic and biochemical basis of the technological characters that one wishes to change, and the necessity for initiating research is obvious. However, one should not neglect the importance of a certain amount of actual breeding work at such early points, even if it has to be carried out on the basis of modest knowledge. Not only does this prepare for rapid progress when more knowledge is available, but the breeding work will inevitably itself yield important knowledge.

In other cases it will appear that a breeding programme can be based on a large amount of coherent biochemical and genetic knowledge. Also here, however, a balanced view is important. Without ongoing research in the particular area one will be in an unfavourable position to tackle the unforeseen problems that always appear in the actual breeding work.

In short, although the balance between research and direct breeding must vary as one goes through a project aimed at strain improvement, neither of the two should at any time be neglected.

2. Model systems

2.1 Fermenter models

In most breeding projects it is necessary to screen a large number of strains for a set of characteristics, including those characteristics that are to be changed. When a broad genetic variation is foreseen, typically in mutation-breeding or cross-breeding, it is usually expedient to screen in several steps. In early steps many strains are screened, while fermenter volumes and the extent of analysis are modest as compared with later steps. The last step is always a series of full-scale production trials with one or a few selected strains.

Even in fairly early screening steps it can be important to approximate the fermentation conditions to those of production. In the case of stirred fermentation, sufficient approximation can often be reached in commercial small-volume fermenters. Included in these cases are those where the process is highly aerobic, such as the manufacture of bakers' yeast; here an adequate simulation and control of oxygen transfer to the medium is necessary.

In unstirred fermentations, efficient contact between the medium and yeast cells depends on natural convection to keep the cells suspended. Carbon dioxide bubbles are the main source of this convection. The natural convection that occurs in large tanks is, however, difficult to simulate in small volumes. In brewing, a practical simulation of the main beer fermentation in a volume of 2 litres has become a European standard. In this standard (1) the problems created by the small volume are partially solved by using a fermenter with the shape of a tall tube (diameter 50 mm, height of the column of fermenting beer about 1 m). Semi-automatic equipment, which accommodates 60 such fermentations at a time, has been described (2). It includes facilities for automatic daily sampling and storage of the samples for a few days.

2.2 Biological models

Because of the rather close phylogenetic relationship between the genetic reference strains of *S. cerevisiae* and other *Saccharomyces* yeasts, the vast amount of biochemical and genetic information existing on the former group of strains is highly relevant in breeding of industrial *Saccharomyces* yeasts. As already mentioned, however, it is often necessary to carry out additional research to evaluate particular opportunities for breeding. Regulatory effects of the amounts of end products or intermediates of a metabolic pathway may, for instance, be insufficiently known. In such cases, a physiological study of mutants blocked at various points in the pathway may be instrumental. Because of the relative ease of obtaining and studying mutants in genetic reference strains of *S. cerevisiae*, it is highly recommended that such strains are used as biological model systems for industrial *Saccharomyces* strains. This recommendation is made despite the fact that important physiological differences exist between the strains, which of course dictates caution in the use of models. For instance, there are several differences between a typical lager brewing strain (*Saccharomyces carlsbergensis*) and genetic reference strains (*S. cerevisiae*) that can be directly related to transport or metabolism of small molecules:

(a) The brewing yeast does not decarboxylate ferulic acid, while the non-brewing yeast does so enzymatically. In beer fermentation this process would lead to the presence of 4-vinyl guaiacol, which has an undesirable flavour in lager beer.

(b) The brewing yeast takes up maltotriose (a constituent of brewers' wort) better than the non-brewing yeast.

(c) The brewing yeast is worse at taking up valine than the non-brewing yeast.

(d) The lager brewing yeast can cleave and utilize melibiose, while ale brewing yeasts and most genetic reference strains cannot. The responsible enzyme (α-galactosidase) is mostly present in the cell wall.

As recently reviewed (3), the *S. cerevisiae* biological model system has been important in current breeding efforts, such as control of the formation of diacetyl, monoterpenes, and sulphite.

3. Genetic analysis and cross-breeding

Haploid *Saccharomyces* cells usually occur only rather transiently outside laboratories. Many wine yeasts, for instance, are homothallic diploids. If immediate mitotic progeny of a haploid ascospore do not mate with cells derived from another ascospore, it will on a solid substrate soon mate with cousins that have changed mating type.

Industrial *Saccharomyces* strains are in general polyploid or diploid and do not ordinarily mate, being heterozygous (*MAT***a**/*MAT*α) at the mating type locus. There are various ways of changing the genotype of such a yeast:

(a) The chromosome number ('ploidy') can be reduced by sporulation, which includes meiosis. If a reasonable amount of genetic work is going to be carried out, it is highly recommended to emphasize this step.

(b) The barrier to mating can be overridden. This can be done by protoplast fusion or rare-mating. Both techniques depend on the selection of rare progeny.

(c) Direct mutation can be used. One can expect to find only non-Mendelian (e.g. respiration-deficient) or dominant mutations at reasonable frequency.

(d) The genotype can be altered by genetic transformation.

3.1 Sporulation

Reduction of the chromosome number by sporulation is not only the main entrance to genetic analysis and cross-breeding, it is also extremely important in mutation-breeding. Many industrial yeasts, however, do not sporulate well when subjected to regimens that are standard for genetic reference strains of *S. cerevisiae*. Since good conditions for sporulation are strain-specific, it is highly recommended that a great deal of trial and error is used, but various points are worth noting:

(a) Some sporulation media contain a small amount of glucose and yeast extract (see *Protocol 1*) to allow a gradual shift from conditions supporting mitotic growth to conditions of nitrogen starvation and energy supply through a non-fermentable carbon source, i.e. conditions that promote sporulation. Growth conditions preceding transfer to such sporulation media are probably not critical. Transfer from growth media to other sporulation media, such as simple 0.5% or 1% potassium acetate, represents a much more abrupt change, and the presporulation medium can be quite critical. In this situation it has been reported (4) that an *S. carlsbergensis* strain sporulated much more efficiently if the presporulation medium was based on acetate (see *Protocol 2*) rather than on glucose. Also the growth phase (exponential or stationary) is important when the abrupt shift is used. Strain differences make it important to experiment to find the optimum procedure.

(b) Some yeasts, including strains of *S. carlsbergensis*, do not sporulate at 30°C, while they sporulate at 22°C (5) or lower temperatures (see *Protocols 1* and *2*).

(c) If the frequency of sporulation is low, or if many asci contain only one spore, it may be expedient to monitor spore formation by staining (6). The staining procedure has been adopted for replica plating (see *Protocol 3*), allowing monitoring of the efficiency of sporulation of many strains at a time.

(d) Trace elements, such as zinc (25 p.p.m.), and other compounds, such as *S*-lactoylglutathione (5 mM) have been reported to stimulate sporulation and/or spore viability when added to presporulation and/or sporulation media. The effects of various conditions on the sporulation of *S. cerevisiae* have been comprehensively reviewed (7).

3.2 Spore germination

It is highly recommended that ascus dissection is performed if at all possible. With industrial strains, however, the percentage of spores that can germinate is often low, and one frequently needs many meiotic segregants, so mass spore isolation (see *Protocol 4*) may be necessary.

Protocol 1. An example of gradual shift to sporulation conditions (5)

Materials

- plates of YPD (see Appendix 1)
- Plates of sporulation medium (see Appendix 1)

Method

1. Grow yeast for 2 days at 22°C on a YPD plate.
2. Streak[a] cells on to a plate of sporulation medium.
3. Incubate at 22°C for 7 days or more.
4. Scrape or wash cells and asci off the plate.

[a] The density of seeding influences how gradual the shift will be; for new strains it is recommended that variations are tried. Replica transfer with sterile velvet can give a rather reproducible density.

Protocol 2. An example of abrupt shift to sporulation conditions (4)

Materials

- PSM medium (0.35% Difco Bacto Peptone, 0.3% Difco Bacto Yeast Extract, 4.1 mM magnesium sulphate (e.g. as heptahydrate), 0.1% ammonium sulphate, 1.5% potassium acetate)

Method

1. Inoculate cells into PSM at 10^6 cells/ml.
2. Grow at 21°C for 2–3 days with vigorous aeration, e.g. in a volume of 50 ml in a baffled 300 ml Erlenmeyer flask on a gyratory shaker at 150 r.p.m.
3. Harvest cells by centrifugation and suspend at 10^7 cells/ml in 0.5% potassium acetate.
4. Aerate vigorously at 21°C for 3 days or more, e.g. by shaking 10 ml in a baffled 300 ml Erlenmeyer flask on a gyratory shaker at 150 r.p.m.

Protocol 3. Staining[a] of a replica to assay efficiency of sporulation (5)

1. With sterile velvet, replicate patches of cells/asci on a plate of sporulation medium on to a 10 cm ×10 cm glass plate.

Protocol 3. *Continued*

2. Allow to dry at room temperature for a few minutes.
3. Fix by heating gently on a hot plate. Slight experimentation with heat and time is needed the first time this is done.
4. Stain in a 5% filtered solution of malachite green oxalate at 70–80 °C for 1 min.
5. Rinse with tap water.
6. Counter-stain in 0.5% safranin O at room temperature for 1 min.
7. Rinse with tap water.
8. Inspect the wet glass plate for spores (green) by the naked eye, and under the microscope using coverslips to cover all patches, within 10 min.[b]

[a] Note that this procedure kills the spores and cells.
[b] If the plate is dried, rather than covered with coverslips, the staining keeps longer.

Protocol 4. An example[a] of mass spore isolation

1. After sporulation (see *Protocols 1* and *2*), harvest the cells and asci and wash them with distilled water. The following volumes are adequate for the amount of material from one Petri plate of sporulation medium according to *Protocol 1*:
2. Resuspend in 5 ml of 75 mM mercaptoacetic acid, 100 mM Tris–HCl, pH 8.8 and incubate for 15 min at 22°C.
3. Spin and wash twice with distilled water.
4. Resuspend in 2.5 ml of 2% (w/v) β-glucuronidase[b] (Sigma no. G-0751) in 1 M sorbitol[c] and incubate for 30 min at 22°C.
5. Spin and wash twice with 1 M sorbitol.
6. Resuspend in 10 ml of distilled water.
7. Freeze at −18°C overnight.
8. Allow to thaw at room temperature.
9. Add one drop of Triton X-100.
10. Sonicate until no more separation of spores is evident by phase contrast microscopy. Spores are ready for plating.

[a] This procedure has been used for *S. carlsbergensis* (5). Most procedures commonly used for *S. cerevisiae* will probably work in other *Saccharomyces* species.
[b] Other glucanase-containing enzyme preparations commonly used for degrading cell walls of *S. cerevisiae* are probably generally applicable.
[c] An unbuffered solution has always worked well. Experimentation with buffering might in principle yield a more robust procedure.

In addition to this problem of low spore viability, industrial *Saccharomyces* strains may exhibit the difficulty presented by homothallism due to an active *HO* gene. In this case, spore germination becomes critical if meiotic segregants are desired for subsequent mating. When a homothallic spore with mating ability (say, of mating type α) germinates alone on the surface of an agar plate, the young colony will soon contain cells in which **a** mating type information has been transferred from a silent mating type cassette to the *MAT* locus. If the α spore was haploid, the shifted cells will have **a** mating ability, and diploid cells resulting from mating will soon overgrow any haploid cells. If the α spore was diploid, polyploid, or disomic for chromosome III (e.g. *MATα*/*MATα*), a non-mater (*MATα*/*MAT***a**) might arise directly by *HO* action. In both cases, non-mating progeny are recovered. Non-mating meiotic segregants can be caused by characteristics of the parent strain other than homothallism, e.g. polyploidy, and a single sporulation with usual analysis may well be insufficient to decide why non-maters have arisen.

One way of dealing with homothallism in crosses and genetic analysis is to let spores germinate in close contact with the mating strain to which a cross is desired. At some point, the *HO* gene can then be crossed-out. In another approach, one copy of the *HO* gene is disrupted by integrative transformation, so that some spores will be heterothallic. A convenient tool for disruption of *HO* has been described (8).

3.3 Mating

Two phenomena encountered in industrial strains should be emphasized.

(a) Among meiotic segregants one can find bisexual maters (5). They could possibly be strains that are in principle non-maters but mitotically give off at moderate frequency variants (genetical or non-genetical) of either mating type.
(b) Zygotes, isolated from a single cross of two meiotic segregants of opposite mating type, do not always give identical and uniform colonies (9). Inefficient karyogamy is one possible explanation, since addition of incomplete genomes (see Section 3.4) could have this effect.

When recognized, these phenomena do not necessarily present serious obstacles in cross breeding and may even be used to advantage. However, the nature of both phenomena deserves more study.

Possibly related to the first of the two phenomena is the technique called rare-mating, where cells of a non-mating strain are mixed with cells of a mater under favourable conditions, e.g. close contact on a YPD plate (see Appendix 1). Mating between the two strains will often take place at low frequency. Because of the low frequency, it is necessary to use efficient selection in order to recover the progeny. In genetically marked *S. cerevisiae* strains, rare-mating can usually be ascribed to a low frequency of mitotic recombination or non-disjunction of chromosome III in the non-mater, i.e. genetic changes resulting in mating cells. However, rare-mating could conceivably also be caused by other factors, such as epigenetic phenomena in *MAT*, analogous with those known in certain mutants of *HML* and *HMR*.

Another way of extending the possibilities of bringing genomes together is protoplast fusion. In this case selection must also be established. Both rare-mating and protoplast fusion may be an entrance to genetics of strains from which meiotic progeny have not been obtained. However, one drawback is that the complexity of the genome is increased, rather than decreased, by these methods. The possibility that only parts of genomes contribute to the progeny should be considered in both rare-mating and protoplast fusion.

3.4 Genetic characterization of individual chromosomes

Detailed genetic characterization of industrial *Saccharomyces* strains is usually hampered by insufficient recovery of viable tetrads. However, Nilsson-Tillgren and co-workers (e.g. ref. 10) have devised an approach to study the genetics of individual chromosomes from yeast strains that can mate with *S. cerevisiae*. The principle is to transfer the chromosome to *S. cerevisiae* and study it there.

(a) Choose at least one convenient, dominant, selectable marker to define the chromosome to be analysed, typically a wild-type marker.

(b) Construct a haploid *S. cerevisiae* strain ('recipient strain') carrying the following genes:

  i. mating type opposite to the strain ('donor strain') that carries the chromosome to be analysed

  ii. *kar1-1*, which reduces the efficiency of karyogamy. Crosses involving a *kar1-1* parent lead to the formation of transient heterokaryons, in which there is a rare but significant transfer of single chromosomes from one nucleus to another

  iii. a recessive allele of the dominant wild-type marker chosen to define the chromosome to be analysed

  iv. preferably more mutant markers of the same linkage group

  v. two or more recessive mutant markers that belong to other linkage groups and which confer a selectable or easily screened trait, such as *can1* (confers canavanine resistance), *ura3* (fluoro-orotic acid resistance), *cyh2* (cycloheximide resistance), *lys2* (aminoadipate resistance), or *ade1* or *ade2* (red colour)

  vi. at least one more recessive nutritional marker to ease subsequent crosses

(c) Mate the donor and recipient and plate on a medium that:

  i. allows growth only of cells with the dominant wild-type marker of the chromosome to be analysed *ii.* contains one or several inhibitors of growth of cells with certain other chromosomes present in the donor strain. If, for instance, the recipient carries *lys2*, aminoadipate will not inhibit growth of progeny with the genome of the recipient, unless this progeny has received a chromosome with the wild-type *LYS2* gene from the donor (chromosome II if linkage is as in *S. cerevisiae*). Neither the donor strain nor progeny produced by complete nuclear fusion will grow on the aminoadipate-containing medium.

(d) The selected progeny will mainly have the genotype of the recipient strain but, importantly, will carry the chromosome to be analysed, while few, if any, other chromosomes from the donor will be present. Most progeny will be chromosome addition strains, i.e. contain the full complement of chromosomes of the recipient, with the additional, transferred chromosome. In some cases, however, progeny strains are directly selected which have lost a recipient chromosome corresponding to the transferred chromosome (chromosome substitution strains). Test the selected progeny by replica plating on relevant selective or diagnostic media.

(e) Analyse progeny, for example by:

i. mitotic segregation. In chromosome addition strains, mitotic recombination or chromosome loss/non-disjunction may take place and be seen by uncovering of recessive markers or disappearance of instabilities. One should attempt to isolate substitution strains for further analysis

ii. Southern analysis with probes representative of the chromosome to be analysed. Low stringency is often necessary to detect divergent regions

iii. PCR

iv. electrophoretic karyotyping

v. meiotic analysis. As *kar1-1* does not completely block karyogamy, crosses are generally possible, and good spore viability can be expected. The use of many markers belonging to the relevant linkage group is recommended

As reviewed (11), the *kar1-1* mediated chromosome transfer technique has allowed genetic and molecular characterization of several linkage groups of *S. carlsbergensis* brewing yeast. These studies have revealed the alloploidy of this yeast.

4. Genetic transformation

Industrial *Saccharomyces* yeasts can generally be transformed by standard procedures, although frequencies of recovery of transformants per μg of DNA may vary widely. The obvious difference from the situation in genetically marked strains of *S. cerevisiae* is the choice of selection marker.

Standard techniques in transformation of *S. cerevisiae* are described in Chapter 8 and elsewhere (12). Use of dominant wild-type *S. cerevisiae* genes, which is standard in genetically marked strains, is possible in industrial yeasts if natural polymorphism or copy number effects allow for it, which can be the case for *CUP1*, *MEL1*, *KILk1*, and *POF1*, or if suitable recessive markers have been introduced in the industrial strain (13). Markers that can in principle be used more directly or generally are dominant mutant *S. cerevisiae* genes, such as *ILV2* alleles which give resistance to sulphonyl urea herbicides, or prokaryotic drug resistance markers, such as *neo* from Tn*601*, which when expressed under the control of a yeast promoter gives resistance to geneticin (trade mark of Gibco, Life Technologies, Inc.; also called G418) (14). The appropriate levels of sulphonyl urea herbicides and geneticin are highly strain-dependent, and, as for any other drug resistance selection system, concentrations should be adjusted to both yeast strain and vector characteristics. Geneticin is available from several companies. At the Carlsberg Laboratory, we have used geneticin sulphate (Sigma cat. no. G5013) in YPD plates (see Appendix 1) at concentrations from 30 p.p.m. (*S. carlsbergensis*) to 500 p.p.m. (*S. cerevisiae* distillers' yeast). Use of various dominant selection markers in industrial *Saccharomyces* strains has been reviewed (15).

4.1 Chromosomal integration

With autonomously replicating plasmids, questions on gene expression and plasmid maintenance are in most cases related to regulatory systems, selection conditions, and the plasmids themselves. When it comes to chromosomal integration, on the other hand, it is quite important to consider genetic structure, such as whether the strain is polyploid, alloploid, or both.

The considerations depend to some extent on whether the gene to be introduced is recessive or dominant. In both cases, standard procedures (12) for gene replacement (one-step or two-step) can be used.

(a) A desired recessive gene can, for example, be a deletion in a gene for a step in a biosynthetic pathway (such as *ilv2*, see refs 3 and 16). Note that:
   i. if the strain is not haploid, it should be investigated whether a pair of meiotic segregants can be found which by mating re-establish a useful approximation of the original strain. The reduced ploidy simplifies any mutagenesis, including directed replacement
   ii. if the strain is not haploid, it is attractive to use the two-step gene replacement technique ('pop-in/pop-out'), since 'pop-out' of the vector allows reuse of the same selective marker in the one or more additional rounds of replacement necessary for replacing all wild-type genes for the function in question. However, it will usually be necessary to screen for pop-outs by replica plating, since positive selection for marker loss is not possible with most of the selective markers used in industrial strains

iii. if the strain is not haploid, one round of two-step gene replacement does not give an altered phenotype. It is therefore expedient to use genomic Southern analysis to find those pop-outs in which the desired gene replacement has happened

iv. Southern analysis is also useful if a second round of two-step gene replacement is to be carried out on a highly homologous chromosome. The analysis is done after the first step (transformation) of the second round, in order to distinguish integration at the desired place from integration where gene replacement has already been carried out. The latter, unwanted integration can sometimes be avoided by deleting, in the first gene replacement, the sequences that are used as integration targets in the second round

v. multiple, tandem integration is a well-known complication in plasmid integration (the 'pop-in' step). For most efficient pop-out, single integrations are desired. With some selective markers there is a tendency to select multiple integrations. This is true for geneticin. The remedy is to use a low drug concentration and pick the least resistant transformants. Single and multiple integrations can be distinguished by Southern analysis

vi. successive gene replacement in homologous and divergent chromosomes is not only a tool for the introduction of recessive mutations—it is also a powerful way of analysing the ploidy of a strain

vii. if the strain is alloploid, it is useful, and perhaps even necessary, to construct the recessive change in both wild-type alleles of the gene in question (see ref. 16), since there is a preference for integration where most homology exists

(b) A desired dominant gene could be a cassette for an antisense RNA with a goal similar to that mentioned above, or it could be a gene from a different strain or species, intended to contribute a specific function to the industrial yeast, such as the ability of a brewing yeast to secrete glucoamylase (14). Note that:

i. depending on the expression system and the type of desired function, a single copy of the introduced gene may be functionally sufficient

ii. if the strain is not haploid, a single copy of the integrated gene is potentially subject to loss, e.g. by mitotic crossing-over between the locus and the centromere. If the integrated gene exerts a load on the strain, even a modest rate of loss can have consequences. Under conditions designed to simulate industrial production, such instability has been found to be insignificant in some systems, while significant in others (14). In a polyploid yeast, integration of a second copy of the gene on a homologous chromosome was found to give sufficient stability (14)

iii. in an alloploid yeast, there is less mitotic recombination between divergent chromosomes than between highly homologous chromosomes. Unless extreme stability is needed, it is probably sufficient to integrate the desired gene in all copies of one of the two divergent types

5. Concluding remarks

Research into the molecular genetics of *S. cerevisiae* has obviously provided traditional *Saccharomyces* industries with remarkable strain construction techniques, of which gene replacement is quite noteworthy, in comparison with opportunities in, for example, plants. In view of this, the following advice and reminders should be considered.

(a) Molecular genetics is useful not only in direct construction of new industrial strains but also in model systems, genetic characterization, and other research that will give information relevant for strategies of the desired breeding.

(b) If one regards the plasmid-based techniques just as a convenient substitute for classical mutagenesis and cross-breeding in troublesome strains, important opportunities are lost. Combination of DNA technology with shifts in ploidy and utilization of natural polymorphisms will be much more powerful. The availability of adequate pilot fermentation facilities is of key importance in taking advantage of this potential, as it is in other strategies of cross-breeding.

References

1. Enari, T.-M. (1977). *J. Inst. Brew.*, **83**, 109.
2. Sigsgaard, P. and Rasmussen, J. N. (1985). *Am. Soc. Brew. Chem. J.*, **43**, 104.
3. Kielland-Brandt, M. C. In *Proceedings of the 6th European congress on biotechnology* (Florence) (ed. L. Alberghina, L. Frontali, and P. Sensi). Elsevier Science Publishers B.V., Amsterdam. (In press).
4. Bilinski, C. A., Russell, I., and Stewart, G. G. (1987). *J. Inst. Brew.*, **93**, 216.
5. Gjermansen, C. and Sigsgaard, P. (1981). *Carlsberg Res. Commun.*, **46**, 1.
6. Schaeffer, A. B. and Fulton, M. D. (1933). *Science*, **77**, 194.
7. Miller, J. J. (1989). In *The yeasts*, 2nd edn, Vol. 3 (ed. A. H. Rose and J. S. Harrison), pp. 489–550. Academic Press, London.
8. Van Zyl, W. H., Lodolo, E. J., and Gericke, M. (1993). *Curr. Genet.*, **23**, 290.
9. Gjermansen, C. and Sigsgaard, P. (1987). In *European brewery convention symposium on brewers' yeast* (Helsinki), pp. 156–68. Verlag Hans Carl (Brauwelt-Verlag), Nürnberg.
10. Nilsson-Tillgren, T., Gjermansen, C., Holmberg, S., Petersen, J. G. L., and Kielland-Brandt, M. C. (1986). *Carlsberg Res. Commun.*, **51**, 309.
11. Kielland-Brandt, M. C., Nilsson-Tillgren, T., Gjermansen, C., Holmberg, S., and Pedersen, M. B. In *The yeasts*, 2nd edn, Vol. 6 (ed. A. H. Rose, J. S. Harrison, and A. E. Wheals). Academic Press, London (In press).

12. Guthrie, C. and Fink, G. R. (ed.) (1991). *Methods in enzymology*, Vol. 194. Academic Press, San Diego.
13. Gjermansen, C. (1983). *Carlsberg Res. Commun.*, **48**, 557.
14. Yocum, R. R. (1986). In *BioExpo 86 Proceedings* (Boston), pp. 171–80. Butterworth Publishers, Stoneham, USA.
15. Hinchliffe, E. (1991). In *Applied molecular genetics of fungi* (ed. J. F. Peberdy, C. E. Caten, J. E. Ogden, and J. W. Bennett), pp. 129–45. Cambridge University Press, Cambridge, UK.
16. Gjermansen, C., Nilsson-Tillgren, T., Petersen, J. G. L., Kielland-Brandt, M. C., Sigsgaard, P., and Holmberg, S. (1988) *J. Basic Microbiol.*, **28**, 175.

A1

More common media and buffers

Relevant chapter numbers are given in **bold**, in parentheses.

1. Media

For solid media, agar is added to the recipes below at 2% (w/v) unless noted otherwise.

1.1 Growth media

1.1.1 Extract media (incompletely defined)

(a) YPD (or YEP, YEPD)

• yeast extract	10 g
• glucose (or dextrose)	20 g
• Bacto-Peptone	10 g (**2, 11, 12**) or 20 g (**1, 3, 8, 10, 16**)
• Distilled water to 1 litre	

(Bacto-Peptone is often increased from 10 g to 20 g in solid medium)

(b) YPAD (**8**)

• YPD	as (a)
• adenine hemisulphate	100 mg

Adjust to pH 6.0 with NaOH.

(c) YPGE (**11**)

• yeast extract	10 g
• Bacto-Peptone	10 g (or 20 g for solid medium)
• glycerol	20 g
• ethanol	10 g

(d) YM5 (**12**)

• yeast extract	1 g

• glucose	20 g
• Bacto-Peptone	2 g
• Succinic acid	10 g
• Yeast nitrogen base (with amino acids)	6.7 g
• NaOH	6.0 g
• distilled water to 1 litre	

(e) Halvorson-YEPD (**14**)

	Liquid	**Solid**
• Yeast extract	5 g	10 g
• Bacto-Peptone	5 g	20 g
• Halvorson 5 × salts (see below)	200 ml	200 ml
• Glucose[c]	20 g	20 g
• distilled water to 1 litre		

Notes: the glucose should be sterilized and added separately. For solid medium, autoclave 2.5 × Halvorson salts *separately* from the agar (the agar will not set properly otherwise).

The composition of **Halvorson 5 × salts** (pH 4.7) is as follows

• ammonium sulphate ($(NH_4)_2SO_4$)	20 g
• potassium phosphate, dibasic (K_2HPO_4)	43.5 g
• Succinic acid ($C_4H_6O_4$)	29 g
• calcium chloride ($CaCl_2.2H_2O$)	1.99 g
• magnesium sulphate ($MgSO_4.7H_2O$)	5.11 g
• trace element solution[a]	5 ml
• distilled water to 1 litre	

Dissolve each component in order before adding the next. This solution can be stored at 4°C for several weeks.

An alternative to Halvorson salts is **succinic acid buffer** (pH 4.7), which contains 10 g of succinic acid and 3 g of NaOH per litre of solution.

[a] Trace element solution:

• ferric sulphate	307 mg
• manganese sulphate	280 mg
• zinc sulphate	445 mg
• cupric sulphate	391 mg
• distilled water to 500 ml.	

(f) L broth (LB) (**2–4**)

• yeast extract	5 g
• Bacto-tryptone	10 g

- NaCl 10 g
- distilled water to 1 litre

(for solid medium, add 1.2% agar)

1.1.2 Synthetic (defined) media

(a) SC (synthetic complete) (also called SD or GYNB) (**7**, **8**, **10**, **11**, **13**)

- yeast nitrogen base (Difco) (**without** amino acids) 6.5–6.7 g
- glucose 20 g

These components constitute a *minimal* medium, but 'complete' media require the addition of amino acids and, usually, purines and pyrimidines (mostly adenine and uracil). The concentration of these supplements frequently varies quite considerably in different laboratories, e.g. from identical concentrations for each (50 μg/ml, **10**) to considerably different concentrations, e.g. see Chapter **7**, *Protocol 3*. The pH is often adjusted to 5.6–6.0 (**8**).

(b) M9 (minimal, **6**)

- Na_2HPO_4 6 g
- KH_2PO_4 3 g
- NaCl 0.5 g
- NH_4Cl 1 g
- distilled water to 988 ml

Adjust to pH 7.4, autoclave, cool, then add:

- 1 M $MgSO_4$ 2 ml
- 20% (w/v) glucose 10 ml
- 1 M $CaCl_2$ 0.1 ml

These three solutions should be autoclaved separately. Supplemented M9 media contain the following:

- leucine 40 μg/ml
- tryptophan 30 μg/ml
- uracil 20 μg/ml

(c) Halvorson minimal medium (MM) (**14**)

- Halvorson 5 × salts (see 1.1.1(e)) 200 ml
- distilled water to 1 litre

Autoclave, then add:

- glucose (sterile, 40%) 50 ml
- vitamin solution[a] 0.5 ml
- amino acid supplements, as required

[a] Vitamin stock:

• D-biotin	100 mg
• calcium pantothenate	30 mg
• folic acid	30 mg
• myo-inositol	330 mg
• pyridoxine-HCl	330 mg
• nicotinic acid	330 mg
• *p*-aminobenzoic acid	330 mg
• thiamine-HCl	330 mg
• sodium carbonate-H_2O	2 g
• distilled water to 100 ml.	

Autoclave; this solution can be stored at 4°C, in the dark, for months.

(d) AHC[a] (**5**)

• yeast nitrogen base (without amino acids)	6.7 g
• ammonium sulphate	5 g
• casein hydrolysate	10 g
• glucose	20 g
• adenine hemisulphate	20 mg
• distilled water to 1 litre; adjust to pH 5.8	

[a] Denoted as AHC$^{=}$ (double minus, i.e. lacking uracil and tryptophan) (see Gnirke, A. and Huxley, C. (1991). *Somat. Cell Mol. Genet.*, **17**, 573).

1.2 Sporulation medium (16)

• potassium acetate	10 g
• yeast extract (Difco)	1 g
• glucose	0.5 g
• agar	20 g
• distilled water to 1 litre	

(also see PSM, Chapter **16**).

1.3 Sphaeroplast-stabilizing medium (12) (YM5–sorbitol)

- YM5 medium (see 1.1.1 (d))
- 1 M sorbitol

1.4

Authors of some chapters in this book cite media as defined by Sherman, F. (1991) (In *Methods in enzymology*, Vol. 194 (ed. C. Guthrie and G. R. Fink), pp. 3–21. Academic Press, San Diego and London).

2. Buffers

(a) TE buffer

- 10 mM Tris–HCl
- 1 mM Na_2 EDTA

pH 8.0 (**1, 3, 4, 6–8**), or 7.5 (**5, 9, 10**), or 7.6 (**2**), or 7.4 (**13**).

(b) TEN buffer (**13**)

- 10 mM Tris–HCl, pH 7.4
- 2 mM EDTA
- 140 mM NaCl

(c) TBE buffer (**3, 5**)[a]

Usually kept as 10 × stock solution:

• Tris base	108 g
• boric acid	55 g
• EDTA (Na_2)	9.3 g
• distilled water to 1 litre	

Adjust to pH 8.3.

[a] **Note:** a different composition of TBE buffer is used in Chapter 9.

(d) TAE buffer (**5**)

Usually kept as 5 × stock solution:

• Tris base	24.2 g
• acetic acid (glacial)	5.71 ml
• EDTA (0.5 M)	20 ml
• distilled water to 1 litre	

Adjust to pH 8.0

A2

Addresses of suppliers

Aldrich Chemical Co. Inc., 1001 W. St Paul Avenue, PO Box 355, Milwaukee, WI 53201, USA.

Amersham Corporation, 2636 S. Clearbrook Drive, Arlington Heights, IL 60005, USA.

Amersham International plc, Amersham Place, Little Chalfont, Buckinghamshire, HP7 9NA, UK.

Amicon Division, W. R. Grace & Co., 72 Cherry Hill Drive, Beverly, MA 01915, USA.

ATCC–American Type Culture Collection, 12301 Parklawn Drive, Rockville, MD 20852–1776, USA.

Barnstead/Thermolyne Company, 2555 Kerper Boulevard, Dubuque, IA 52001, USA.

BDH, Merck House, Poole, Dorset, BH15 ITD, UK.

Beckman (a Smithkline Beecham company), Progress Road, Sands Industrial Estate, High Wycombe, HP12 4JL, UK.

Beckman Instruments, Inc., Spinco Division, 1050 Page Mill Road, Palo Alto, CA 94304, USA.

BIO 101 Inc., PO Box 2284, La Jolla, CA 92038–2284, USA.

Biocent b.v., Heereweg 441 B, 2161 DB Lisse; PO Box 280, 2160 AG Lisse, The Netherlands.

Bio-Rad, Microscience Division, Bio-Rad House, Maylands Avenue, Hemel Hempstead, Hertfordshire, HP2 7TD, UK.

Bio-Rad Laboratories GmbH, Heidenmannstrasse 164, D-8000 München 45, Germany

Bio-Rad Laboratories, Chemicals Group, 3300 Regatta Boulevard, Richmond, CA 94804, USA.

Biosupplies Australia Pty Ltd, PO Box 835, Parkville, Victoria 3052, Australia.

Boehringer Mannheim Canada Ltd, 200 Micro Boulevard, Laval, Quebec, Canada H7V 3Z9.

Boehringer Mannheim UK, Bell Lane, Lewes, East Sussex, BN7 1LG, UK.

B. Braun Biotech, 999 Postal Road, Allentown, PA 18103, USA.

B. Braun Biotech International GmbH, Schwarzenberger Weg 73–79, D-3508, Melsungen, Germany.

Collaborative Research, 1365 Main Street, Waltham, MA 02154, USA.

Corning Science Products, Corning, NY 14830, USA.

Diagen GmbH, Niederheider Strasse 3, D-4000 Düsseldorf 13, Germany.
Diagen GmbH/Quiagen Inc. (Quiagen Inc.), Chatsworth, CA, USA.
Difco Laboratories, PO Box 14B, Central Avenue, East Molesey, Surrey, UK.
Difco Laboratories, PO Box 1058, Detroit, MI 48232, USA.
DNAstar Ltd, Link House, 565–569 Chiswick High Road, London W4 3AY, UK.
DNAstar Inc., 1801 University Avenue, Madison, WI 53705, USA.
Dupont (UK) Ltd, Wedgewood Way, Stevenage, Herts, SG1 4QN, UK.
Dupont Co., 1007 Market Street, Wilmington, DE 19898, USA.
E-Y Laboratories Inc., 127 N. Amphlett Boulevard, San Mateo, CA 94401, USA.
FMC Bioproducts, 191 Thomaston, Rockland, ME 04841, USA.
FMC Bioproducts Europe, Risingevej 1, DK-2665, Vallensbaek Strand, Denmark.
The Genetics Society of America, 9650 Rockville Pike, Bethesda, MD 20814, USA.
Gibco BRL, Division of Life Technologies Inc., 3175 Staley Road, Grand Island, NY 14072–0068, USA.
ICN Biomedicals Inc., 3300 Hyland Avenue, PO Box 5023, Costa Mesa, CA 92626, USA.
ISCO Inc., Instrument Division, PO Box 5347, Lincoln, NB 68505–9987, USA.
Jencons Scientific Ltd, Cherrycourt Way Industrial Estate, Stanbridge Road, Leighton Buzzard, Beds, LU7 8UA, UK.
Lablogic, St Johns House, 131 Psalter Lane, Sheffield S11 8UX, UK.
Mallinckrodt Chemicals, Science Products Division, Paris, KY 40361, USA.
Miles Diagnostics, PO Box 2000, Elkhart, IN 46515, USA.
Millipore (UK) Ltd, The Boulevard, Blackmoor Lane, Watford, Herts, WD1 8YW, UK.
Millipore Corporation, 80 Ashby Road, Bedford, MA 01730, USA.
Nalgene Company (Sybron Corporation), PO Box 20365, Rochester, NY 14602–0365, USA.
New England Biolabs, 32 Tozer Road, Beverley, MA 01915–5599, USA.
New England Nuclear Corp. (NEN), 549 Albany Street, Boston, MA 01915, USA.
New England Nuclear Corp. (NEN), Du Pont de Nemours (Deutschland) GmbH, Biotechnology Systems Division, NEN Research Products, Postfach 401240, D-6072, Dreieich 4, Germany.
Nunc A/S, Postbox 280, Kamstrup, DK-4000 Roskilde, Denmark.
Nunc Inc., 2000 North Aurora Road, Naperville, IL 60566, USA.
Nunc Inc., Gibco Ltd, PO Box 35, Trident House, Renfrew Road, Paisley, PA3 4EF, UK.
Pall, Postfach 102120, Philipp-Reis-Strasse 6, D-6072, Dreieich 1, Germany.
Pall BioSupport Div., Glen Cove, NY 11542, USA.

Perkin Elmer Corporation, 12300 Parcreast Drive, Suite 1000, Stafford, TX 77477, USA.

Pfizer Limited, Groton Plant & Research Center, Eastern Point Road, Groton, CT 06340–5146, USA.

Pharmacia Ltd, Davey Avenue, Knowlhill, Milton Keynes MK5 8PH, UK.

Pharmacia LKB Biotechnology, 800 Centennial Avenue, PO Box 1327, Piscataway, NJ 08855–1327, USA.

Polysciences GmbH, Ulmenhof 28, 5401 St Goar, Germany.

Polysciences Inc., 400 Valley Road, Warrington, PA 18976, USA.

Polysciences Ltd, 24 Low Farm Place, Moulton Park, Northampton NN3 1HY, UK.

Promega, Delta House, Enterprise Road, Southampton SO1 7NS, UK.

Promega, 2800 Woods Hollow Road, Madison, WI 53711–5399, USA.

Sartorius AG, 3400 Göttingen, Germany.

Savant Instruments Inc., 110–103 Bi-County Boulevard, Farmingdale, NY 11735, USA.

Schleicher & Schull, PO Box 4, D-3354, Dassel, Germany.

Schleicher & Schull Inc., 10 Optical Avenue, Keene, NH 03431, USA.

SCM Speciality Chemicals, PO Box 1466, Gainesville, FL 32602, USA.

Seikagaku America, 30 West Gude Drive, Suite 2600, Rockville, MD 20850, USA.

Sigma Chemical Co., Fancy Road, Poole, Dorset, BH17 7BR, UK.

Sigma Chemical Co., PO Box 14508, St Louis, MO 63178, USA.

Sigma Chemie, PO Box 260, CH-9470 Buchs, Switzerland.

Sigma Chimie, L'Isle D'Abeau Chesnes, BP 701, 38297 St Quentin Fallavier Cedex, France.

Sigma Chemie GmbH, Grunwalder Weg 30, W-8024, Deisenhofen, Germany.

Spectrum Medical Industries Inc., 60916 Terminal Annex, Los Angeles, CA 90060, USA.

SPI Supplies, (division of **Structure Probe Inc.**), 535 E. Gay Street, PO Box 656, West Chester, PA 19381–0656, USA.

Stratagene, 11099 North Torrey Pines Road, La Jolla, CA 92037, USA.

Stratagene Ltd, Cambridge Innovation Centre, Cambridge Science Park, Milton Road, Cambridge CB4 4GF, UK.

Ted Pella Inc. (Agar Scientific Ltd), 4595 Mountain Lakes Boulevard, Redding, CA 96003, USA.

Whatman Labsales, PO Box 1359, Hillsboro, OR 97123, USA.

Whatman Scientific Ltd, St Leonard's Road, 20/20 Maidstone, Kent ME16 0LS, UK.

Wheaton, 1301 N. Tenth St., Millville, NJ 08332, USA.

Worthington Biochemicals, PO Box 650, Halls Mill Road, Freehold, NJ 07728, USA.

Yeast Genetic Stock Center, Donner Laboratory, University of California, Berkeley, CA 94720, USA.

Index

Index

ORDER OTHER TITLES OF INTEREST TODAY

Price list for: UK, Europe, Rest of World (excluding US and Canada)

138. Plasmids (2/e) Hardy, K.G. (Ed)
.... Spiralbound hardback 0-19-963445-9 **£30.00**
.... Paperback 0-19-963444-0 **£19.50**
136. RNA Processing: Vol. II Higgins, S.J. & Hames, B.D. (Eds)
.... Spiralbound hardback 0-19-963471-8 **£30.00**
.... Paperback 0-19-963470-X **£19.50**
135. RNA Processing: Vol. I Higgins, S.J. & Hames, B.D. (Eds)
.... Spiralbound hardback 0-19-963344-4 **£30.00**
.... Paperback 0-19-963343-6 **£19.50**
134. NMR of Macromolecules Roberts, G.C.K. (Ed)
.... Spiralbound hardback 0-19-963225-1 **£32.50**
.... Paperback 0-19-963224-3 **£22.50**
133. Gas Chromatography Baugh, P. (Ed)
.... Spiralbound hardback 0-19-963272-3 **£40.00**
.... Paperback 0-19-963271-5 **£27.50**
132. Essential Developmental Biology Stern, C.D. & Holland, P.W.H. (Eds)
.... Spiralbound hardback 0-19-963423-8 **£30.00**
.... Paperback 0-19-963422-X **£19.50**
131. Cellular Interactions in Development Hartley, D.A. (Ed)
.... Spiralbound hardback 0-19-963391-6 **£30.00**
.... Paperback 0-19-963390-8 **£18.50**
129 Behavioural Neuroscience: Volume II Sahgal, A. (Ed)
.... Spiralbound hardback 0-19-963458-0 **£32.50**
.... Paperback 0-19-963457-2 **£22.50**
128 Behavioural Neuroscience: Volume I Sahgal, A. (Ed)
.... Spiralbound hardback 0-19-963368-1 **£32.50**
.... Paperback 0-19-963367-3 **£22.50**
127. Molecular Virology Davison, A.J. & Elliott, R.M. (Eds)
.... Spiralbound hardback 0-19-963358-4 **£35.00**
.... Paperback 0-19-963357-6 **£25.00**
126. Gene Targeting Joyner, A.L. (Ed)
.... Spiralbound hardback 0-19-963407-6 **£30.00**
.... Paperback 0-19-9634036-8 **19.50**
125. Glycobiology Fukuda, M. & Kobata, A. (Eds)
.... Spiralbound hardback 0-19-963372-X **£32.50**
.... Paperback 0-19-963371-1 **£22.50**
124. Human Genetic Disease Analysis (2/e) Davies, K.E. (Ed)
.... Spiralbound hardback 0-19-963309-6 **£30.00**
.... Paperback 0-19-963308-8 **£18.50**
122. Immunocytochemistry Beesley, J. (Ed)
.... Spiralbound hardback 0-19-963270-7 **£35.00**
.... Paperback 0-19-963269-3 **£22.50**
123. Protein Phosphorylation Hardie, D.G. (Ed)
.... Spiralbound hardback 0-19-963306-1 **£32.50**
.... Paperback 0-19-963305-3 **£22.50**
121. Tumour Immunobiology Gallagher, G., Rees, R.C. & others (Eds)
.... Spiralbound hardback 0-19-963370-3 **£40.00**
.... Paperback 0-19-963369-X **£27.50**
120. Transcription Factors Latchman, D.S. (Ed)
.... Spiralbound hardback 0-19-963342-8 **£30.00**
.... Paperback 0-19-963341-X **£19.50**
119. Growth Factors McKay, I. & Leigh, I. (Eds)
.... Spiralbound hardback 0-19-963360-6 **£30.00**
.... Paperback 0-19-963359-2 **£19.50**
118. Histocompatibility Testing Dyer, P. & Middleton, D. (Eds)
.... Spiralbound hardback 0-19-963364-9 **£32.50**
.... Paperback 0-19-963363-0 **£22.50**
117. Gene Transcription Hames, B.D. & Higgins, S.J. (Eds)
.... Spiralbound hardback 0-19-963292-8 **£35.00**
.... Paperback 0-19-963291-X **£25.00**
116. Electrophysiology Wallis, D.I. (Ed)
.... Spiralbound hardback 0-19-963348-7 **£32.50**
.... Paperback 0-19-963347-9 **£22.50**
115. Biological Data Analysis Fry, J.C. (Ed)
.... Spiralbound hardback 0-19-963340-1 **£50.00**
.... Paperback 0-19-963339-8 **£27.50**
114. Experimental Neuroanatomy Bolam, J.P. (Ed)
.... Spiralbound hardback 0-19-963326-6 **£32.50**
.... Paperback 0-19-963325-8 **£22.50**
113. Preparative Centrifugation Rickwood, D. (Ed)
.... Spiralbound hardback 0-19-963208-1 **£45.00**
.... Paperback 0-19-963211-1 **£25.00**
.... Paperback 0-19-963099-2 **£25.00**
112. Lipid Analysis Hamilton, R.J. & Hamilton, Shiela (Eds)
.... Spiralbound hardback 0-19-963098-4 **£35.00**
.... Paperback 0-19-963099-2 **£25.00**
111. Haemopoiesis Testa, N.G. & Molineux, G. (Eds)
.... Spiralbound hardback 0-19-963366-5 **£32.50**
.... Paperback 0-19-963365-7 **£22.50**
110. Pollination Ecology Dafni, A.
.... Spiralbound hardback 0-19-963299-5 **£32.50**
.... Paperback 0-19-963298-7 **£22.50**
109. In Situ Hybridization Wilkinson, D.G. (Ed)
.... Spiralbound hardback 0-19-963328-2 **£30.00**
.... Paperback 0-19-963327-4 **£18.50**
108. Protein Engineering Rees, A.R., Sternberg, M.J.E. & others (Eds)
.... Spiralbound hardback 0-19-963139-5 **£35.00**
.... Paperback 0-19-963138-7 **£25.00**
107. Cell-Cell Interactions Stevenson, B.R., Gallin, W.J. & others (Eds)
.... Spiralbound hardback 0-19-963319-3 **£32.50**
.... Paperback 0-19-963318-5 **£22.50**
106. Diagnostic Molecular Pathology: Volume I Herrington, C.S. & McGee, J. O'D. (Eds)
.... Spiralbound hardback 0-19-963237-5 **£30.00**
.... Paperback 0-19-963236-7 **£19.50**
105. Biomechanics-Materials Vincent, J.F.V. (Ed)
.... Spiralbound hardback 0-19-963223-5 **£35.00**
.... Paperback 0-19-963222-7 **£25.00**
104. Animal Cell Culture (2/e) Freshney, R.I. (Ed)
.... Spiralbound hardback 0-19-963212-X **£30.00**
.... Paperback 0-19-963213-8 **£19.50**
103. Molecular Plant Pathology: Volume II Gurr, S.J., McPherson, M.J. & others (Eds)
.... Spiralbound hardback 0-19-963352-5 **£32.50**
.... Paperback 0-19-963351-7 **£22.50**
102 Signal Transduction Milligan, G. (Ed)
.... Spiralbound hardback 0-19-963296-0 **£30.00**
.... Paperback 0-19-963295-2 **£18.50**
101. Protein Targeting Magee, A.I. & Wileman, T. (Eds)
.... Spiralbound hardback 0-19-963206-5 **£32.50**
.... Paperback 0-19-963210-3 **£22.50**
100. Diagnostic Molecular Pathology: Volume II: Cell and Tissue Genotyping Herrington, C.S. & McGee, J.O'D. (Eds)
.... Spiralbound hardback 0-19-963239-1 **£30.00**
.... Paperback 0-19-963238-3 **£19.50**
99. Neuronal Cell Lines Wood, J.N. (Ed)
.... Spiralbound hardback 0-19-963346-0 **£32.50**
.... Paperback 0-19-963345-2 **£22.50**

98. Neural Transplantation Dunnett, S.B. & Björklund, A. (Eds)
.... Spiralbound hardback 0-19-963286-3 **£30.00**
.... Paperback 0-19-963285-5 **£19.50**

97. Human Cytogenetics: Volume II: Malignancy and Acquired Abnormalities (2/e) Rooney, D.E. & Czepulkowski, B.H. (Eds)
.... Spiralbound hardback 0-19-963290-1 **£30.00**
.... Paperback 0-19-963289-8 **£22.50**

96. Human Cytogenetics: Volume I: Constitutional Analysis (2/e) Rooney, D.E. & Czepulkowski, B.H. (Eds)
.... Spiralbound hardback 0-19-963288-X **£30.00**
.... Paperback 0-19-963287-1 **£22.50**

95. Lipid Modification of Proteins Hooper, N.M. & Turner, A.J. (Eds)
.... Spiralbound hardback 0-19-963274-X **£32.50**
.... Paperback 0-19-963273-1 **£22.50**

94. Biomechanics-Structures and Systems Biewener, A.A. (Ed)
.... Spiralbound hardback 0-19-963268-5 **£42.50**
.... Paperback 0-19-963267-7 **£25.00**

93. Lipoprotein Analysis Converse, C.A. & Skinner, E.R. (Eds)
.... Spiralbound hardback 0-19-963192-1 **£30.00**
.... Paperback 0-19-963231-6 **£19.50**

92. Receptor-Ligand Interactions Hulme, E.C. (Ed)
.... Spiralbound hardback 0-19-963090-9 **£35.00**
.... Paperback 0-19-963091-7 **£27.50**

91. Molecular Genetic Analysis of Populations Hoelzel, A.R. (Ed)
.... Spiralbound hardback 0-19-963278-2 **£32.50**
.... Paperback 0-19-963277-4 **£22.50**

90. Enzyme Assays Eisenthal, R. & Danson, M.J. (Eds)
.... Spiralbound hardback 0-19-963142-5 **£35.00**
.... Paperback 0-19-963143-3 **£25.00**

89. Microcomputers in Biochemistry Bryce, C.F.A. (Ed)
.... Spiralbound hardback 0-19-963253-7 **£30.00**
.... Paperback 0-19-963252-9 **£19.50**

88. The Cytoskeleton Carraway, K.L. & Carraway, C.A.C. (Eds)
.... Spiralbound hardback 0-19-963257-X **£30.00**
.... Paperback 0-19-963256-1 **£19.50**

87. Monitoring Neuronal Activity Stamford, J.A. (Ed)
.... Spiralbound hardback 0-19-963244-8 **£30.00**
.... Paperback 0-19-963243-X **£19.50**

86. Crystallization of Nucleic Acids and Proteins Ducruix, A. & Giegé, R. (Eds)
.... Spiralbound hardback 0-19-963245-6 **£35.00**
.... Paperback 0-19-963246-4 **£25.00**

85. Molecular Plant Pathology: Volume I Gurr, S.J., McPherson, M.J. & others (Eds)
.... Spiralbound hardback 0-19-963103-4 **£30.00**
.... Paperback 0-19-963102-6 **£19.50**

84. Anaerobic Microbiology Levett, P.N. (Ed)
.... Spiralbound hardback 0-19-963204-9 **£32.50**
.... Paperback 0-19-963262-6 **£22.50**

83. Oligonucleotides and Analogues Eckstein, F. (Ed)
.... Spiralbound hardback 0-19-963280-4 **£32.50**
.... Paperback 0-19-963279-0 **£22.50**

82. Electron Microscopy in Biology Harris, R. (Ed)
.... Spiralbound hardback 0-19-963219-7 **£32.50**
.... Paperback 0-19-963215-4 **£22.50**

81. Essential Molecular Biology: Volume II Brown, T.A. (Ed)
.... Spiralbound hardback 0-19-963112-3 **£32.50**
.... Paperback 0-19-963113-1 **£22.50**

80. Cellular Calcium McCormack, J.G. & Cobbold, P.H. (Eds)
.... Spiralbound hardback 0-19-963131-X **£35.00**
.... Paperback 0-19-963130-1 **£25.00**

79. Protein Architecture Lesk, A.M.
.... Spiralbound hardback 0-19-963054-2 **£32.50**
.... Paperback 0-19-963055-0 **£22.50**

78. Cellular Neurobiology Chad, J. & Wheal, H. (Eds)
.... Spiralbound hardback 0-19-963106-9 **£32.50**
.... Paperback 0-19-963107-7 **£22.50**

77. PCR McPherson, M.J., Quirke, P. & others (Eds)
.... Spiralbound hardback 0-19-963226-X **£30.00**
.... Paperback 0-19-963196-4 **£19.50**

76. Mammalian Cell Biotechnology Butler, M. (Ed)
.... Spiralbound hardback 0-19-963207-3 **£30.00**
.... Paperback 0-19-963209-X **£19.50**

75. Cytokines Balkwill, F.R. (Ed)
.... Spiralbound hardback 0-19-963218-9 **£35.00**
.... Paperback 0-19-963214-6 **£25.00**

74. Molecular Neurobiology Chad, J. & Wheal, H. (Eds)
.... Spiralbound hardback 0-19-963108-5 **£30.00**
.... Paperback 0-19-963109-3 **£19.50**

73. Directed Mutagenesis McPherson, M.J. (Ed)
.... Spiralbound hardback 0-19-963141-7 **£30.00**
.... Paperback 0-19-963140-9 **£19.50**

72. Essential Molecular Biology: Volume I Brown, T.A. (Ed)
.... Spiralbound hardback 0-19-963110-7 **£32.50**
.... Paperback 0-19-963111-5 **£22.50**

71. Peptide Hormone Action Siddle, K. & Hutton, J.C.
.... Spiralbound hardback 0-19-963070-4 **£32.50**
.... Paperback 0-19-963071-2 **£22.50**

70. Peptide Hormone Secretion Hutton, J.C. & Siddle, K. (Eds)
.... Spiralbound hardback 0-19-963068-2 **£35.00**
.... Paperback 0-19-963069-0 **£25.00**

69. Postimplantation Mammalian Embryos Copp, A.J. & Cockroft, D.L. (Eds)
.... Spiralbound hardback 0-19-963088-7 **£15.00**
.... Paperback 0-19-963089-5 **£12.50**

68. Receptor-Effector Coupling Hulme, E.C. (Ed)
.... Spiralbound hardback 0-19-963094-1 **£30.00**
.... Paperback 0-19-963095-X **£19.50**

67. Gel Electrophoresis of Proteins (2/e) Hames, B.D. & Rickwood, D. (Eds)
.... Spiralbound hardback 0-19-963074-7 **£35.00**
.... Paperback 0-19-963075-5 **£25.00**

66. Clinical Immunology Gooi, H.C. & Chapel, H. (Eds)
.... Spiralbound hardback 0-19-963086-0 **£32.50**
.... Paperback 0-19-963087-9 **£22.50**

65. Receptor Biochemistry Hulme, E.C. (Ed)
.... Paperback 0-19-963093-3 **£25.00**

64. Gel Electrophoresis of Nucleic Acids (2/e) Rickwood, D. & Hames, B.D. (Eds)
.... Spiralbound hardback 0-19-963082-8 **£32.50**
.... Paperback 0-19-963083-6 **£22.50**

63. Animal Virus Pathogenesis Oldstone, M.B.A. (Ed)
.... Spiralbound hardback 0-19-963100-X **£15.00**
.... Paperback 0-19-963101-8 **£12.50**

62. Flow Cytometry Ormerod, M.G. (Ed)
.... Paperback 0-19-963053-4 **£22.50**

61. Radioisotopes in Biology Slater, R.J. (Ed)
.... Spiralbound hardback 0-19-963080-1 **£32.50**
.... Paperback 0-19-963081-X **£22.50**

60. Biosensors Cass, A.E.G. (Ed)
.... Spiralbound hardback 0-19-963046-1 **£30.00**
.... Paperback 0-19-963047-X **£19.50**

59. Ribosomes and Protein Synthesis Spedding, G. (Ed)
.... Spiralbound hardback 0-19-963104-2 **£15.00**
.... Paperback 0-19-963105-0 **£12.50**

58. Liposomes New, R.R.C. (Ed)
.... Spiralbound hardback 0-19-963076-3 **£35.00**
.... Paperback 0-19-963077-1 **£22.50**

57. Fermentation McNeil, B. & Harvey, L.M. (Eds)
.... Spiralbound hardback 0-19-963044-5 **£30.00**
.... Paperback 0-19-963045-3 **£19.50**

56. Protein Purification Applications Harris, E.L.V. & Angal, S. (Eds)
.... Spiralbound hardback 0-19-963022-4 **£30.00**
.... Paperback 0-19-963023-2 **£18.50**

55. Nucleic Acids Sequencing Howe, C.J. & Ward, E.S. (Eds)
.... Spiralbound hardback 0-19-963056-9 **£30.00**
.... Paperback 0-19-963057-7 **£19.50**

54. Protein Purification Methods Harris, E.L.V. & Angal, S. (Eds)
.... Spiralbound hardback 0-19-963002-X **£30.00**
.... Paperback 0-19-963003-8 **£22.50**

53. Solid Phase Peptide Synthesis Atherton, E. & Sheppard, R.C.
.... Spiralbound hardback 0-19-963066-6 **£15.00**
.... Paperback 0-19-963067-4 **£12.50**

52. Medical Bacteriology Hawkey, P.M. & Lewis, D.A. (Eds)
.... Paperback 0-19-963009-7 **£25.00**

51. Proteolytic Enzymes Beynon, R.J. & Bond, J.S. (Eds)
.... Spiralbound hardback 0-19-963058-5 **£30.00**
.... Paperback 0-19-963059-3 **£19.50**

50. Medical Mycology Evans, E.G.V. & Richardson, M.D. (Eds)
.... Spiralbound hardback 0-19-963010-0 **£37.50**
.... Paperback 0-19-963011-9 **£25.00**

49. Computers in Microbiology Bryant, T.N. & Wimpenny, J.W.T. (Eds)
.... Paperback 0-19-963015-1 **£12.50**

48. **Protein Sequencing** Findlay, J.B.C. & Geisow, M.J. (Eds)
.... Spiralbound hardback 0-19-963012-7 **£15.00**
.... Paperback 0-19-963013-5 **£12.50**

47. **Cell Growth and Division** Baserga, R. (Ed)
.... Spiralbound hardback 0-19-963026-7 **£15.00**
.... Paperback 0-19-963027-5 **£12.50**

46. **Protein Function** Creighton, T.E. (Ed)
.... Spiralbound hardback 0-19-963006-2 **£32.50**
.... Paperback 0-19-963007-0 **£22.50**

45. **Protein Structure** Creighton, T.E. (Ed)
.... Spiralbound hardback 0-19-963000-3 **£32.50**
.... Paperback 0-19-963001-1 **£22.50**

44. **Antibodies: Volume II** Catty, D. (Ed)
.... Spiralbound hardback 0-19-963018-6 **£30.00**
.... Paperback 0-19-963019-4 **£19.50**

43. **HPLC of Macromolecules** Oliver, R.W.A. (Ed)
.... Spiralbound hardback 0-19-963020-8 **£30.00**
.... Paperback 0-19-963021-6 **£19.50**

42. **Light Microscopy in Biology** Lacey, A.J. (Ed)
.... Spiralbound hardback 0-19-963036-4 **£30.00**
.... Paperback 0-19-963037-2 **£19.50**

41. **Plant Molecular Biology** Shaw, C.H. (Ed)
.... Paperback 1-85221-056-7 **£12.50**

40. **Microcomputers in Physiology** Fraser, P.J. (Ed)
.... Spiralbound hardback 1-85221-129-6 **£15.00**
.... Paperback 1-85221-130-X **£12.50**

39. **Genome Analysis** Davies, K.E. (Ed)
.... Spiralbound hardback 1-85221-109-1 **£30.00**
.... Paperback 1-85221-110-5 **£18.50**

38. **Antibodies: Volume I** Catty, D. (Ed)
.... Paperback 0-947946-85-3 **£19.50**

37. **Yeast** Campbell, I. & Duffus, J.H. (Eds)
.... Paperback 0-947946-79-9 **£12.50**

36. **Mammalian Development** Monk, M. (Ed)
.... Hardback 1-85221-030-3 **£15.00**
.... Paperback 1-85221-029-X **£12.50**

35. **Lymphocytes** Klaus, G.G.B. (Ed)
.... Hardback 1-85221-018-4 **£30.00**

34. **Lymphokines and Interferons** Clemens, M.J., Morris, A.G. & others (Eds)
.... Paperback 1-85221-035-4 **£12.50**

33. **Mitochondria** Darley-Usmar, V.M., Rickwood, D. & others (Eds)
.... Hardback 1-85221-034-6 **£32.50**
.... Paperback 1-85221-033-8 **£22.50**

32. **Prostaglandins and Related Substances** Benedetto, C., McDonald-Gibson, R.G. & others (Eds)
.... Hardback 1-85221-032-X **£15.00**
.... Paperback 1-85221-031-1 **£12.50**

31. **DNA Cloning: Volume III** Glover, D.M. (Ed)
.... Hardback 1-85221-049-4 **£15.00**
.... Paperback 1-85221-048-6 **£12.50**

30. **Steroid Hormones** Green, B. & Leake, R.E. (Eds)
.... Paperback 0-947946-53-5 **£19.50**

29. **Neurochemistry** Turner, A.J. & Bachelard, H.S. (Eds)
.... Hardback 1-85221-028-1 **£15.00**
.... Paperback 1-85221-027-3 **£12.50**

28. **Biological Membranes** Findlay, J.B.C. & Evans, W.H. (Eds)
.... Hardback 0-947946-84-5 **£15.00**
.... Paperback 0-947946-83-7 **£12.50**

27. **Nucleic Acid and Protein Sequence Analysis** Bishop, M.J. & Rawlings, C.J. (Eds)
.... Hardback 1-85221-007-9 **£35.00**
.... Paperback 1-85221-006-0 **£25.00**

26. **Electron Microscopy in Molecular Biology** Sommerville, J. & Scheer, U. (Eds)
.... Hardback 0-947946-64-0 **£15.00**
.... Paperback 0-947946-54-3 **£12.50**

25. **Teratocarcinomas and Embryonic Stem Cells** Robertson, E.J. (Ed)
.... Paperback 1-85221-004-4 **£19.50**

24. **Spectrophotometry and Spectrofluorimetry** Harris, D.A. & Bashford, C.L. (Eds)
.... Hardback 0-947946-69-1 **£15.00**
.... Paperback 0-947946-46-2 **£12.50**

23. **Plasmids** Hardy, K.G. (Ed)
.... Paperback 0-947946-81-0 **£12.50**

22. **Biochemical Toxicology** Snell, K. & Mullock, B. (Eds)
.... Paperback 0-947946-52-7 **£12.50**

19. **Drosophila** Roberts, D.B. (Ed)
.... Hardback 0-947946-66-7 **£32.50**
.... Paperback 0-947946-45-4 **£22.50**

17. **Photosynthesis: Energy Transduction** Hipkins, M.F. & Baker, N.R. (Eds)
.... Hardback 0-947946-63-2 **£15.00**
.... Paperback 0-947946-51-9 **£12.50**

16. **Human Genetic Diseases** Davies, K.E. (Ed)
.... Hardback 0-947946-76-4 **£15.00**
.... Paperback 0-947946-75-6 **£12.50**

14. **Nucleic Acid Hybridisation** Hames, B.D. & Higgins, S.J. (Eds)
.... Hardback 0-947946-61-6 **£15.00**
.... Paperback 0-947946-23-3 **£12.50**

13. **Immobilised Cells and Enzymes** Woodward, J. (Ed)
.... Hardback 0-947946-60-8 **£15.00**

12. **Plant Cell Culture** Dixon, R.A. (Ed)
.... Paperback 0-947946-22-5 **£19.50**

11a. **DNA Cloning: Volume I** Glover, D.M. (Ed)
.... Paperback 0-947946-18-7 **£12.50**

11b. **DNA Cloning: Volume II** Glover, D.M. (Ed)
.... Paperback 0-947946-19-5 **£12.50**

10. **Virology** Mahy, B.W.J. (Ed)
.... Paperback 0-904147-78-9 **£19.50**

9. **Affinity Chromatography** Dean, P.D.G., Johnson, W.S. & others (Eds)
.... Paperback 0-904147-71-1 **£19.50**

7. **Microcomputers in Biology** Ireland, C.R. & Long, S.P. (Eds)
.... Paperback 0-904147-57-6 **£18.00**

6. **Oligonucleotide Synthesis** Gait, M.J. (Ed)
.... Paperback 0-904147-74-6 **£18.50**

5. **Transcription and Translation** Hames, B.D. & Higgins, S.J. (Eds)
.... Paperback 0-904147-52-5 **£12.50**

3. **Iodinated Density Gradient Media** Rickwood, D. (Ed)
.... Paperback 0-904147-51-7 **£12.50**

Sets

Essential Molecular Biology: 2 vol set Brown, T.A. (Ed)
.... Spiralbound hardback 0-19-963114-X **£58.00**
.... Paperback 0-19-963115-8 **£40.00**

Antibodies: 2 vol set Catty, D. (Ed)
.... Paperback 0-19-963063-1 **£33.00**

Cellular and Molecular Neurobiology: 2 vol set Chad, J. & Wheal, H. (Eds)
.... Spiralbound hardback 0-19-963255-3 **£56.00**
.... Paperback 0-19-963254-5 **£38.00**

Protein Structure and Protein Function: 2 vol set Creighton, T.E. (Ed)
.... Spiralbound hardback 0-19-963064-X **£55.00**
.... Paperback 0-19-963065-8 **£38.00**

DNA Cloning: 2 vol set Glover, D.M. (Ed)
.... Paperback 1-85221-069-9 **£30.00**

Molecular Plant Pathology: 2 vol set Gurr, S.J., McPherson, M.J. & others (Eds)
.... Spiralbound hardback 0-19-963354-1 **£56.00**
.... Paperback 0-19-963353-3 **£37.00**

Protein Purification Methods, and Protein Purification Applications: 2 vol set Harris, E.L.V. & Angal, S. (Eds)
.... Spiralbound hardback 0-19-963048-8 **£48.00**
.... Paperback 0-19-963049-6 **£32.00**

Diagnostic Molecular Pathology: 2 vol set Herrington, C.S. & McGee, J. O'D. (Eds)
.... Spiralbound hardback 0-19-963241-3 **£54.00**
.... Paperback 0-19-963240-5 **£35.00**

RNA Processing: 2 vol set Higgins, S.J. & Hames, B.D. (Eds)
.... Spiralbound hardback 0-19-963473-4 **£54.00**
.... Paperback 0-19-963472-6 **£35.00**

Receptor Biochemistry; Receptor-Effector Coupling; Receptor-Ligand Interactions: 3 vol set Hulme, E.C. (Ed)
.... Paperback 0-19-963097-6 **£62.50**

Human Cytogenetics: 2 vol set (2/e) Rooney, D.E. & Czepulkowski, B.H. (Eds)
.... Hardback 0-19-963314-2 **£58.50**
.... Paperback 0-19-963313-4 **£40.50**

Behavioural Neuroscience: 2 vol set Sahgal, A. (Ed)
.... Spiralbound hardback 0-19-963460-2 **£58.00**
.... Paperback 0-19-963459-9 **£40.00**

Peptide Hormone Secretion/Peptide Hormone Action: 2 vol set Siddle, K. & Hutton, J.C. (Eds)
.... Spiralbound hardback 0-19-963072-0 **£55.00**
.... Paperback 0-19-963073-9 **£38.00**

ORDER FORM for UK, Europe and Rest of World

(Excluding USA and Canada)

Qty	ISBN	Author	Title	Amount
			P&P	
			*VAT	
			TOTAL	

Please add postage and packing: £1.75 for UK orders under £20; £2.75 for UK orders over £20; overseas orders add 10% of total.
* EC customers please note that VAT must be added (excludes UK customers)

Name ....

Address ....

....

.... Post code ....

[ ] Please charge £ .... to my credit card
Access/VISA/Eurocard/AMEX/Diners Club (circle appropriate card)

Card No .... Expiry date ....

Signature ....

Credit card account address if different from above:

....

.... Postcode ....

[ ] I enclose a cheque for £....................

Please return this form to: OUP Distribution Services, Saxon Way West, Corby, Northants NN18 9ES, UK

OR ORDER BY CREDIT CARD HOTLINE: Tel +44-(0)536-741519 or Fax +44-(0)536-746337

ORDER OTHER TITLES OF INTEREST TODAY

Price list for: USA and Canada

128. Behavioural Neuroscience: Volume I Sahgal, A. (Ed)
.... Spiralbound hardback 0-19-963368-1 **$57.00**
.... Paperback 0-19-963367-3 **$37.00**

127. Molecular Virology Davison, A.J. & Elliott, R.M. (Eds)
.... Spiralbound hardback 0-19-963358-4 **$49.00**
.... Paperback 0-19-963357-6 **$32.00**

126. Gene Targeting Joyner, A.L. (Ed)
.... Spiralbound hardback 0-19-963407-6 **$49.00**
.... Paperback 0-19-9634036-8 **$34.00**

124. Human Genetic Disease Analysis (2/e) Davies, K.E. (Ed)
.... Spiralbound hardback 0-19-963309-6 **$54.00**
.... Paperback 0-19-963308-8 **$33.00**

123. Protein Phosphorylation Hardie, D.G. (Ed)
.... Spiralbound hardback 0-19-963306-1 **$65.00**
.... Paperback 0-19-963305-3 **$45.00**

122. Immunocytochemistry Beesley, J. (Ed)
.... Spiralbound hardback 0-19-963270-7 **$62.00**
.... Paperback 0-19-963269-3 **$42.00**

121. Tumour Immunobiology Gallagher, G., Rees, R.C. & others (Eds)
.... Spiralbound hardback 0-19-963370-3 **$72.00**
.... Paperback 0-19-963369-X **$50.00**

120. Transcription Factors Latchman, D.S. (Ed)
.... Spiralbound hardback 0-19-963342-8 **$48.00**
.... Paperback 0-19-963341-X **$31.00**

119. Growth Factors McKay, I. & Leigh, I. (Eds)
.... Spiralbound hardback 0-19-963360-6 **$48.00**
.... Paperback 0-19-963359-2 **$31.00**

118. Histocompatibility Testing Dyer, P. & Middleton, D. (Eds)
.... Spiralbound hardback 0-19-963364-9 **$60.00**
.... Paperback 0-19-963363-0 **$41.00**

117. Gene Transcription Hames, B.D. & Higgins, S.J. (Eds)
.... Spiralbound hardback 0-19-963292-8 **$72.00**
.... Paperback 0-19-963291-X **$50.00**

116. Electrophysiology Wallis, D.I. (Ed)
.... Spiralbound hardback 0-19-963348-7 **$56.00**
.... Paperback 0-19-963347-9 **$39.00**

115. Biological Data Analysis Fry, J.C. (Ed)
.... Spiralbound hardback 0-19-963340-1 **$80.00**
.... Paperback 0-19-963339-8 **$60.00**

114. Experimental Neuroanatomy Bolam, J.P. (Ed)
.... Spiralbound hardback 0-19-963326-6 **$59.00**
.... Paperback 0-19-963325-8 **$39.00**

113. Preparative Centrifugation Rickwood, D. (Ed)
.... Spiralbound hardback 0-19-963208-1 **$78.00**
.... Paperback 0-19-963211-1 **$44.00**

111. Haemopoiesis Testa, N.G. & Molineux, G. (Eds)
.... Spiralbound hardback 0-19-963366-5 **$59.00**
.... Paperback 0-19-963365-7 **$39.00**

110. Pollination Ecology Dafni, A.
.... Spiralbound hardback 0-19-963299-5 **$56.95**
.... Paperback 0-19-963298-7 **$39.95**

109. In Situ Hybridization Wilkinson, D.G. (Ed)
.... Spiralbound hardback 0-19-963328-2 **$58.00**
.... Paperback 0-19-963327-4 **$36.00**

108. Protein Engineering Rees, A.R., Sternberg, M.J.E. & others (Eds)
.... Spiralbound hardback 0-19-963139-5 **$64.00**
.... Paperback 0-19-963138-7 **$44.00**

107. Cell-Cell Interactions Stevenson, B.R., Gallin, W.J. & others (Eds)
.... Spiralbound hardback 0-19-963319-3 **$55.00**
.... Paperback 0-19-963318-5 **$38.00**

106. Diagnostic Molecular Pathology: Volume I Herrington, C.S. & McGee, J. O'D. (Eds)
.... Spiralbound hardback 0-19-963237-5 **$50.00**
.... Paperback 0-19-963236-7 **$33.00**

105. Biomechanics-Materials Vincent, J.F.V. (Ed)
.... Spiralbound hardback 0-19-963223-5 **$70.00**
.... Paperback 0-19-963222-7 **$50.00**

104. Animal Cell Culture (2/e) Freshney, R.I. (Ed)
.... Spiralbound hardback 0-19-963212-X **$55.00**
.... Paperback 0-19-963213-8 **$35.00**

103. Molecular Plant Pathology: Volume II Gurr, S.J., McPherson, M.J. & others (Eds)
.... Spiralbound hardback 0-19-963352-5 **$65.00**
.... Paperback 0-19-963351-7 **$45.00**

102. Signal Transduction Milligan, G. (Ed)
.... Spiralbound hardback 0-19-963296-0 **$60.00**
.... Paperback 0-19-963295-2 **$38.00**

101. Protein Targeting Magee, A.I. & Wileman, T. (Eds)
.... Spiralbound hardback 0-19-963206-5 **$75.00**
.... Paperback 0-19-963210-3 **$50.00**

100. Diagnostic Molecular Pathology: Volume II: Cell and Tissue Genotyping Herrington, C.S. & McGee, J.O'D. (Eds)
.... Spiralbound hardback 0-19-963239-1 **$60.00**
.... Paperback 0-19-963238-3 **$39.00**

99. Neuronal Cell Lines Wood, J.N. (Ed)
.... Spiralbound hardback 0-19-963346-0 **$68.00**
.... Paperback 0-19-963345-2 **$48.00**

98. Neural Transplantation Dunnett, S.B. & Bj‹148›rklund, A. (Eds)
.... Spiralbound hardback 0-19-963286-3 **$69.00**
.... Paperback 0-19-963285-5 **$42.00**

97. Human Cytogenetics: Volume II: Malignancy and Acquired Abnormalities (2/e) Rooney, D.E. & Czepulkowski, B.H. (Eds)
.... Spiralbound hardback 0-19-963290-1 **$75.00**
.... Paperback 0-19-963289-8 **$50.00**

96. Human Cytogenetics: Volume I: Constitutional Analysis (2/e) Rooney, D.E. & Czepulkowski, B.H. (Eds)
.... Spiralbound hardback 0-19-963288-X **$75.00**
.... Paperback 0-19-963287-1 **$50.00**

95. Lipid Modification of Proteins Hooper, N.M. & Turner, A.J. (Eds)
.... Spiralbound hardback 0-19-963274-X **$75.00**
.... Paperback 0-19-963273-1 **$50.00**

94. Biomechanics-Structures and Systems Biewener, A.A. (Ed)
.... Spiralbound hardback 0-19-963268-5 **$85.00**
.... Paperback 0-19-963267-7 **$50.00**

93. Lipoprotein Analysis Converse, C.A. & Skinner, E.R. (Eds)
.... Spiralbound hardback 0-19-963192-1 **$65.00**
.... Paperback 0-19-963231-6 **$42.00**

92. Receptor-Ligand Interactions Hulme, E.C. (Ed)
.... Spiralbound hardback 0-19-963090-9 **$75.00**
.... Paperback 0-19-963091-7 **$50.00**

91. Molecular Genetic Analysis of Populations Hoelzel, A.R. (Ed)
.... Spiralbound hardback 0-19-963278-2 **$65.00**
.... Paperback 0-19-963277-4 **$45.00**

90. **Enzyme Assays** Eisenthal, R. & Danson, M.J. (Eds)
.... Spiralbound hardback 0-19-963142-5 **$68.00**
.... Paperback 0-19-963143-3 **$48.00**
89. **Microcomputers in Biochemistry** Bryce, C.F.A. (Ed)
.... Spiralbound hardback 0-19-963253-7 **$60.00**
.... Paperback 0-19-963252-9 **$40.00**
88. **The Cytoskeleton** Carraway, K.L. & Carraway, C.A.C. (Eds)
.... Spiralbound hardback 0-19-963257-X **$60.00**
.... Paperback 0-19-963256-1 **$40.00**
87. **Monitoring Neuronal Activity** Stamford, J.A. (Ed)
.... Spiralbound hardback 0-19-963244-8 **$60.00**
.... Paperback 0-19-963243-X **$40.00**
86. **Crystallization of Nucleic Acids and Proteins** Ducruix, A. & Giegé, R. (Eds)
.... Spiralbound hardback 0-19-963245-6 **$60.00**
.... Paperback 0-19-963246-4 **$50.00**
85. **Molecular Plant Pathology: Volume I** Gurr, S.J., McPherson, M.J. & others (Eds)
.... Spiralbound hardback 0-19-963103-4 **$60.00**
.... Paperback 0-19-963102-6 **$40.00**
84. **Anaerobic Microbiology** Levett, P.N. (Ed)
.... Spiralbound hardback 0-19-963204-9 **$75.00**
.... Paperback 0-19-963262-6 **$45.00**
83. **Oligonucleotides and Analogues** Eckstein, F. (Ed)
.... Spiralbound hardback 0-19-963280-4 **$65.00**
.... Paperback 0-19-963279-0 **$45.00**
82. **Electron Microscopy in Biology** Harris, R. (Ed)
.... Spiralbound hardback 0-19-963219-7 **$65.00**
.... Paperback 0-19-963215-4 **$45.00**
81. **Essential Molecular Biology: Volume II** Brown, T.A. (Ed)
.... Spiralbound hardback 0-19-963112-3 **$65.00**
.... Paperback 0-19-963113-1 **$45.00**
80. **Cellular Calcium** McCormack, J.G. & Cobbold, P.H. (Eds)
.... Spiralbound hardback 0-19-963131-X **$75.00**
.... Paperback 0-19-963130-1 **$50.00**
79. **Protein Architecture** Lesk, A.M.
.... Spiralbound hardback 0-19-963054-2 **$65.00**
.... Paperback 0-19-963055-0 **$45.00**
78. **Cellular Neurobiology** Chad, J. & Wheal, H. (Eds)
.... Spiralbound hardback 0-19-963106-9 **$73.00**
.... Paperback 0-19-963107-7 **$43.00**
77. **PCR** McPherson, M.J., Quirke, P. & others (Eds)
.... Spiralbound hardback 0-19-963226-X **$55.00**
.... Paperback 0-19-963196-4 **$40.00**
76. **Mammalian Cell Biotechnology** Butler, M. (Ed)
.... Spiralbound hardback 0-19-963207-3 **$60.00**
.... Paperback 0-19-963209-X **$40.00**
75. **Cytokines** Balkwill, F.R. (Ed)
.... Spiralbound hardback 0-19-963218-9 **$64.00**
.... Paperback 0-19-963214-6 **$44.00**
74. **Molecular Neurobiology** Chad, J. & Wheal, H. (Eds)
.... Spiralbound hardback 0-19-963108-5 **$56.00**
.... Paperback 0-19-963109-3 **$36.00**
73. **Directed Mutagenesis** McPherson, M.J. (Ed)
.... Spiralbound hardback 0-19-963141-7 **$55.00**
.... Paperback 0-19-963140-9 **$35.00**
72. **Essential Molecular Biology: Volume I** Brown, T.A. (Ed)
.... Spiralbound hardback 0-19-963110-7 **$65.00**
.... Paperback 0-19-963111-5 **$45.00**
71. **Peptide Hormone Action** Siddle, K. & Hutton, J.C.
.... Spiralbound hardback 0-19-963070-4 **$70.00**
.... Paperback 0-19-963071-2 **$50.00**
70. **Peptide Hormone Secretion** Hutton, J.C. & Siddle, K. (Eds)
.... Spiralbound hardback 0-19-963068-2 **$70.00**
.... Paperback 0-19-963069-0 **$50.00**
69. **Postimplantation Mammalian Embryos** Copp, A.J. & Cockroft, D.L. (Eds)
.... Spiralbound hardback 0-19-963088-7 **$70.00**
.... Paperback 0-19-963089-5 **$50.00**
68. **Receptor-Effector Coupling** Hulme, E.C. (Ed)
.... Spiralbound hardback 0-19-963094-1 **$70.00**
.... Paperback 0-19-963095-X **$45.00**
67. **Gel Electrophoresis of Proteins (2/e)** Hames, B.D. & Rickwood, D. (Eds)
.... Spiralbound hardback 0-19-963074-7 **$75.00**
.... Paperback 0-19-963075-5 **$50.00**
66. **Clinical Immunology** Gooi, H.C. & Chapel, H. (Eds)
.... Spiralbound hardback 0-19-963086-0 **$69.95**
.... Paperback 0-19-963087-9 **$50.00**
65. **Receptor Biochemistry** Hulme, E.C. (Ed)
.... Paperback 0-19-963093-3 **$50.00**
64. **Gel Electrophoresis of Nucleic Acids (2/e)** Rickwood, D. & Hames, B.D. (Eds)
.... Spiralbound hardback 0-19-963082-8 **$75.00**
.... Paperback 0-19-963083-6 **$50.00**
63. **Animal Virus Pathogenesis** Oldstone, M.B.A. (Ed)
.... Spiralbound hardback 0-19-963100-X **$68.00**
.... Paperback 0-19-963101-8 **$40.00**
62. **Flow Cytometry** Ormerod, M.G. (Ed)
.... Paperback 0-19-963053-4 **$50.00**
61. **Radioisotopes in Biology** Slater, R.J. (Ed)
.... Spiralbound hardback 0-19-963080-1 **$75.00**
.... Paperback 0-19-963081-X **$45.00**
60. **Biosensors** Cass, A.E.G. (Ed)
.... Spiralbound hardback 0-19-963046-1 **$65.00**
.... Paperback 0-19-963047-X **$43.00**
59. **Ribosomes and Protein Synthesis** Spedding, G. (Ed)
.... Spiralbound hardback 0-19-963104-2 **$75.00**
.... Paperback 0-19-963105-0 **$45.00**
58. **Liposomes** New, R.R.C. (Ed)
.... Spiralbound hardback 0-19-963076-3 **$70.00**
.... Paperback 0-19-963077-1 **$45.00**
57. **Fermentation** McNeil, B. & Harvey, L.M. (Eds)
.... Spiralbound hardback 0-19-963044-5 **$65.00**
.... Paperback 0-19-963045-3 **$39.00**
56. **Protein Purification Applications** Harris, E.L.V. & Angal, S. (Eds)
.... Spiralbound hardback 0-19-963022-4 **$54.00**
.... Paperback 0-19-963023-2 **$36.00**
55. **Nucleic Acids Sequencing** Howe, C.J. & Ward, E.S. (Eds)
.... Spiralbound hardback 0-19-963056-9 **$59.00**
.... Paperback 0-19-963057-7 **$38.00**
54. **Protein Purification Methods** Harris, E.L.V. & Angal, S. (Eds)
.... Spiralbound hardback 0-19-963002-X **$60.00**
.... Paperback 0-19-963003-8 **$40.00**
53. **Solid Phase Peptide Synthesis** Atherton, E. & Sheppard, R.C.
.... Spiralbound hardback 0-19-963066-6 **$58.00**
.... Paperback 0-19-963067-4 **$39.95**
52. **Medical Bacteriology** Hawkey, P.M. & Lewis, D.A. (Eds)
.... Paperback 0-19-963009-7 **$50.00**
51. **Proteolytic Enzymes** Beynon, R.J. & Bond, J.S. (Eds)
.... Spiralbound hardback 0-19-963058-5 **$60.00**
.... Paperback 0-19-963059-3 **$39.00**
50. **Medical Mycology** Evans, E.G.V. & Richardson, M.D. (Eds)
.... Spiralbound hardback 0-19-963010-0 **$69.95**
.... Paperback 0-19-963011-9 **$50.00**
49. **Computers in Microbiology** Bryant, T.N. & Wimpenny, J.W.T. (Eds)
.... Paperback 0-19-963015-1 **$40.00**
48. **Protein Sequencing** Findlay, J.B.C. & Geisow, M.J. (Eds)
.... Spiralbound hardback 0-19-963012-7 **$56.00**
.... Paperback 0-19-963013-5 **$38.00**
47. **Cell Growth and Division** Baserga, R. (Ed)
.... Spiralbound hardback 0-19-963026-7 **$62.00**
.... Paperback 0-19-963027-5 **$38.00**
46. **Protein Function** Creighton, T.E. (Ed)
.... Spiralbound hardback 0-19-963006-2 **$65.00**
.... Paperback 0-19-963007-0 **$45.00**
45. **Protein Structure** Creighton, T.E. (Ed)
.... Spiralbound hardback 0-19-963000-3 **$65.00**
.... Paperback 0-19-963001-1 **$45.00**
44. **Antibodies: Volume II** Catty, D. (Ed)
.... Spiralbound hardback 0-19-963018-6 **$58.00**
.... Paperback 0-19-963019-4 **$39.00**
43. **HPLC of Macromolecules** Oliver, R.W.A. (Ed)
.... Spiralbound hardback 0-19-963020-8 **$54.00**
.... Paperback 0-19-963021-6 **$45.00**
42. **Light Microscopy in Biology** Lacey, A.J. (Ed)
.... Spiralbound hardback 0-19-963036-4 **$62.00**
.... Paperback 0-19-963037-2 **$38.00**
41. **Plant Molecular Biology** Shaw, C.H. (Ed)
.... Paperback 1-85221-056-7 **$38.00**
40. **Microcomputers in Physiology** Fraser, P.J. (Ed)
.... Spiralbound hardback 1-85221-129-6 **$54.00**
.... Paperback 1-85221-130-X **$36.00**
39. **Genome Analysis** Davies, K.E. (Ed)
.... Spiralbound hardback 1-85221-109-1 **$54.00**
.... Paperback 1-85221-110-5 **$36.00**
38. **Antibodies: Volume I** Catty, D. (Ed)
.... Paperback 0-947946-85-3 **$38.00**
37. **Yeast** Campbell, I. & Duffus, J.H. (Eds)
.... Paperback 0-947946-79-9 **$36.00**

36. **Mammalian Development** Monk, M. (Ed)
.... Hardback 1-85221-030-3 **$60.00**
.... Paperback 1-85221-029-X **$45.00**
35. **Lymphocytes** Klaus, G.G.B. (Ed)
.... Hardback 1-85221-018-4 **$54.00**
34. **Lymphokines and Interferons** Clemens, M.J., Morris, A.G. & others (Eds)
.... Paperback 1-85221-035-4 **$44.00**
33. **Mitochondria** Darley-Usmar, V.M., Rickwood, D. & others (Eds)
.... Hardback 1-85221-034-6 **$65.00**
.... Paperback 1-85221-033-8 **$45.00**
32. **Prostaglandins and Related Substances** Benedetto, C., McDonald-Gibson, R.G. & others (Eds)
.... Hardback 1-85221-032-X **$58.00**
.... Paperback 1-85221-031-1 **$38.00**
31. **DNA Cloning: Volume III** Glover, D.M. (Ed)
.... Hardback 1-85221-049-4 **$56.00**
.... Paperback 1-85221-048-6 **$36.00**
30. **Steroid Hormones** Green, B. & Leake, R.E. (Eds)
.... Paperback 0-947946-53-5 **$40.00**
29. **Neurochemistry** Turner, A.J. & Bachelard, H.S. (Eds)
.... Hardback 1-85221-028-1 **$56.00**
.... Paperback 1-85221-027-3 **$36.00**
28. **Biological Membranes** Findlay, J.B.C. & Evans, W.H. (Eds)
.... Hardback 0-947946-84-5 **$54.00**
.... Paperback 0-947946-83-7 **$36.00**
27. **Nucleic Acid and Protein Sequence Analysis** Bishop, M.J. & Rawlings, C.J. (Eds)
.... Hardback 1-85221-007-9 **$66.00**
.... Paperback 1-85221-006-0 **$44.00**
26. **Electron Microscopy in Molecular Biology** Sommerville, J. & Scheer, U. (Eds)
.... Hardback 0-947946-64-0 **$54.00**
.... Paperback 0-947946-54-3 **$40.00**
24. **Spectrophotometry and Spectrofluorimetry** Harris, D.A. & Bashford, C.L. (Eds)
.... Hardback 0-947946-69-1 **$56.00**
.... Paperback 0-947946-46-2 **$39.95**
23. **Plasmids** Hardy, K.G. (Ed)
.... Paperback 0-947946-81-0 **$36.00**
22. **Biochemical Toxicology** Snell, K. & Mullock, B. (Eds)
.... Paperback 0-947946-52-7 **$40.00**
19. **Drosophila** Roberts, D.B. (Ed)
.... Hardback 0-947946-66-7 **$67.50**
.... Paperback 0-947946-45-4 **$46.00**
17. **Photosynthesis: Energy Transduction** Hipkins, M.F. & Baker, N.R. (Eds)
.... Hardback 0-947946-63-2 **$54.00**
.... Paperback 0-947946-51-9 **$36.00**
16. **Human Genetic Diseases** Davies, K.E. (Ed)
.... Hardback 0-947946-76-4 **$60.00**
.... Paperback 0-947946-75-6 **$34.00**
14. **Nucleic Acid Hybridisation** Hames, B.D. & Higgins, S.J. (Eds)
.... Hardback 0-947946-61-6 **$60.00**
.... Paperback 0-947946-23-3 **$36.00**
12. **Plant Cell Culture** Dixon, R.A. (Ed)
.... Paperback 0-947946-22-5 **$36.00**
11a. **DNA Cloning: Volume I** Glover, D.M. (Ed)
.... Paperback 0-947946-18-7 **$36.00**
11b. **DNA Cloning: Volume II** Glover, D.M. (Ed)
.... Paperback 0-947946-19-5 **$36.00**
10. **Virology** Mahy, B.W.J. (Ed)
.... Paperback 0-904147-78-9 **$40.00**
9. **Affinity Chromatography** Dean, P.D.G., Johnson, W.S. & others (Eds)
.... Paperback 0-904147-71-1 **$36.00**
7. **Microcomputers in Biology** Ireland, C.R. & Long, S.P. (Eds)
.... Paperback 0-904147-57-6 **$36.00**
6. **Oligonucleotide Synthesis** Gait, M.J. (Ed)
.... Paperback 0-904147-74-6 **$38.00**
5. **Transcription and Translation** Hames, B.D. & Higgins, S.J. (Eds)
.... Paperback 0-904147-52-5 **$38.00**
3. **Iodinated Density Gradient Media** Rickwood, D. (Ed)
.... Paperback 0-904147-51-7 **$36.00**

Sets

Essential Molecular Biology: 2 vol set Brown, T.A. (Ed)
.... Spiralbound hardback 0-19-963114-X **$118.00**
.... Paperback 0-19-963115-8 **$78.00**
Antibodies: 2 vol set Catty, D. (Ed)
.... Paperback 0-19-963063-1 **$70.00**
Cellular and Molecular Neurobiology: 2 vol set Chad, J. & Wheal, H. (Eds)
.... Spiralbound hardback 0-19-963255-3 **$133.00**
.... Paperback 0-19-963254-5 **$79.00**
Protein Structure and Protein Function: 2 vol set Creighton, T.E. (Ed)
.... Spiralbound hardback 0-19-963064-X **$114.00**
.... Paperback 0-19-963065-8 **$80.00**
DNA Cloning: 2 vol set Glover, D.M. (Ed)
.... Paperback 1-85221-069-9 **$92.00**
Molecular Plant Pathology: 2 vol set Gurr, S.J., McPherson, M.J. & others (Eds)
.... Spiralbound hardback 0-19-963354-1 **$110.00**
.... Paperback 0-19-963353-3 **$75.00**
Protein Purification Methods, and Protein Purification Applications: 2 vol set Harris, E.L.V. & Angal, S. (Eds)
.... Spiralbound hardback 0-19-963048-8 **$98.00**
.... Paperback 0-19-963049-6 **$68.00**
Diagnostic Molecular Pathology: 2 vol set Herrington, C.S. & McGee, J. O'D. (Eds)
.... Spiralbound hardback 0-19-963241-3 **$105.00**
.... Paperback 0-19-963240-5 **$69.00**
Receptor Biochemistry; Receptor-Effector Coupling; Receptor-Ligand Interactions: 3 vol set Hulme, E.C. (Ed)
.... Paperback 0-19-963097-6 **$130.00**
Human Cytogenetics: (2/e): 2 vol set Rooney, D.E. & Czepulkowski, B.H. (Eds)
.... Hardback 0-19-963314-2 **$130.00**
.... Paperback 0-19-963313-4 **$90.00**
Peptide Hormone Secretion/Peptide Hormone Action: 2 vol set Siddle, K. & Hutton, J.C. (Eds)
.... Spiralbound hardback 0-19-963072-0 **$135.00**
.... Paperback 0-19-963073-9 **$90.00**

ORDER FORM for USA and Canada

Qty	ISBN	Author	Title	Amount
			S&H	
			CA and NC residents add appropriate sales tax	
			TOTAL	

Please add shipping and handling: US $2.50 for first book, (US $1.00 each book thereafter)

Name ....

Address ....

....

.... Zip ....

[ ] Please charge $ .... to my credit card
Mastercard/VISA/American Express (circle appropriate card)

Acct. .... Expiry date ....

Signature ....

Credit card account address if different from above:

....

.... Zip ....

[ ] I enclose a cheque for US $...........

Mail orders to: Order Dept. Oxford University Press, 2001 Evans Road, Cary, NC 27513